PROCEEDINGS

OF

THE JOHNS HOPKINS WORKSHOP

ON

CURRENT PROBLEMS IN PARTICLE THEORY

23

PREVIOUS JOHNS HOPKINS WORKSHOP PROCEEDINGS

1	BALTIMORE	1974	....
2	BALTIMORE	1978	....
3	FLORENCE	1979	....
4	BONN	1980	Lattice Gauge Theories. Integrable Systems.
5	BALTIMORE	1981	Unified Field Theories and Beyond.
6	FLORENCE	1982	Lattice Gauge Theories. Supersymmetry and Grand Unification.
7	BONN	1983	Lattice Gauge Theories. Supersymmetry and Grand Unification.
8	BALTIMORE	1984	Particles and Gravity.
9	FLORENCE	1985	New Trends in Particle Theory.
10	BONN	1986	Infinite Lie Algebras and Conformal Invariance in Condensed Matter and Particle Physics.
11	LANZHOU	1987	Frontiers in Particle Theory.
12	BALTIMORE	1988	TeV Physics.
13	FLORENCE	1989	Knots, Topology and Quantum Field Theory.
14	HUNGARY	1990	Nonperturbative Methods in Low Dimensional Quantum Field Theories.
15	BALTIMORE	1991	Particle Physics from Underground to Heaven.
16	GÖTEBORG	1992	Pathways to Fundamental Theories.
17	BUDAPEST	1993	Particles and the Universe.
18	FLORENCE	1994	Theory Meets Experiment.
19	BALTIMORE	1995	Particles, Strings and Cosmology.
20	HEIDELBERG	1996	Non-Perturbative Particle Theory and Experimental Tests.
21	LANZHOU	1997	Topology in Particle and Field Theory.
22	GÖTEBORG	1998	Novelties in String Theory.

PROCEEDINGS
OF
THE JOHNS HOPKINS WORKSHOP
ON
CURRENT PROBLEMS IN PARTICLE THEORY 23

BALTIMORE, 1999
(June 10 – 12)

NEUTRINOS IN THE NEW MILLENNIUM

Edited by

G. DOMOKOS & S. KOVESI-DOMOKOS
Johns Hopkins University, Baltimore, USA

World Scientific
Singapore • New Jersey • London • Hong Kong

Published by

World Scientific Publishing Co. Pte. Ltd.

P O Box 128, Farrer Road, Singapore 912805

USA office: Suite 1B, 1060 Main Street, River Edge, NJ 07661

UK office: 57 Shelton Street, Covent Garden, London WC2H 9HE

Library of Congress Cataloging-in-Publication Data

Johns Hopkins Workshop on Current Problems in Particle Theory : (23rd : 1999 :
Baltimore, Md.)

Neutrinos in the new millennium : proceedings of the Johns Hopkins Workshop Current Problems in Particle Theory 23, Baltimore, 1999, (June 10–12) / edited by G. Domokos, S. Kovesi-Domokos.

p. cm.

ISBN 9810242522

1. Neutrinos--Congresses. I. Domokos, G. II. Kövesi-Domokos, S. III. Title.

QC793.5.N42 J64 1999 00-031993

539.7'215--dc21

British Library Cataloguing-in-Publication Data

A catalogue record for this book is available from the British Library.

This book is printed on acid and chlorine free paper.

Printed in Singapore by World Scientific Printers

INTERNATIONAL ADVISORY COMMITTEE

THE JOHNS HOPKINS WORKSHOPS ON CURRENT PROBLEMS IN PARTICLE THEORY ARE ORGANIZED BY THE FOLLOWING INSTITUTIONS:

EÖTVÖS LORÁND UNIVERSITY
UNIVERSITY OF FLORENCE
UNIVERSITY OF GÖTEBORG
UNIVERSITY OF HEIDELBERG
THE JOHNS HOPKINS UNIVERSITY
UNIVERSITY OF LANZHOU

ORGANIZING COMMITTEE

Lars BRINK (Göteborg)
Roberto CASALBUONI (Florence)
Gabor DOMOKOS (Johns Hopkins)
Yi-Shi DUAN (Lanzhou)
Zalán HORVÁTH (Eötvös)
Susan KOVESI-DOMOKOS (Johns Hopkins)
Luca LUSANNA (Florence)
Robert MARNELIUS (Göteborg)
Otto NACHTMANN (Heidelberg)
András PATKÓS (Eötvös)
Berthold STECH (Heidelberg)

LOCAL ORGANIZERS

J. BAGGER
G. DOMOKOS
A. FALK
S. KOVESI-DOMOKOS

FOREWORD

The 23^d *Johns Hopkins Workshop on Current Problems in Particle Theory* took place at Johns Hopkins University, June 10th thru 12th, 1999. The entire workshop was devoted to neutrino physics. The recent past brought astonishingly beautiful results in this field: there is now evidence for flavor oscillations, hence, apparently, neutrinos (at least some of them) are massive. Also, new experiments are now available in order to study the interactions of neutrinos at high energies – and others are under construction. One hopes that the first decade of the twenty first century will bring more results and some surprizes. Therefore, it was felt appropriate for the workshop taking place in the last year of the twentieth century to survey and summarize the state of the art in neutrino physics.

For that reason, we deviated somewhat from the usual format of the workshops. Traditionally, the *Johns Hopkins Workshops* were mostly concerned with theoretical results in the field chosen as the subject of any given workshop; new experimental developments have been summarized by one speaker. In the field of neutrino physics, however, there is a much closer relationship between theory and experiment. To a large extent, this is due to the fact neutrino physics provides a probe of the *"terra incognita"*, the physics beyond the well established Standard Model. As a result, more experimental talks were scheduled than usual. The ensuing lively discussions are, to a certain extent, reflected in the written version of the talks printed in this volume. We hope that in this way, the workshop stimulated new ideas about future experiments and theoretical investigations.

It is always a pleasant duty of the organizers to thank all those who contributed to the success of the workshop. First and foremost, we wish to thank the members of the International Advisory Committee for sharing their wisdom with the organizers: their suggestions contributed substantially to the success of the workshop.

Financial support by The Johns Hopkins University, the U.S. Department of Energy and the Space Telescope Science Institute is gratefully acknowledged. Particular thanks are due to Dean Herbert Kessler, to Ron Allen at STScI and to Holland Ford, director of CAS at Johns Hopkins University for their support of this workshop.

Last but not least, we thank members of the staff of the *Henry A. Rowland Department of Physics and Astronomy* for their invaluable help in organizing and running the workshop. In particular, Walt Earls and Ann Evans have been instrumental in taking care of all the details of the organizational work. In addition, Walt has lent competent and devoted help in putting together this Proceedings.

The Organizing Committee

CONTENTS

Foreword		vii
F. HALZEN	High Energy Neutrino Astronomy	1
R. GANDHI	The Interactions and Detection of Ultra High Energy Neutrinos: A Brief Review	25
T. MILLER *et al.*	From AMANDA to IceCube: Current and Future High Energy Neutrino Telescopes at the South Pole	47
D. SECKEL	Detecting Ultra High Energy Neutrinos ($E_\nu > 10^{18}$ EV) with a Large Radio Array	63
S. COUTU, X. BERTOU & P. BILLOIR	Ultra-High Energy Neutrinos with Auger	73
J. N. BAHCALL	Solar Neutrinos	91
J. R. KLEIN	The Sudbury Neutrino Observatory at Turn-On	113
J. G. LEARNED	The Discovery of Neutrino Oscillations and Mass: The Disappearance of Muon Neutrinos	125
J. M. LOSECCO	Atmospheric Neutrinos: Past, Present and ...	151
T. STANEV	Atmospheric Neutrino Oscillations and Cosmic Ray Leptons in the Atmosphere	161
Y. F. WANG	Recent Results and Future Prospects on Neutrino Oscillations from Reactors	179
D. B. CLINE	The Neutrino Physics Potential of the ICARUS Detector Program	189
B. J. KING	High Rate Physics at Neutrino Factories	201

B. KAYSER	Neutrino Mass: What Has Been Accomplished, and What Do We Need to Do?	223
J. C. PATI	The Link Between Neutrino Masses and Proton Decay in Supersymmetric Unification	225
P. RAMOND	Neutrino Masses and Lepton-Quark Symmetries	265
F. FERUGLIO	Models of Neutrino Masses from Oscillations with Large Mixings	279
B. STECH	Are the Neutrino Masses and Mixings Closely Related to the Masses and Mixings of Quarks?	295
C. GIUNTI	Four-Neutrino Mixing	305
A. A. PETROV	QCD and Electroweak Studies at Neutrino Factories	321
D. O. CALDWELL	The Hot Dark Matter	331
G. M. FULLER	Implications of Neutrino Flavor Conversion for Cosmology and Supernovae	341
J. F. BEACOM	Neutrinos from the Next Galactic Supernova	357
L. WOLFENSTEIN	Neutrinos in the Next Millennium – Summary Talk	385

HIGH ENERGY NEUTRINO ASTRONOMY

FRANCIS HALZEN
Department of Physics, University of Wisconsin, Madison, WI 53706

With the completion of the first-generation Lake Baikal and AMANDA detectors, the past year represents a milestone in neutrino astronomy. Neutrino signals from the Lake Baikal detector bode well for the flurry of R&D activities leading to the deployment of similar instruments in the Mediterranean. The AMANDA telescope announced first light, neutrinos actually, and produced results that validate plans for commissioning a kilometer-scale detector in Antarctic ice. Although high energy neutrino astronomy is a multidisciplinary science, gamma ray bursts have, with supermassive black holes, become the theoretical focus since recent astronomical observations revealed their potential as cosmic particle accelerators. This spotlight is shared with investigations of the potential of high energy telescopes to observe oscillating neutrinos. The Superkamiokande results have boosted atmospheric neutrinos from a calibration tool and a background for doing astronomy, to an opportunity to confirm the evidence for neutrino mass.

1 Neutrino Astronomy: Multidisciplinary Science

Using optical sensors deployed in deep ocean and lake waters or buried in the deep clear Antarctic ice sheet, neutrino astronomers are attacking major problems in astronomy, astrophysics, cosmic ray physics and particle physics by commissioning a first generation of neutrino telescopes.[1] According to estimates covering a wide range of scientific objectives, a neutrino telescope with effective telescope area of 1 kilometer squared is required to address the most fundamental questions.[2] Planning is already underway to instrument a cubic volume of ice, 1 kilometer on the side, as a neutrino detector. This infrastructure provides unique opportunities for yet more interdisciplinary science covering the geosciences and biology.

The commissioning of neutrino telescopes is primarily motivated by their discovery potential. Among the many scientific problems we nevertheless anticipate high energy neutrino telescopes to address are the origin of cosmic rays, the nature of gamma ray bursts (GRB), the physical processes in the engines which power active galaxies (AGN), the search for the annihilation products of halo cold dark matter and, possibly, even the structure of the Earth's interior. In burst mode they scan the sky for galactic supernovae and, more speculatively, for the birth of supermassive black holes.[3] Coincident experiments with Earth- and space-based gamma ray observatories, cosmic ray telescopes and gravitational wave detectors such as LIGO can be contemplated. With high-energy neutrino astrophysics we are poised to open a new window into space and back in time to the highest-energy processes in the Universe.

Multi-wavelength Astronomy: Gamma Ray Bursts

Gamma ray bursts have, arguably, become the best motivated source for high energy neutrinos.[4] Evidence has been steadily accumulating that GRB emission is the result of a relativistically expanding fireball energized by a process involving neutron stars or black holes.[5] In the early stage, the fireball cannot emit efficiently because the radiation is trapped due to the very large optical depth. The fireball's energy is dissipated in kinetic energy until it becomes optically thin and produces the observed display. This scenario can accommodate the observed energy- and time scales provided the bulk Lorentz factor γ of the expanding shock is ≥ 300.

The production of high energy neutrinos is anticipated: protons, accelerated in the kinetic phase of the shock, interact with photons producing charged pions which are the parents of neutrinos. Standard particle physics and fireball phenomenology are sufficient to compute the neutrino energy and event rate:[4]

$$N_\nu = 9\ \mathrm{km}^{-2}\,\mathrm{yr}^{-1}\left[\frac{\mathrm{f}_\pi}{15\%}\right]\left[\frac{\dot{\mathrm{E}}}{4\times 10^{44}\,\mathrm{Mpc}^{-3}\,\mathrm{yr}^{-1}}\right]\left[\frac{\mathrm{E}_\nu}{700\ \mathrm{TeV}}\right]^{-1}, \qquad (1)$$

i.e., we expect on average 10^{-2} detected neutrino events per burst, or about 10 events in a km^2 detector in one year assuming 10^3 bursts per year. This average is however not relevant to observation as we will see further on. $\dot{E}$ is the injection rate of energy into protons (rather than electrons) in the kinetic fireball. This is a critical parameter in the calculations; it can be fixed, for instance, by assuming that GRBs are the source of the highest energy cosmic rays[6] beyond the ankle in the spectrum, or $\dot{E} \simeq 4 \times 10^{44}$ erg Mpc^{-3} yr^{-1}.[7] The factor f_π represents the fraction of total energy going into pion production. It is calculated by known particle physics and of order 15%.

One should keep in mind that neutrinos can also be produced in other stages of the fireball, e.g. when it expands through the interstellar medium.[8] More importantly from an observational point of view, the relevant observed neutrino rates are determined by the fluctuations, and not by the average burst computed above.[9] The neutrino rates of individual bursts fluctuate with i) the square of the distance which is assumed to follow a cosmological distribution, ii) with the energy, assuming that the neutrino rate depends linearly on energy and that ten percent of GRB produce more energy than average by a factor of ten, and one percent by a factor of 100, and, most importantly, iii) fluctuations in the γ factor around its average value of 300. The fluctuations in γ affect the value of f_π which varies approximately as γ^{-4} ([4]) as well as the position of the break energy which varies as γ^2. Both effects are taken into account. Clearly a factor of ten variation of γ leads to a change in flux by roughly 4 orders of magnitude.

The origin of the large fluctuations with γ should be a general property of boosted accelerators. With high luminosities emitted over short times, the large photon density renders the GRB opaque unless γ is very large. Only transparent sources with large boost factors emit photons. They are however relatively weak neutrino sources because the actual photon target density in the fireball is diluted by the large Lorentz factor. The assumption of a smaller γ-factor results, on the other hand, in a higher photon density in the source reference frame. Because of the large photon target density seen by the accelerated protons, it turns into a spectacular neutrino source. Some have argued[10] for $\langle\gamma\rangle \sim 30$ on the basis that the unusual fluctuations in the morphology of GRBs can only be produced in a relatively dense, turbulent medium. Lower γ sources which radiate lower energy photons over longer time-scales could have been missed by selection effects.

Even in the absence of the dominant fluctuations in γ, the rate of about 10 detected neutrinos per year over 10^3 bursts increases by an order of magnitude, even if only fluctuations in energy and distance are taken into account. Not knowing the distribution of γ-factors around its average, one can parametrize it as a Gaussian with width σ which is, in the end, the most critical parameter. Neither theory nor experiment provides compelling information. Note that we require $\sigma > 70$ in order to allow a significant part of the bursts to have γ-factors less than 10^2, or one third the average value. For $\sigma = 70$, the observed rate is $\sim$ 600 per year in a kilometer square detector, with 23 individual bursts yielding more than 4 detected neutrinos, and 7 yielding more than 10! Absorption inside the source eventually reduces the overall rates when γ is much smaller than average. Absorption is included in the rates quoted.[9]

Even though bursts with somewhat lower γ produce neutrinos with reduced energy, compared to 700 TeV for $\gamma = 300$, and over a longer time-scale than 1 second, the signatures are spectacular. A single GRB with high energy, close proximity and a relatively low Lorentz factor could produce more detectable neutrino events than all other GRB over several years time. It is far more likely that neutrino detectors detect one GRB with favorable characteristics, than hundreds with average values. Because these spectacular events arrive with the GRB time-stamp of order 1 second and with an unmistakable high-energy signature, background rejection is greatly simplified. Although one cannot say with total confidence that GRB are a guaranteed source of neutrinos, it is clear that **not observing them over several years of IceCube live-time will challenge our understanding of the best-buy fireball model of GRB** by requiring a severe limit on the number of accelerated protons in the kinetic phase of the burst, or complete exclusion of all but the highest values of the Lorentz boost factor.

Clearly, the primary importance of fluctuations is a key consideration for other GRB models, as well as for blazars and any other boosted sources. Fluctuations may also be a key factor in the production of photons. Note that we expect no direct correlation between neutrino and high energy gamma sources because cosmic events with abundant neutrino production are almost certainly opaque to high energy photons.

The observation of GRB neutrinos over a cosmological baseline has scientific potential beyond testing the "best-buy" fireball model:[4] GRB neutrinos can test with unmatched precision special relativity and the equivalence principle, and permit study of oscillating neutrino flavors over the ultimate baseline of $z \simeq 1$.

Cosmic Accelerators

Before itemizing other potential sources of high energy neutrinos, one has to keep in mind that Nature has been more imaginative than scientists in anticipating the rewards of building discovery instruments studying heavens with novel techniques. We nevertheless speculate on potential extra-galactic and galactic cosmic accelerators of cosmic rays. There is no concensus in this discussion but one cannot forget that Nature does accelerate particles to 10^{20} eV, somewhere.

Besides GRB, AGN are the other bright gamma ray sources in the Universe. Their engines must not only be powerful, but also extremely compact because their luminosities are observed to flare by over an order of magnitude over time periods as short as hours.[11] Only sites in the vicinity of black holes which are a billion times more massive than our sun will do. It is generally believed that the electromagnetic spectrum at all wavelengths, from radio waves to TeV gamma rays, is produced in the interactions of the accelerated electrons with the magnetic fields and ambient photons in the source. If protons are accelerated along with electrons to PeV–EeV energy, they will produce high energy photons by photoproduction of neutral pions on the ubiquitous UV thermal background. Although the relative merits of the electron and proton acceleration are hotly debated, it is here relevant that the issues can be settled experimentally. Proton acceleration turns AGN into sources of high energy cosmic rays and neutrinos, not just gamma rays. Weakly interacting neutrinos can, unlike high energy gamma-rays and high energy cosmic rays, reach us from more distant and much more powerful AGN. The opportunities for high energy neutrino astronomy are wonderfully obvious.

Models have mostly focussed on acceleration in the jets beamed perpendicular to the accretion disc. It is assumed that particles are accelerated by Fermi shocks in blobs of matter traveling along the jet with a bulk Lorentz factor of $\gamma \sim 10$, possibly higher, relative to the observer. In order to accommodate bursts lasting a day, or less, in the observer's frame, the size of the blob must be of order $\gamma c \Delta t \sim 10^{-2}$ parsecs or less. The blobs are actually more like sheets, thinner than the jet's size of roughly 1 parsec. The transformation between blob and observer is $R' = \gamma R$ and $E' = \frac{1}{\gamma} E$ for distances and energies, respectively. Here primes refer to a frame attached to the blob. The observed radiation at all wavelengths is produced by the interaction of the accelerated particles in the blob and the ambient radiation in the AGN.

From the observed photon luminosity L_γ the energy density of photons in the shocked region can be inferred:

$$U'_\gamma = \frac{L_\gamma \Delta t}{\frac{4}{3}\pi R'^3} \sim \frac{L_\gamma \Delta t}{\gamma} \frac{1}{(\gamma c \Delta t)^3} \sim \frac{L_\gamma}{\gamma^4 \Delta t^2} \,. \tag{2}$$

With high luminosities L_γ emitted over short Δt, the large photon density renders the blob opaque to photons of 10 TeV energy and above unless γ is very large. A large γ factor, typically larger than 10, is the agent that dilutes the blob until 10 TeV γ's fall below the $\gamma\gamma \to e^+e^-$ threshold in the blob. Only transparent sources with large boost factors emit TeV photons, e.g. Markarian 421 and 501. They are weak neutrino sources because, even if the TeV photon is of π^0 origin, one expects only one neutrino per photon, and this translates into observation of only a few events per year in a kilometer-scale detector.[12] They should still be within reach of IceCube, especially because fluctuations will increase rates and sharpen observational signatures, as in the case of GRB. More importantly, a source with the same morphology, but $\gamma \simeq 1$, is completely opaque to high energy photons and protons. It is a "hidden" source, with reduced or extinguished emission of high energy particles, but abundant neutrino production by protons on the high density photon target. Nature hopefully made AGN with a distribution of boost factors; *observed* boost factors are actually typically less than 10.

While uninteresting for high energy gamma ray astronomy, hidden sources have the potential to be powerful neutrino emitters with fluxes exceeding those of TeV gamma ray sources by several orders of magnitude; a representative range of models is shown in Fig. 1a. Highlighted is the pioneering calculation by Stecker and collaborators who integrated the neutrino flux from single generic AGN to obtain a diffuse flux from all cosmological AGN. In their model protons are accelerated at a shock in the accretion flow which is formed where its ram pressure balances the radiation pressure near the central black hole. In Fig. 1b we show the energy dependence of neutrino-induced muons observed with the AMANDA for a diffuse neutrino flux with a typical E^{-2} spectrum. Also shown is the background from lower energy atmospheric neutrinos. Although energy measurement in the present detectors is challenging, it is sufficiently accurate to exclude the excess of high energy events above 1 TeV predicted in Fig. 1a.[13] A preliminary AMANDA analysis[13] provides a limit on a diffuse cosmic neutrino flux roughly at the level of the Stecker *et al.* prediction:

$$E^2 \Phi_\nu \leq 3.6 \times 10^{-6} \, (\mathrm{cm}^2 \, \mathrm{sec \ sr})^{-1} \, \mathrm{GeV} \,. \tag{3}$$

Models in Fig. 1a exceeding this highlighted flux are ruled out. AMANDA also obtains a preliminary limit on the neutrino flux from individual AGN from the absence of hotspots in the skymap. After a data selection which optimizes the detector response to a E^{-2} spectrum, AMANDA obtains a limit

$$\Phi_\nu(> 1 \ \mathrm{TeV}) \leq 3 \times 10^{-9} \, (\mathrm{cm}^2 \, \mathrm{sec})^{-1} \tag{4}$$

at the 90% CL.

The recent detection of TeV gamma ray emission from supernova remnants such as IC 443 and γ-Cygni has provided further indirect support for the idea that the bulk of the cosmic rays may be born in supernova blast waves. The cosmic ray energy spectrum can be understood, up to perhaps 1000 TeV, in terms of acceleration by supernova shocks exploding into the interstellar medium of our galaxy. If some or all of the gamma rays are indeed decay products of neutral pions produced in the cosmic accelerator, the concomitant flux of neutrinos should be detectable. Although the precise flux is difficult to compute,[14] the aggregate flux of the cosmic ray producing supernovae could be a copious source of neutrinos.[1]

X-ray binaries, mini-quasars such as SS 433, and pulsars have also been identified as potential nearby sources of high energy neutrinos. The Crab supernova remnant has been identified[15] as a promising source of high energy ions accelerated in the wind blowing off the pulsar. The ions drift through the remnant producing pions, and therefore neutrinos, in interactions with the thermal gas.

Neutrino Telescopes: Their More Routine Capabilities

Neutrino telescopes are observing muons and neutrinos with energies exceeding those produced at accelerators, or observed in previous underground experiments. They have the potential to search for prompt cosmic ray muons,

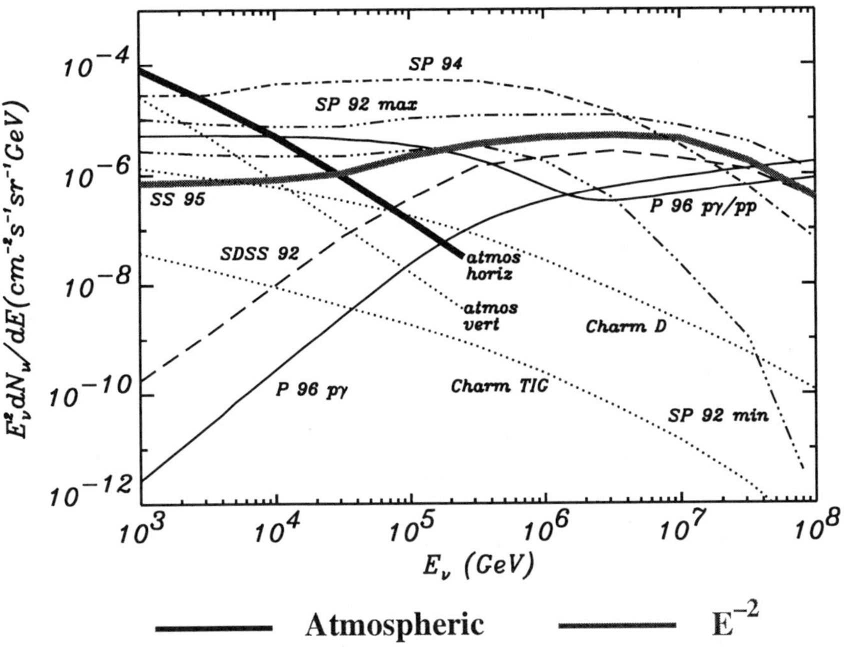

Figure 1a: Predictions for the diffuse neutrino background from AGN are confronted with the atmospheric neutrino background.

for instance, those produced by the decay of heavy quarks produced in high energy cosmic ray intereactions in the atmosphere. Hints of a prompt charm component in the muon spectrum may have been identified in both Gran Sasso experiments.[16]

The atmosphere is not the only source of cosmic neutrinos guaranteed by the existence of the cosmic ray beam. Cosmic rays produce neutrinos in interactions with the sun and with the interstellar gas in our galaxy.[1] It takes kilometer-scale detectors to observe them. The same is true for the neutrinos made in interactions of extragalactic cosmic rays with microwave photons. A search for them would be of renewed interest with the accumulating evidence that there is no Greisen cutoff in the cosmic ray spectrum. If extra-galactic cosmic rays are not truly universal, but confined to clusters of galaxies, the rate of Greisen neutrinos will be reduced and, most probably, unobservable.[1]

Detection of high energy neutrino emission by the galactic plane will be more than routine. The expected event rate, of order 10 per year depending

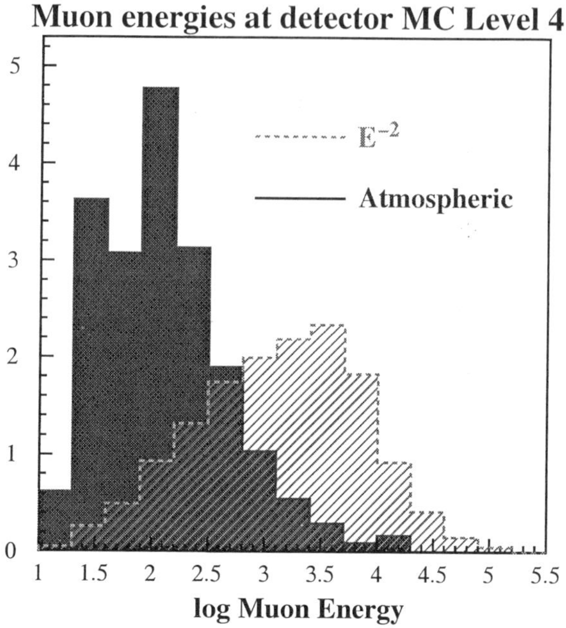

Figure 1b: Response of the AMANDA detector to signal and background in Fig. 1a.

on the detector threshold, can be directly related to the cosmic ray flux and, therefore, sensitive to its spectrum outside our galactic neighbourhood where it can not be directly observed. The neutrino rate would, for instance, be increased by an order of magnitude if it were characterized by a harder cosmic ray spectrum in the outer galaxy. Satellite experiments have been accumulating persistent evidence for this.

In the extremely low-noise detector environment provided by sterile ice, the simultaneous short-time increase of the counting rate in all photomultipliers resulting from the production of electrons by supernova $\bar{\nu}_e$, would provide a statistically significant detection out to the Magelanic Cloud. IceCube will continue, with increased sensitivity, the ongoing AMANDA search for the bursts of low energy neutrinos of order 10 MeV originating from galactic supernova collapses.[3] Because of its South Pole location IceCube plays a pivotal role in the supernova network which will attempt to triangulate the neutrino arrival times with northern hemisphere detectors. The goal is to reach the ~10 degree resolution of a single Schmidt telescope. The potential of IceCube, operated in burst mode, for observing the birth of galaxies or the low energy neutrino fluxes from GRB has also been investigated.[17]

Finally, a relatively shallow neutrino detector can be operated as a TeV gamma ray telescope. It is possible to observe the gamma rays of TeV sources, by detecting the muon secondaries in the air showers produced in the interaction of the very high energy photons in the Earth's atmosphere. The muon flux can be unambiguously derived from the γ-ray flux — there is no model dependence. In the IceCube detector the VELA pulsar produces almost 10^5 events per year on a background of 4×10^7 cosmic ray muons in a $1°$ bin around the source. The signal to square root of noise is 15. For a Markarian type southern hemisphere source in "steady state", and at the same zenith as VELA (45 degrees), we predict close to 10^4 events per year. The signal to noise would be excellent because of the low background that would be accumulated in the detector during the short time of a bursting source. This is especially true for TeV gamma ray observations of GRB whose observation should be a high priority with the recent discovery that the energy in TeV emission may exceed that inferred from keV-MeV observations by over one order of magnitude.[18]

The potential of neutrino telescopes discussed in this section is not a matter of speculation; the fluxes associated with its more routine capabilities discussed above are calculable and guaranteed.

Particle Physics

Neutrino masses provide a window beyond the Standard Model, which is known to be incomplete. Because cosmic sources of neutrinos produce ν_e, ν_μ beams, IceCube can perform the ultimate long-baseline experiment: ν_τ appearance studies from cosmological sources such as a GRB can probe the ν_τ mass with a sensitivity of $\Delta m^2 \gtrsim 10^{-17}\,\mathrm{eV}^2$. PeV τ-neutrinos can be identified by unique signatures: i) the "double-bang event" consisting of a pair of separated showers associated with the charged-current production of a τ-lepton followed by its decay,[19] and ii) its zenith angle distribution which is different from ν_e and ν_μ because high energy ν_τ's can penetrate the Earth at all zenith angles; they do not range out.[20]

Above 1 PeV ν_e and ν_μ are absorbed by charged-current interactions in the Earth before reaching a detector at the opposite surface. In contrast, the Earth never becomes opaque to ν_τ since the τ produced in a charged-current ν_τ interaction decays back into a ν_τ with smaller energy and interaction probability. The process repeats until the ν_τ can penetrate Earth.[20] This penetration of tau neutrinos through the Earth above 10^2 TeV provides an experimental signature for neutrino oscillations. The appearance of a ν_τ component in a pure $\nu_{e,\mu}$ beam would be signalled by a flat angular dependence of a source intensity at the highest neutrino energies. Such a flat zenith angle dependence for the farthest sources is a signature for tau neutrino mixing with a sensitivity in Δm^2 as low as $10^{-17}\,\mathrm{eV}^2$.

Arguably the most urgent question in neutrino physics is whether SuperKamiokande has indeed observed ν_μ to μ_τ oscillations. In the 10 GeV energy region, ν_τ charged-current interactions signaling the appearance of ν_τ in the atmospheric neutrino flux, can be identified by the additional energy released by the prompt τ decay.[21] Such measurements probably require modifications of the planned detectors in order to achieve good energy measurement at low threshold; see the later discussion of ANTARES however.

Finally, where the search for the particle nature of the dark matter is concerned, neutrino telescopes occupy a unique place in the diverse landscape of dark matter searches.[22] This is usually illustrated in a framework where cold dark matter particles, e.g. neutralinos, constitute the non-luminous halo of our galaxy. For neutralino masses above 0.5 TeV, IceCube's sensitivity peaks because of the high energy of the neutrinos produced by neutralino annihilation. It roughly increases linearly with the mass up to 10 TeV. Higher masses are inconsistent with standard cosmology. Direct searches whose sensitivity decreases with mass are unlikely ever to cover this mass range. Neither will the Large Hadron Collider, which operates below the threshold for producing neutralinos of 0.5 TeV mass and higher. The rate of neutrino events from dark matter annihilation is calculable; therefore non-observation can be translated into constraints on models, supersymmetry or other.

With neutrino telescopes we will also search for ultrahigh-energy neutrino signatures from topological defects predicted by grand unified theories, for magnetic monopoles[23] and for clues to entirely new physical phenomena.

2 Large Natural Cherenkov Detectors

The study of GRBs is one more example of a science mission that requires kilometer-scale neutrino detectors. This is not a real surprise. The probability to detect a PeV neutrino is roughly 10^{-3}. This is easily computed from the requirement that, in order to be detected, the neutrino has to interact within a distance of the detector which is shorter than the range of the muon it produces.[1] At PeV energy the cosmic ray flux is of order 1 per m^2 per year and the probability to detect a neutrino of this energy is of order 10^{-3}. A neutrino flux equal to the cosmic ray flux will therefore yield only a few events per day in a kilometer squared detector. At EeV energy the situation is worse. With a cosmic ray rate of 1 per km^2 per year and a detection probability of 0.1, one can only detect several events per year in a kilometer squared detector provided the neutrino flux exceeds the proton flux by 2 orders of magnitude or more. For the neutrino flux generated by cosmic rays interacting with CMBR photons and for sources like active galaxies and topological defects,[24] this is indeed the case. All above estimates are however conservative and the rates should be higher because absorption of protons in the source is expected, and the neutrinos escape the source with a flatter energy spectrum than the protons. In summary, at least where cosmic rays are part of the beam dump, their flux and the neutrino cross section and muon range define the size of a neutrino telescope. A telescope with kilometer squared effective area represents a neutrino detector of kilometer cubed volume.

The first generation of neutrino telescopes, launched by the bold decision of the DUMAND collaboration over 25 years ago to construct such an instrument, are designed to reach a large telescope area and detection volume for a neutrino threshold of order 10 GeV. This relatively low threshold permits calibration of the novel instrument on the known flux of atmospheric neutrinos. The architecture is optimized for reconstructing the Cherenkov light front radiated by an up-going, neutrino-induced muon. Only up-going muons made by neutrinos reaching us through the Earth can be successfully detected. The Earth is used as a filter to screen the fatal background of cosmic ray muons. This makes neutrino detection possible over the lower hemisphere of the detector. Up-going muons must be identified in a background of down-going, cosmic ray muons which are more than 10^5 times more frequent for a depth of 1~2 kilometers. The method is sketched in Fig. 2.

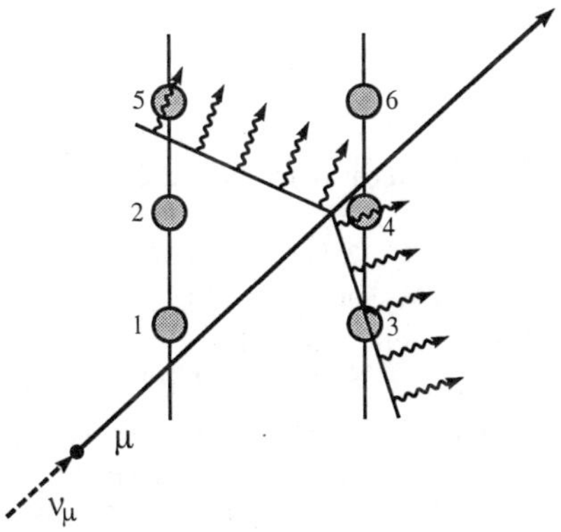

Figure 2: The arrival times of the Cherenkov photons in 6 optical sensors determine the direction of the muon track.

The optical requirements of the detector medium are severe. A large absorption length is required because it determines the spacings of the optical sensors and, to a significant extent, the cost of the detector. A long scattering length is needed to preserve the geometry of the Cherenkov pattern. Nature has been kind and offered ice and water as adequate natural Cherenkov media. Their optical properties are, in fact, complementary. Water and ice have similar attenuation length, with the role of scattering and absorption reversed; see Table 1. Optics seems, at present, to drive the evolution of ice and water detectors in predictable directions: towards very large telescope area in ice exploiting the large absorption length, and towards lower threshold and good muon track reconstruction in water exploiting the large scattering length.

Table 1: Optical properties of South Pole ice at 1750 m, Lake Baikal water at 1 km, and the range of results from measurements in ocean water below 4 km.

$\lambda = 385$ nm *	(1700 m) AMANDA	BAIKAL	OCEAN
attenuation	~ 30 m **	$\sim$8 m	25–30 m ***
absorption	95 ± 5 m	8 m	—
scattering length	24 ± 2 m	150–300 m	—

* peak PMT efficiency

** same for bluer wavelengths

*** smaller for bluer wavelengths

3 Baikal, ANTARES, Nestor and NEMO: Northern Water

Whereas the science is compelling, the real challenge is to develop a reliable, expandable and affordable detector technology. With the termination of the pioneering DUMAND experiment, the efforts in water are, at present, spearheaded by the Baikal experiment.[25] The Baikal Neutrino Telescope is deployed in Lake Baikal, Siberia, 3.6 km from shore at a depth of 1.1 km. An umbrella-like frame holds 8 strings, each instrumented with 24 pairs of 37-cm diameter *QUASAR* photomultiplier tubes (PMT). Two PMTs in a pair are switched in coincidence in order to suppress background from natural radioactivity and bioluminescence. Operating with 144 optical modules since April 1997, the *NT-200* detector has been completed in April 1998 with 192 optical modules

(OM). The Baikal detector is well understood and the first atmospheric neutrinos have been identified.

The Baikal site is competitive with deep oceans although the smaller absorption length of Cherenkov light in lake water requires a somewhat denser spacing of the OMs. This does however result in a lower threshold which is a definite advantage, for instance for oscillation measurements and WIMP searches. They have shown that their shallow depth of 1 kilometer does not represent a serious drawback. By far the most significant advantage is the site with a seasonal ice cover which allows reliable and inexpensive deployment and repair of detector elements.

With data taken with 96 OMs only, they have shown that atmospheric muons can be reconstructed with sufficient accuracy to identify atmospheric neutrinos; see Fig. 3. The neutrino events are isolated from the cosmic ray muon background by imposing a restriction on the chi-square of the Cherenkov fit, and by requiring consistency between the reconstructed trajectory and the spatial locations of the OMs reporting signals. In order to guarantee a minimum lever arm for track fitting, they were forced to reject events with a projection of the most distant channels on the track smaller than 35 meters. This does, of course, result in a higher threshold.

In the following years, *NT-200* will be operated as a neutrino telescope with an effective area between $10^3 \sim 5 \times 10^3\,\mathrm{m}^2$, depending on energy. Presumably too small to detect neutrinos from extraterrestrial sources, *NT-200* will serve as the prototype for a larger telescope. For instance, with 2000 OMs, a threshold of $10 \sim 20\,\mathrm{GeV}$ and an effective area of $5 \times 10^4 \sim 10^5\,\mathrm{m}^2$, an expanded Baikal telescope would fill the gap between present underground detectors and planned high threshold detectors of cubic kilometer size. Its key advantage would be low threshold.

The Baikal experiment represents a proof of concept for deep ocean projects. These have the advantage of larger depth and optically superior water. Their challenge is to find reliable and affordable solutions to a variety of technological challenges for deploying a deep underwater detector. Several groups are confronting the problem, both NESTOR and ANTARES are developing rather different detector concepts in the Mediterranean.

The NESTOR collaboration,[26] as part of a series of ongoing technology tests, is testing the umbrella structure which will hold the OMs. They have already deployed two aluminum “floors”, 34 m in diameter, to a depth of 2600 m. Mechanical robustness was demonstrated by towing the structure, submerged below 2000 m, from shore to the site and back. These test should soon be repeated with fully instrumented floors. The actual detector will consist of a tower of 12 six-legged floors vertically separated by 30 m. Each floor con-

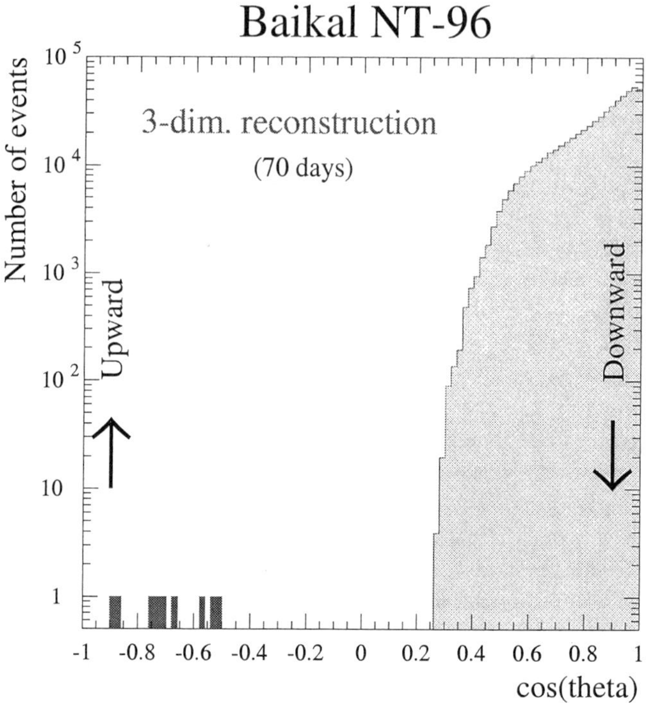

Figure 3: Angular distribution of muon tracks in the Lake Baikal experiment after the cuts described in the text.

tains 14 OMs with four times the photocathode area of the commercial 8 inch photomultipliers used by AMANDA and ANTARES.

The detector concept is patterned along the Baikal design. The symmetric up/down orientation of the OMs will result in uniform angular acceptance and the relatively close spacings in a low threshold. NESTOR does have the advantage of a superb site, possibly the best in the Mediterranean. The detector can be deployed below 3.5 km relatively close to shore. With the attenuation length peaking at 55 m near 470 nm the site is optically superior to that of all deep water sites investigated for neutrino astronomy.

The ANTARES collaboration[27] is investigating the suitability of a 2400 m deep Mediterranean site off Toulon, France. The site is a trade-off between acceptable optical properties of the water and easy access to ocean technology. Their detector concept requires indeed remotely operated vehicles for making underwater connections. First results on water quality are very encouraging with an attenuation length of 40 m at 467 nm and a scattering length exceeding 100 m. Random noise exceeding 50 khz per OM is eliminated by requiring coincidences between neighboring OMs, as is done in the Lake Baikal design. Unlike other water experiments, they will point all photomultipliers sideways or down in order to avoid the effects of biofouling. The problem is significant at the Toulon site, but only affects the upper pole region of the OM. Relatively weak intensity and long duration bioluminescence results in an acceptable deadtime of the detector. They have demonstrated their capability to deploy and retrieve a string.

With the study of atmospheric neutrino oscillations as a top priority, they had planned to deploy in 2001-2003 10 strings instrumented over 400 m with 100 OMs. After study of the underwater currents they decided that they can space the strings by 100 m, and possibly by 60 m. The large photocathode density of the array will allow the study of oscillations in the range $255 < L/E < 2550$ km GeV^{-1} with neutrinos in the energy range $5 < E_\nu < 50$ GeV. More recent plans call for a different deployment of the 1000 OMs, on 15 strings separated by 80 m. Each string is instrumented with triplets of PMTs spaced by 8-16 m.

A new R&D initiative based in Catania, Sicily has been mapping Mediterranean sites, studying mechanical structures and low power electronics. One must hope that with a successful pioneering neutrino detector of 10^{-3} km^3 in Lake Baikal, a forthcoming 10^{-2} km^3 detector near Toulon, the Mediterranean effort will converge on a 10^{-1} km^3 detector at the NESTOR site.[28] For neutrino astronomy to become a viable science several of these, or other, projects will have to succeed besides AMANDA. Astronomy, whether in the optical or in any other wave-band, thrives on a diversity of complementary instruments, not on "a single best instrument". When, for instance, the Soviet government tried out the latter method by creating a national large mirror project, it virtually annihilated the field.

4 AMANDA: Southern Ice

Construction of the first-generation AMANDA detector was completed in the austral summer 96–97. It consists of 300 optical modules deployed at a depth of 1500–2000 m; see Fig. 4. Here the optical module consists of an 8 inch photomultiplier tube and nothing else. It is connected to the surface by a cable which transmits the HV as well as the anode current of a triggered photomultiplier. The instrumented volume and the effective telescope area of this instrument matches those of the ultimate DUMAND Octogon detector which, unfortunately, could not be completed.

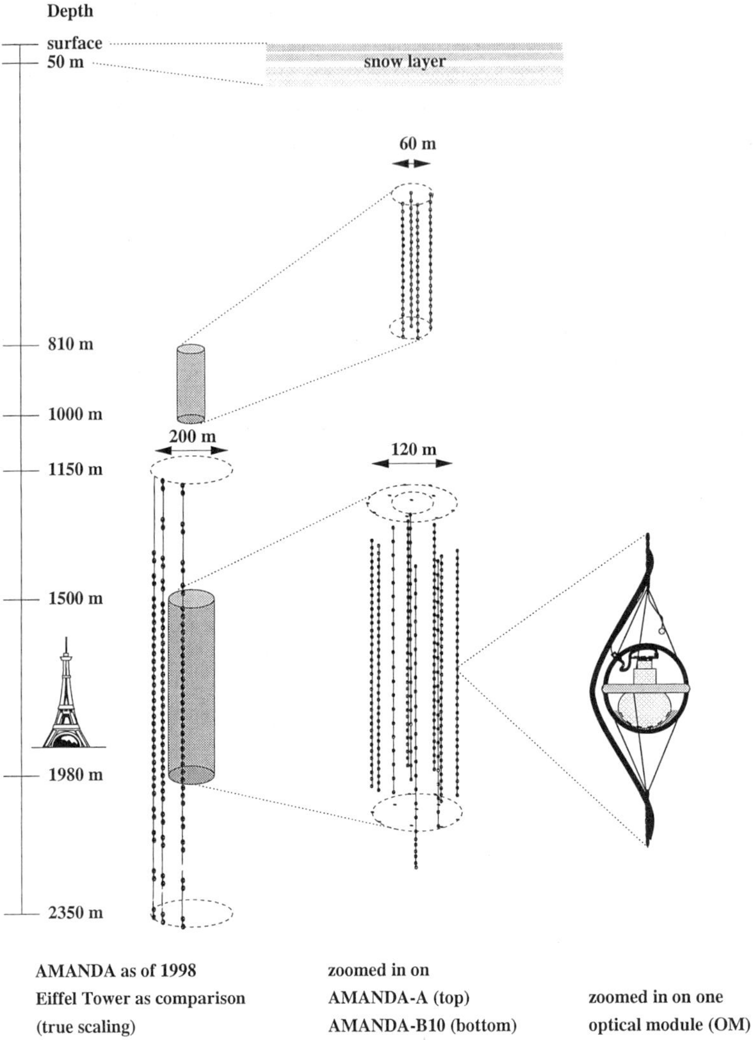

Figure 4: The Antarctic Muon And Neutrino Detector Array (AMANDA).

As predicted from transparency measurements performed with strings near 1 km depth,[29] it was found that ice is bubble-free below 1400 m. Its optical properties were a surprise, nevertheless the bubble-free ice turned out to be an adequate Cherenkov medium. We explain this next.

The AMANDA detector was antecedently proposed on the premise that inferior properties of ice as a particle detector with respect to water could be compensated by additional optical modules. The technique was supposed to be a factor 5~10 more cost-effective and, therefore, competitive. The design was based on then current information:[30]

- the absorption length at 370 nm, the wavelength where photomultipliers are maximally efficient, had been measured to be 8 m;
- the scattering length was unknown;
- the AMANDA strategy was to use a large number of closely spaced OM's to overcome the short absorption length. Simulations indicated that muon tracks triggering 6 or more OM's were reconstructed with degree accuracy. Taking data with a simple majority trigger of 6 OM's or more at 100 Hz resulted in an average effective telescope area of 10^4 m^2, somewhat smaller for atmospheric neutrinos and significantly larger for the high energy signals.

The reality is that:

- the absorption length is 100 m or more, depending on depth;[29]
- the scattering length is ~25 m (preliminary, this number represents an average value which may include the combined effects of deep ice and the refrozen ice disturbed by the hot water drilling);
- because of the large absorption length, OM spacings are similar, actually larger, than those of proposed water detectors. Also, in a trigger 20 OM's report, not 6. Of these more than 6 photons are, on average, "not scattered." A "direct" photon is typically required to arrive within 25 nsec of the time predicted by the Cherenkov fit. This allows for a small amount of scattering and includes the dispersion of the anode signals over the 2 km cable. In the end, reconstruction is therefore as before, although additional information can be extracted from scattered photons by minimizing a likelihood function which matches their measured and expected time delays.

The most striking demonstration of the quality of natural ice as a Cherenkov detector medium is the observation of atmospheric neutrino candidates with the partially deployed AMANDA detector which consisted of only eighty 8-inch photomultiplier tubes.[31] The up-going muons are separated from the down-going cosmic ray background once a sufficient number of direct photons and a minimum track length guarantee adequate reconstruction of the Cherenkov cone. For details, see Ref. 31. The analysis methods were verified by reconstructing cosmic ray muon tracks registered in coincidence with a surface air shower array.

After completion of the AMANDA detector with 300 OMs, a similar analysis led to a first calibration of the instrument using the atmospheric neutrino beam. The separation of signal and background are shown in Fig. 5 after requiring, sequentially, 5 direct photons, a minimum 100 m track length, and 6 direct photons per event. The details are somewhat more complicated; see Ref. 32. A neutrino event is shown in Fig. 6. By requiring the long muon track the events are gold-plated, but the threshold high, roughly $E_\nu \geq 50$ GeV. This type of analysis will allow AMANDA to harvest of order 100 atmospheric neutrinos per year, adequate for calibration.

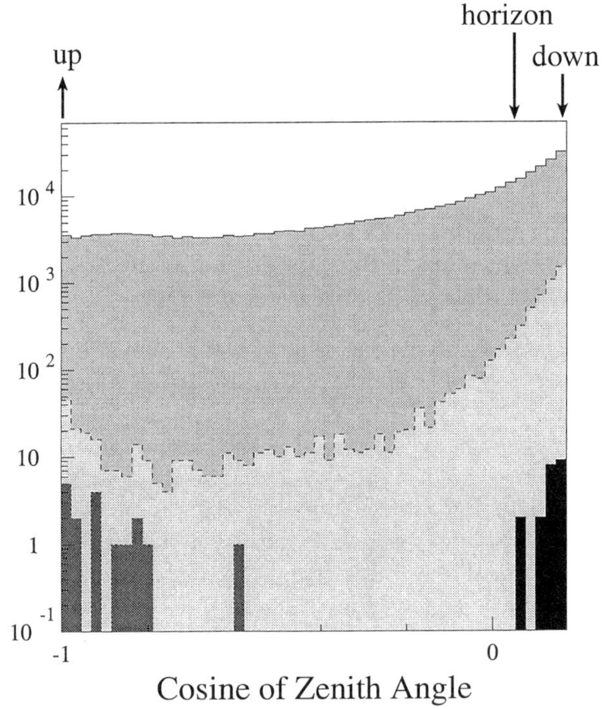

Figure 5: Angular distribution of muon tracks in AMANDA at three levels of quality cuts. Roughly, $N_{\text{direct}} \geq 5$, muon track longer than 100 m, and $N_{\text{direct}} \geq 6$. Details in Ref. 32.

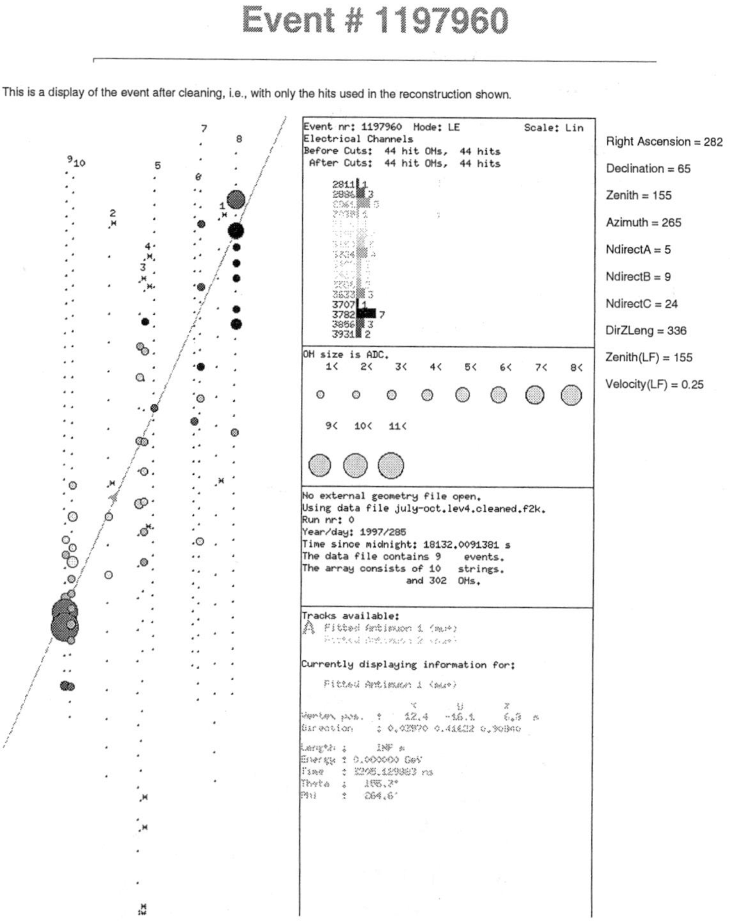

Figure 6: Neutrino candidate in AMANDA. The shading (size) of the dots represents time (amplitude) of triggered photomultipliers. The reconstructed muon track moves upward over more than 300 m.

While water detectors exploit large scattering length to achieve sub-degree angular resolution, ice detectors can more readily achieve large telescope area because of the long absorption length of blue light. Currently, the only proven technology to build a neutrino detector in a cost-effective way is to look for flashes of light associated with secondary electrons and muons in enormous amounts of ice. In early 1998 three 2,400 m deep strings, with optical modules spread over the lowest kilometer, were deployed. They form part of an intermediate detector, AMANDA II which will be completed in 99-00 with the addition of six more strings. Construction of IceCube will be staged over approximately five deployments (possibly 01-02 to 05-06; reducing this time period by improved drilling techniques is under study) of 16 strings per year for a total of ~5000 optical modules. A straw-man design for IceCube calls for strings with 60 PMTs in 80 holes spaced laterally by 100 m, or more.

Extrapolating from the performance of AMANDA, achieving degree resolution of long muon tracks in ICECUBE has been demonstrated by Monte Carlo. Doing nothing more sophisticated than adjusting the direct hit and track length cuts used in AMANDA, simulation establishes the strawman IceCube design as a kilometer-scale detector over most of the solid angle; see Fig. 7. It should be noted that the nominal resolution per muon track of 2.5° for AMANDA and 1° for IceCube is already smaller than the average neutrino-muon angle at the respective energy thresholds of these two detectors.

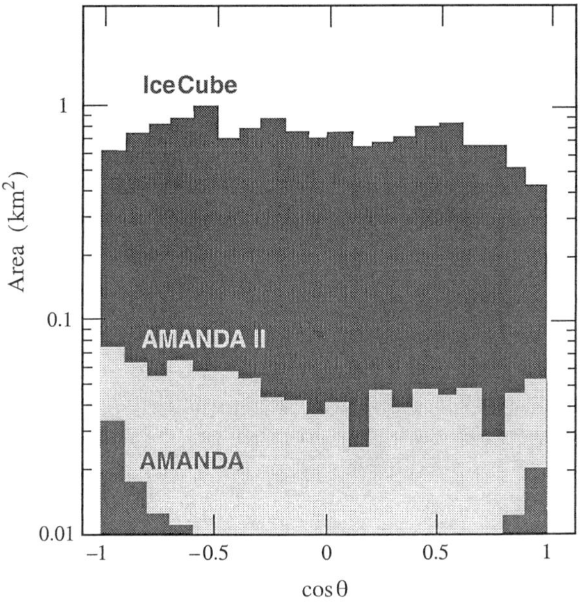

Figure 7: Effective area of a strawman IceCube detector described in the text as a function of zenith angle. We required six or more photons delayed by no more than 75 seconds and a muon track-length exceeding 150 m. We assumed the expected doubling of the collection area based on the use of larger PMTs with wavelength shifting coating.

IceCube will offer great advantages over AMANDA and AMANDA II beyond its larger size: it will have a much higher efficiency to reconstruct tracks, map showers from electron- and tau-neutrinos (events where both the production and decay of a τ produced by a ν_τ can be identified) and, most importantly, measure neutrino energy. Initial simulation of a strawman ICECUBE design indicate that the direction of showers can be reconstructed to better than $10°$ in both θ, ϕ above 10 TeV. Energy reconstruction is expected to be superior in ice, in part because of the scattering. Energy resolution is critical because, once one establishes that the energy exceeds 100 TeV, there is, in practice, no background in a kilometer-scale detector. Simulations predict a linear response in energy of about 25%. This has to be contrasted with the logarithmic energy resolution of first-generation detectors.

Acknowledgments

This work was supported in part by the University of Wisconsin Research Committee with funds granted by the Wisconsin Alumni Research Foundation, and in part by the U.S. Department of Energy under Grant No. DE-FG02-95ER40896.

1. For a review, see T.K. Gaisser, F. Halzen and T. Stanev, *Phys. Rep.* **258**(3), 173 (1995); R. Gandhi, C. Quigg, M. H. Reno and I. Sarcevic, *Astropart. Phys.*, **5**, 81 (1996).
2. F. Halzen, *The case for a kilometer-scale neutrino detector*, in Nuclear and Particle Astrophysics and Cosmology, Proceedings of Snowmass 94, R. Kolb and R. Peccei, eds.; *The Case for a Kilometer-Scale Neutrino Detector: 1996*, in Proc. of the Sixth International Symposium on Neutrino Telescopes, ed. by M. Baldo-Ceolin, Venice (1996).
3. R. Wischnewski for the AMANDA collaboration, Proceedings of the 26th International Cosmic Ray Conference, Salt Lake City, Utah, USA (1999).
4. E. Waxman and J. N. Bahcall, Phys. Rev. Lett. **78**, 2292 (1997); M. Vietri, Phys. Rev. Lett. **80**, 3690 (1998); M. Boettcher and C. D. Dermer, astro-ph/9801027.
5. For a recent review, see Piran, T. 1999, astro-ph/9810256, Phys. Rep., **314**, 575 (1999).
6. E. Waxman, Phys. Rev. Lett. **75**, 386 (1995); M. Milgrom, and V. Usov, Astrophy. J. **449**, L37 (1995); M. Vietri, Astrophys. J. **453**, 883 (1995).
7. F. Halzen, Lectures on Neutrino Astronomy: Theory and Experiment, Lectures given at Theoretical Advanced Study Institute in Elementary Particle Physics (TASI 98): Neutrinos in Physics and Astrophysics:

From 10^{-33} to 10^{+28} cm, Boulder, CO, 31 May - 26 June 1998, astro-ph/9810368 (1998).

8. Katz, J.I. 1994, ApJ, 432, L27, Halzen, F. & Jaczko, G. 1996, Phys. Rev. D, 54, 2774, Vietri, M. 1998, Phys. Rev. Lett. 80, 3690, Boettcher, M & Dermer, C. D. 1998, astro-ph/9801027, ApJ, 499, L131, E. Waxman and J. Bahcall, astro-ph/9909286.
9. F. Halzen and D. Hooper, astro-ph/9908138, Astrophy. J. Lett. (in press).
10. Stern, B. 1999, astro-ph/9902203, in "High Energy Processes in Accreting Black Holes", ed. J. Poutanen and R. Svensson (ASP Conf. Series), Vol. 161, 277.
11. R. A. Ong, Phys.Rep. **305**, 93, (1998); T.C. Weekes, High Energy Astrophysics, to appear in Proceedings of DPF'99, UCLA (1999).
12. E. Waxman and J. Bahcall, Phys. Rev. D **59**, 023002, (1999). For earlier references, see P. L. Biermann, & P. A. Strittmatter, Astrophys. J. **322**, 643 (1987); K. Mannheim and P.L. Biermann, *Astron. Astrophys.* **221**, 211 (1989); F. Stecker, C. Done, M. Salamon, and P. Sommers, Phys. Rev. Lett. **66**, 2697 (1991); erratum Phys. Rev. Lett. **69**, 2738 (1992); K. Mannheim, Astron. Astrophys. **269**, 67 (1993); M. Sikora, M.C. Begelman and M.J. Rees, *Ap. J. Lett.* **421**, 153 (1994); K. Mannheim, Astropar. Phys. **3**, 295 (1995); K. Mannheim, S. Westerhoff, H. Meyer and H.H. Fink, Astronomy and Astrophysics, in press (1996); R.J. Protheroe, Gamma Rays and Neutrinos from AGN Jets, Adelaide preprint ADP-AT-96-4 (astro-ph/9607165) (1996); F. Halzen and E. Zas, Astrophys. J. **488**, 669 (1997); J. P. Rachen and P. Mészáros, submitted to Phys. Rev. D (1998) (astro-ph/9802280); F.W. Stecker and M.H. Salamon, Space Sci. Rev. **75**, 341 (1996); G.C. Hill, Astropart. Phys. **6**, 215 (1997), D. Kazanas and A. Mastichiadis, Astrophy. J. Lett., **518**, 17 (1999).
13. J.H. Kim for the AMANDA collaboration, Proceedings of the 26th International Cosmic Ray Conference, Salt Lake City, Utah, USA (1999).
14. T. K. Gaisser, R. Protheroe and T. Stanev, astro-ph/9609044, Astroparticle Physics, in press, (1996).
15. W. Bednarek and R. J. Protheroe, Phys. Rev. Lett. **79**, 2616 (1997).
16. A. Castellina et al., Nuovo Cimento **8C**, 93 (1993); N. P. Il'ina et al., International Cosmic Ray Conference Rome, vol. 1, 524 (1995).
17. X. Shi, G. M. Fuller and F. Halzen, Phys.Rev.Lett. **81**, 5722 (1998).
18. B. Dingus, rapporteur talk at the 26th International Cosmic Ray Conference, Salt Lake City, Utah, USA (1999).
19. J. G. Learned and Sandip Pakvasa, *Astropart. Phys.* **3**, 267 (1995).
20. F. Halzen and D. Saltzberg, Phys. Rev. Lett. **81**, 4305 (1998).

21. T. Stanev, astro-ph/9907018
22. L. Bergstrom, J. Edsjoe and P. Gondolo, astro-ph/9906033
23. Peter Niessen for the AMANDA collaboration, Proceedings of the 26th International Cosmic Ray Conference, Salt Lake City, Utah, USA (1999).
24. G. Sigl, S. Lee, P. Bhattacharjee and D. N. Schramm, Science 270:1977-1980, (1995).
25. I. A. Belolaptikov et al., Astroparticle Physics **7**, 263 (1997).
26. L. Trascatti, in *Procs. of the 5th International Workshop on "Topics in Astroparticle and Underground Physics (TAUP 97)*, Gran Sasso, Italy, 1997, ed. by A. Bottino, A. di Credico, and P. Monacelli, Nucl. Phys. **B70** (Proc. Suppl.), p. 442 (1998).
27. F. Feinstein, in *Procs. of the 5th International Workshop on "Topics in Astroparticle and Underground Physics (TAUP 97)*, Gran Sasso, Italy, 1997, ed. by A. Bottino, A. di Credico, and P. Monacelli, Nucl. Phys. **B70** (Proc. Suppl.), p. 445 (1998).
28. M. Spiro, presentation to AASC Committee of the National Academy of Sciences, Atlanta (1999).
29. The AMANDA collaboration, *Science* **267**, 1147 (1995).
30. S. W. Barwick *et al*, *Proceedings of the 22nd International Cosmic Ray Conference*, Dublin (Dublin Institute for Advanced Studies, 1991), Vol. 4, p. 658.
31. The AMANDA collaboration, Astroparticle Physics, in press (1999).
32. A. Karle, for the AMANDA collaboration, Observation of Atmospheric Neutrinos with the AMANDA Experiment, to be published in *Proceedings of the 17th International Workshop on Weak Interactions and Neutrinos*, Cape Town, South Africa (1999) .

THE INTERACTIONS AND DETECTION OF ULTRA HIGH ENERGY NEUTRINOS: A BRIEF REVIEW

RAJ GANDHI

Mehta Research Institute, Jhusi, Allahabad, India

We review the phenomenology of UHE neutrino detection. Various motivations for looking for such neutrinos, their expected sources and their fluxes are given. Cross-sections with nucleons at energies all the way upto 10^{20} eV and their attenuation in the Earth prior to detection are discussed. Finally, sample event-rates in Water/Ice Cerenkov detectors are provided.

1 Introduction

It is now widely beleived that the recent Super Kamiokande (SK) result [1,2] of an anomaly in the flavour ratios and zenith angle dependance of the atmospheric neutrino flux provides the firmest signal yet of physics beyond the Standard Model (SM). The significance of this can be gauged by the fact that a signal for such physics has been sought for in all the varied and intensive experimental probes that the many sectors of the SM have been subjected to over more than twenty-five years. That this signal comes from the neutrino sector of the theory perhaps assumes increased significance when considered in conjuntion with the fact that two other existing experimental anomalies which persist, the solar deficit [3] and the LSND result [4] are also within the neutrino sector. In particular, the first of these anomalies (the solar neutrino deficit) has its origins in the Chlorine experiment, the first results of which were presented more than thirty years ago. Its existence now stands confirmed by 4 other recent experiments, namely SAGE, GALLEX, Kamiokande and SK [3].

The detailed interpretation of these anomalies and their relation with each other is, however, still a matter of intense and ongoing theoretical and experimental activity, although neutrino masses, mixings and consequent oscillations have provided the simplest framework for understanding the experimental results. What may be said with a reasonable degree of conviction is that once their interpretation is clear, a hitherto unprecedented window into the nature of the physical theory beyond the SM will probably have been opened.

It is within this context, perhaps, that one should view the theoretical and experimental efforts that are ongoing in the area of ultra high energy (UHE) neutrino physics [5,6]. In view of the fact that data collection and further upgradation has just begun for the first generation of UHE neutrino detectors (AMANDA [8] and BAIKAL [9]) and that planning, design, testing and deployment is underway for some others (AUGER [10], NESTOR [11], ICECUBE [12], ANTARES [13], RICE [14]), and NEMO [15]) definitive results of these efforts are still in the future. However, besides the other specific reasons from astrophysics and cosmic ray physics (some of which we discuss below) which provide motivation for such experiments, we note that the energy range covered by these experiments ($E_\nu \simeq 1$ Tev to 10^9 TeV or higher) offers an unprecedented opportunity for particle physics at energies significantly beyond the scope of terrestial accelerators. Although progress is expected to be slow, the potential for serendipitous discovery is undoubtedly high.

2 Why should we search for and expect to detect UHE neutrinos?

In addition to the general particle physics motivations mentioned above, there exist other specific reasons spanning several fields of research (astrophysics, astronomy and cosmic-ray physics) for pursuing the search for UHE neutrinos.

Perhaps the strongest reason comes from the observed cosmic-ray (CR) spectrum. We briefly discuss its observational features prior to making the connection to UHE neutrinos.

Over a period of several decades, a sizeable and impressive body of observations spanning twelve orders of magnitude in energy have been compiled by workers in this field. (For general discussions see [16,17], and for updates on observational efforts and data see [18,19,20].) Fig. 1 (from [30]) shows a compilation of the all-particle CR spectrum spanning the full observational energy range. A power-law behaviour is apparent over almost the entire spectral range, with breaks at what is referred to as the "knee" (corresponding to $E_{primary} \simeq 4 \times 10^{15}$ eV) and the "ankle"(corresponding to $E_{primary} \simeq 10^{19}$ eV). Upto the knee region, the spectrum exhibits a power-law with index -2.7, there onwards steepning to about -3, and flattening again in the ankle region. The steep fall in the flux and the consequent increasing difficulty in detection is reflected in the fact that the number of primaries falls from about one particle per m^2 per sec at energies of about 10^{11} eV to roughly one per km^2 per century around the ankle (10^{19} eV). It is thus necessary to employ very different techniques of detection depending on the energy of the primary. The direct observation of a CR primary is only possible in a detector mounted on a spacecraft or a balloon, due to interactions in the atmosphere. Usually, a combination of mounted nuclear emulsion stacks, spectrometers and calorimeters are used for such detection, examples are the JACEE [21] and EGRET [22] experiments. Such detection, however can collect enough statistics only upto primary energies of about 10^{14} eV. At higher energies, indirect methods involving detection of secondaries produced by interaction of the primaries in the atmosphere is necessary. The most widely used indirect method has been the deployment of Extensive Air Shower (EAS) arrays (for example Yakutsk [24] and AGASA [25]). These sample a lateral cross-section of the multi-particle shower initiated by a CR primary high up in the atmosphere. The arrays typically use a combination of scintillators, water Cerenkov tanks and shielded muon detectors spread over a large ground area to maximize coverage. Clearly, the determination of the energy of the primary from the charged hadrons, electrons, muons and photons detected by the ground-based array is a non-trivial task. However, present Monte Carlo techniques allow this to be accomplished with an accuracy of about 30%.

Still less trivial is the determination of the nature of the primary, which is complicated by fluctuations from shower to shower. In recent years another technique, which focusses on the detection of secondary nitrogen fluorescence radiation excited by shower secondaries has been very succesfully implemented by Fly's Eye [26]. In this method the emitted radiation is isotropic, (unlike, for instance Cerenkov emission) and information about the nature of the primary is conveyed by the atmospheric depth at which the shower attains maximum particle production. The energy of the primary is measured by the total fluoresence, and can be much more reliably determined than in an EAS. However, in contrast to the EAS, the duty cycle is only about 10%, since viewing is restricted to dark and moonless nights.

Using the modes of detection discussed above, a general but incomplete picture of the nature of CR primaries has emerged. Essentially, below primary energies of 10^{14} eV, the composition is fairly accurately known, from a variety of direct detection experiments, to be 98% hadrons, mainly protons. Above these energies, where only indirect detection techniques can be employed, the composition appears to retain a significant hadronic fraction, even though the question as to whether the primaries are protons or heavier nuclei is yet unresolved. For instance, around the knee region, results from DICE [27], a secondary Cerenkov radiation detector, point towards a lighter composition, while evidence from two EAS, KASCADE [28] and HEGRA [29] appear to favour an increasingly heavier one. None of the evidence is statistically firm enough, though, for a conclusive resolution of this question. Similarly, very low statistics hampers conclusions about the composition of CR beyond the knee upto the highest energies (at and above the GKZ cutoff, which is discussed below). However, we stress that all or most of the by now extensive EAS events appear to have a muon content consistent with hadronic primaries.

For our purpose it is sufficient to say that it appears almost certain from the CR observations that there are astophysical sites in the universe which acclerate hadrons upto Eev energies, and there is a reasonable chance that they are accelerated to higher energies, perhaps all the way upto $\approx 10^{20}$ eV. The mechanisms by which this happens are not completely understood, although we discuss some of them below. Hadronic collisions, like $p + p$ or $p + \gamma$, always result in copious pion production. These follow the decay chain $\pi \rightarrow \mu + \nu_\mu, \mu \rightarrow e + \nu_\mu + \nu_e$ resulting in a flux ratio of 2:1 for $\nu_\mu : \nu_e$. The neutrino is thus expected to retain about 20% of the energy of the parent pion on the average, but in general, UHE neutrino production is guaranteed if hadronic collisions play a role in UHE CR production, although the spectrum depends on the nature of the source. Before we discuss specific sources, it is important to emphasize that there exist mechanisms which do not involve the acceleration of matter but can still yeild UHE fluxes of protons, photons and neutrinos. These are the "top-down" (TD) scenarios (for a comprehensive review of these see [30]) in which CR beyond the knee could originate in the decay of GUT scale massive particles produced due to topological defects like monopoles, cosmic strings, etc.

In addition to neutrino production at the source, a UHE CR proton with energy in excess of about 5×10^{19} eV traversing interstellar space will interact with the Cosmic Microwave Background (CMB) to photo-produce pions, and hence neutrinos. The interaction in question, the Δ resonance for single and multiple pion production, sets in at a center-of-mass energy of about 1.5 GeV and is responsible for an expected abrupt fall-off in the proton primary spectrum at and above these energies. This was first pointed out by Greisen, Kuzmin and Zatsepin [34,35] and is called the GKZ cutoff. It is at present associated with one of the most significant puzzles in CR physics, because CR events with energies in excess of 10^{20} eV have been convincingly detected by several experiments employing different detection techniques (Figure 2). Specifically, these are the Volcano Ranch Experiment [31] (1 event), the Haverah Park Array [33] (8 events), the Yakutsk Array [24] (1 event), SUGAR [32] (8 events), Fly's Eye [26] (1 event) and AGASA [25] (6 events). These events are , despite uncertainties of about $10 - 30\%$ in energy determination, convincingly above the GKZ cutoff. We however, do not discuss this puzzle further in this article or the apparent absence of a GKZ cutoff signalled by the highest energy CR while emphasizing that the existence of the cutoff is demanded by well-tested low-energy SM physics.

A diffuse UHE neutrino flux is also expected from interactions of galactic CR with interstellar matter, at energies around the knee region.

Yet another potentially important source of UHE neutrinos could be Gamma-Ray Bursts (GRB). There appears to be observational support for a "fireball" model for GRBs [49]. The physics of this involves an initial merger or collapse of blackholes, neutron stars or some other highly magnetized compact object. The collapse to a small radius object is followed by a very rapid expansion, by about a factor of 10^5 in a time frame of the order of a second. This ultra-relativistic acceleration of the plasma of protons, electrons, positrons and photons leads, at some radius at which the plasma is optically thin to radiation which is detected as the GRB. At the same time, second-order Fermi shocks cause charged particle acceleration, and the ensuing $p - \gamma$ interactions lead to neutrino fluxes, which have been estimated in [50,51] for instance (See figure 2).

Active Galactic Nuclei (AGNs) are also expected to be an important source of UHE neutrinos. These are a class of highly compact bright objects powered presumably by black holes causing acceleration and accretion of matter, characterised usually by high powered jet emmission. High energy gamma-rays (MeV, GeV and TeV) are expected and have been observed from ≈ 40 AGN sites. The acceleration of matter again leads to subsequent $p - \gamma$ (and perhaps $p - p$) interactions, from which neutrinos are expected. (Figure 2).

Finally, as mentioned before, in top-down scenarios, GUT scale particles decay to give fluxes of photons, protons and neutrinos at UHE [30].

The detection of UHE neutrinos from the above sources is expected to help answer many important questions in CR physics and the astrophysics of highly energetic sources. In particular, their detection above $10^{16} - 10^{18}$ eV would provide important support for hadronic blazar (versus electron blazar) models. If GRB are indeed the source of the highest energy CR then a co-related UHE flux is expected, and its non-observation would set important limits on their contribution to these events. Also, the observation of an isotropic neutrino flux beyond GKZ energies would signal that these events are due to universal CR activity rather than a nearby single source and would shed light on the nature of the primaries responsible for these events.

3 UHE Neutrino Fluxes

In Figure 2 we show some of the flux calculations for UHE neutrinos from AGNs, GRBs and TD sources. We refer the reader to the references for details on the models used to obtain the predicted fluxes. The fluxes are labeled as AGN-M95 [52], AGN-SS91 [53], AGN-P96 [54] for Active Galactic Nuclei, GRB-WB [55] for Gamma Ray Bursts and TD-WMB12 [56], TD-WMB16 [56], TD-SLC [57] for top-down models. References and spectra for the remaining fluxes mentioned above may be found in [58].

It is important to stress here that fluxes for all the three sources in Figure 2 are still being modified as our understanding of the nature of the source, the interactions involved and the constraints placed by observations on them improve. For instance, it has been pointed out [59] that observations of the CR spectrum place important constraints on the neutrino flux from sources which are optically thin with respect to neutrons and photons. This argument could significantly restrict several AGN models, but the exact upper bound is presently being debated. In [60], for instance it is argued that this bound depends on the assumption that the overall CR injection spectrum $dN/dE \propto E^{-2}$ upto energies of 10^{19} eV, and that injection spectra for many sources of CR producing shocks may be different, and, for a given source, may not extend all the way upto 10^{19} eV. The importance of using CR and other observations to constrain the physics occuring in the sources, and the neutrino fluxes emmitted as emphasized in [59] cannot, however, be disputed.

4 The Detection of UHE Neutrinos: General Considerations

Having discussed potential sources for UHE neutrinos, we next review some salient points relevant to their detection. The main mode of detection discussed in the literature and implemented at extant detectors has been the observation of long range muons produced via charged current (CC) neutrino-nucleon interactions. The low fluxes necessitate the deployment of large volume water or ice detectors like AMANDA [8], BAIKAL [9], NESTOR [11] and ANTARES [13]. These are shielded from above by several kilometers of water-equivalent (kmwe) rock volume in order to supress the otherwise overwhelming background from muons produced by CR interactions in the atmosphere. Indeed, for a considerable portion of the UHE range (100 TeV and lower), the detection of downward moving muons by Cerenkov radiation in water produced by contained events is less advantageous than that of upward moving ones produced in the rock below the detector. Existing detectors have thus focussed on looking mainly for muons which, after losses due to passage in the earth, still retain sufficient energy to be observed above threshold. In general, a 10 TeV muon will travel several km in rock before its energy is degraded down to 1 TeV, which is a typical detector threshold. [1] The effective detector

[1] A charged particle moving through rock suffers two types of energy losses, continuous and discrete. Continous losses are mainly due to ionization of the medium, and their rate depends weakly (logarithmically) on the energy. Discrete losses stem from bremstrahllung, electromagnetic interactions with nuclei and direct e^+e^- production. In general, such losses are proportional to the energy, but for muons gain importance only at higher energies. The most important loss mechanism, bremstrahllung, is proportional to the inverse square of the

volume for such detection at present capabilities is of the order of 0.05 km^3. Whereas serious neutrino astronomy results are not expected with present detector volumes, the actively underway upgradation of existing facilities and the addition of new ones should yeild significant physics results in the next few years. The present AMANDA detector has seen about 20 UHE ν_μ events, and by Spring 2000 its upgraded version, AMANDA II will be fully deployed. A km^3 extension, ICECUBE, is also planned. BAIKAL has also observed UHE neutrinos and set limits on the UHE fluxes using its prelimnary observation, with analysis of another 3 years of data currently underway. The ANTARES project, off the French coast, is exploring the design and implementation of a km^3 scale deep sea detector for underground muons. Similarly, NESTOR is doing the same southwest of the coast of Greece, while NEMO [15] is yet another planned deep sea detector off the southern Italian coast. The RICE project in the Antarctic [14] will be a pioneering attempt to detect ν_e via radio waves. An UHE electron produced subsequent to a CC interaction in the rock transfers most of its energy to an electromagnetic shower. Positrons produced in the shower annhilate and additional atomic electrons scatter into the shower, causing a charge imbalance which corresponds to a ball of negative charge moving through the rock. The consequent Cerenkov emission gives radio waves ($\lambda \simeq 10$ cm), which are detected by receivers buried in the ice. Finally, the Pierre Auger Observatory [10], primarily designed to detect CR showers, should also be able to detect UHE neutrino induced showers close to the horizon. Rate calculations for this array are given in [7].

An important consideration for calculating event rates for muon neutrinos is the attenuation of neutrinos in the earth due to the rapid rise in the CC cross-section (discussed below) with energy. The interaction length of a neutrino in rock is approximately equal to the diameter of the earth at 40 TeV. A convienient quantity relevant to flux attenuation and event rate calculations is the 'shadow factor', S, which is defined to be an effective solid angle divided by 2π for upward muons and is a function of the energy-dependant cross-section for neutrinos in the earth [5]:

$$S(E_\nu) = \frac{1}{2\pi}\int_{-1}^{0} d\cos\theta \int d\phi \exp\left[-z(\theta)/\mathcal{L}_{int}(E_\nu)\right]. \tag{1}$$

Here $\mathcal{L}$ is the interaction length for the neutrino, defined by

$$\mathcal{L}_{\rm int} = \frac{1}{\sigma_{\nu N}(E_\nu) N_{\rm A}}, \tag{2}$$

---

mass of the charged particle, and hence is highly suppressed for muons relative to electrons.

where $N_A = 6.022 \times 10^{23}$ mol^{-1} $= 6.022 \times 10^{23}$ cm^{-3} (water equivalent) is Avogadro's number. z represents the column-depth as a function of the nadir angle θ. Figure 4 (from Ref [5]) shows that from almost no attenuation at 10^{12} eV, about 93% of the flux is shadowed out at the highest energies at which CR have been observed ($\approx 10^{21}$ eV). It is this effect that makes the (low) downward rate for muons produced within the instrumented volume competetive with the upward rate at energies above 10^{15} eV, where the atmospheric background is essentially absent.

5 The Neutrino-Nucleon Deep Inelastic Scattering Cross-Section at UHE

Of crucial importance to the attenuation and the event-rate calculations is the UHE neutrino-nucleon DIS cross-section. This is given by

$$\nu_\mu N \to \mu^- + \text{anything}, \tag{3}$$

where $N \equiv \dfrac{n+p}{2}$ is an isoscalar nucleon, in the renormalization group-improved parton model. The differential cross section is written in terms of the Bjorken scaling variables $x = Q^2/2M\nu$ and $y = \nu/E_\nu$ as

$$\frac{d^2\sigma}{dxdy} = \frac{2G_F^2 M E_\nu}{\pi}\left(\frac{M_W^2}{Q^2 + M_W^2}\right)^2 \left[xq(x,Q^2) + x\overline{q}(x,Q^2)(1-y)^2\right], \tag{4}$$

where $-Q^2$ is the invariant momentum transfer between the incident neutrino and outgoing muon, $\nu = E_\nu - E_\mu$ is the energy loss in the lab (target) frame, M and M_W are the nucleon and intermediate-boson masses, and $G_F = 1.16632 \times 10^{-5}$ GeV^{-2} is the Fermi constant. The quark distribution functions are

$$q(x,Q^2) = \frac{u_v(x,Q^2) + d_v(x,Q^2)}{2} + \frac{u_s(x,Q^2) + d_s(x,Q^2)}{2} + s_s(x,Q^2) + b_s(x,Q^2) \tag{5}$$

$$\overline{q}(x,Q^2) = \frac{u_s(x,Q^2) + d_s(x,Q^2)}{2} + c_s(x,Q^2) + t_s(x,Q^2),$$

where the subscripts v and s label valence and sea contributions, and u, d, c, s, t, b denote the distributions for various quark flavors in a *proton*. At the energies of interest for neutrino astronomy, perturbative QCD corrections to the cross section formula (4) are insignificant, and can be safely omitted.

A parallel calculation leads to the neutral-current cross section. In this case the differential cross section for the reaction $\nu_\mu N \to \nu +$anything is given by

$$\frac{d^2\sigma}{dxdy} = \frac{G_F^2 M E_\nu}{2\pi}\left(\frac{M_Z^2}{Q^2+M_Z^2}\right)^2 \left[xq^0(x,Q^2) + x\overline{q}^0(x,Q^2)(1-y)^2\right], \quad (6)$$

where M_Z is the mass of the neutral intermediate boson. The quantities involving parton distribution functions are

$$\begin{aligned} q^0(x,Q^2) &= \left[\frac{u_v(x,Q^2)+d_v(x,Q^2)}{2} + \frac{u_s(x,Q^2)+d_s(x,Q^2)}{2}\right](L_u^2+L_d^2) \\ &+ \left[\frac{u_s(x,Q^2)+d_s(x,Q^2)}{2}\right](R_u^2+R_d^2) + \\ &[s_s(x,Q^2)+b_s(x,Q^2)](L_d^2+R_d^2) + [c_s(x,Q^2)+t_s(x,Q^2)](L_u^2+R_u^2) \end{aligned} \quad (7)$$

$$\begin{aligned} \overline{q}^0(x,Q^2) &= \left[\frac{u_v(x,Q^2)+d_v(x,Q^2)}{2} + \frac{u_s(x,Q^2)+d_s(x,Q^2)}{2}\right](R_u^2+R_d^2) \\ &+ \left[\frac{u_s(x,Q^2)+d_s(x,Q^2)}{2}\right](L_u^2+L_d^2) + \\ &[s_s(x,Q^2)+b_s(x,Q^2)](L_d^2+R_d^2) + [c_s(x,Q^2)+t_s(x,Q^2)](L_u^2+R_u^2), \end{aligned} \quad (8)$$

where the chiral couplings are

$$\begin{aligned} L_u &= 1 - \tfrac{4}{3}x_W \qquad & L_d &= -1 + \tfrac{2}{3}x_W \\ R_u &= -\tfrac{4}{3}x_W \qquad & R_d &= \tfrac{2}{3}x_W \end{aligned} \quad (9)$$

and $x_W = \sin^2\theta_W$ is the weak mixing parameter. Again the top-quark sea is negligible.

Recent results from the *ep* collider HERA [36,37,38,39,40] have greatly enhanced our knowledge of parton distributions and importantly, are particularly relevant to the application at hand. Measurements by the ZEUS and H1 collaborations mark the first experimental studies of very small parton momentum fractions x at momentum transfers Q^2 securely in the deeply inelastic regime. The HERA experiments mapped the structure function $F_2(x,Q^2)$ in the interval $10^{-4} \lesssim x \lesssim 10^{-2}$, with $8.5\ \mathrm{GeV}^2 \lesssim Q^2 \lesssim 15\ \mathrm{GeV}^2$. For $x \gtrsim 2\times10^{-2}$, F_2 has been measured over a significant range in Q^2.

For most hard-scattering applications in particle physics, it is straightforward to begin with parametrizations of parton distribution functions tied to data at modest values of Q^2 and evolve them to the desired high scale using the Altarelli-Parisi equations. For UHE neutrino-nucleon scattering, however, the W-boson propagator emphasizes smaller and smaller values of x as the neutrino energy E_ν increases. In the UHE domain, the most important contributions to the νN cross section come from $x \sim M_W^2/2ME_\nu$. Up to $E_\nu \approx 10^5$ GeV, the parton distributions are sampled only at values of x where they have been constrained by experiment. At still higher energies, we require parton distributions at such small values of x that direct experimental constraints are not available, not even at low values of Q^2.

The theoretical uncertainties that enter the evaluation of the UHE neutrino-nucleon cross section arise from the low-Q^2 parametrization, the evolution of the parton distribution functions to large values of $Q^2 \sim M_W^2$, and the extrapolation to small values of x. The greatest uncertainty is due to the small-x extrapolation.

Perturbative QCD gives us two approaches to understanding the Q^2-evolution of small-x parton distributions. The traditional approach, followed in the CTEQ [41] distributions, is to determine parton densities for $Q^2 > Q_0^2$ by solving the next-to-leading-order Altarelli-Parisi [42] evolution equations numerically. The second approach to small-x evolution attempts to solve the Balitskiĭ-Fadin-Kuraev-Lipatov (BFKL) equation [43], which is a leading $\alpha_s \ln(1/x)$ resummation of soft gluon emissions. The BFKL approach predicts a singular behavior in x and a rapid Q^2-variation,

$$xq_s(x, Q^2) \sim \sqrt{Q^2}\, x^{-0.5}. \tag{10}$$

Applying the Altarelli-Parisi equations to singular input distributions $\propto x^{-\frac{1}{2}}$ leads to

$$xq_s(x, Q^2) \sim \ln{(Q^2)} x^{-0.5}, \tag{11}$$

a less rapid growth with Q^2.

The Altarelli-Parisi (AP) approach is applicable in the not-so-small-x and large-Q^2 region, while the BFKL solution applies to the small-x and moderate-Q^2 region. BFKL evolution eventually breaks down at large Q^2, because of the rapid growth exhibited in (10). In the case of ultrahigh-energy neutrino-nucleon interactions, the region of interest is small-x and large-Q^2, which requires a resummation of both $\ln 1/x$ and $\ln Q^2/Q_0^2$ contributions. An attempt to incorporate elements of the BFKL approach into the standard AP approach has recently been made in [44]. That this leads to cross-sections which are comparable (to within 10-15 % at the highest energies) with those obtained by next-to-leading order AP evolution is an indication that barring radically new physics, the cross-section calculation even at the highest energies is under control. Modifications in the cross-sections can occur, of course, as a result of physics beyond the SM. Many of these modifications are strongly constrained by unitarity and for a broad class of plausible extensions are much smaller than the SM cross-section, as shown in [45]. Various possibilities leading to higher neutrino-nucleon cross-sections have been discussed in the literature and some of the mechanisms considered include leptoquark excitations [46], superpartner contributions [47], strong FCNC interactions [48] and finally, higher dimensional gravitational contributions at TeV scales [64,65,66].

In Figure 3 (from [7]) we show the CC, NC and total cross-sections for $\nu - N$ interactions resulting from the SM using the AP approach to small x extrapolation. For 10^{16} eV $\leq E_\nu \leq 10^{21}$ eV, the cross sections are given within 10% by the following convienient analytic parametrisations [7],

$$
\begin{aligned}
\sigma_{\rm CC}(\nu N) &= 5.53 \times 10^{-36}\ {\rm cm}^2 \left(\frac{E_\nu}{1\ {\rm GeV}}\right)^{0.363} \\
\sigma_{\rm NC}(\nu N) &= 2.31 \times 10^{-36}\ {\rm cm}^2 \left(\frac{E_\nu}{1\ {\rm GeV}}\right)^{0.363} \\
\sigma_{\rm tot}(\nu N) &= 7.84 \times 10^{-36}\ {\rm cm}^2 \left(\frac{E_\nu}{1\ {\rm GeV}}\right)^{0.363} \\
\sigma_{\rm CC}(\bar{\nu} N) &= 5.52 \times 10^{-36}\ {\rm cm}^2 \left(\frac{E_\nu}{1\ {\rm GeV}}\right)^{0.363} \\
\sigma_{\rm NC}(\bar{\nu} N) &= 2.29 \times 10^{-36}\ {\rm cm}^2 \left(\frac{E_\nu}{1\ {\rm GeV}}\right)^{0.363} \\
\sigma_{\rm tot}(\bar{\nu} N) &= 7.80 \times 10^{-36}\ {\rm cm}^2 \left(\frac{E_\nu}{1\ {\rm GeV}}\right)^{0.363} .
\end{aligned}
\tag{12}
$$

6 Detection Rates

The event-rate for an underground detector of area A (events/sr/year) is calculable via

$$
Rate = A \int_{E_\mu^{\rm min}}^{E^{\rm max}} dE_\nu\, P_\mu(E_\nu; E_\mu^{\rm min}) S(E_\nu) \frac{dN}{dE_\nu} . \tag{13}
$$

The integrals in (13) are evaluated up to $E^{max} = 10^{11}$ GeV except for AGN-SS91, for which the data files extend only to $E^{max} = 10^{9.8}$ GeV. Here

$P_\mu(E_\nu, E_\mu^{\rm min}$ is the probability that a muon produced in a charged-current interaction arrives in a detector with an energy above the muon energy threshold E_μ^{min} and is given by

$$P_\mu(E_\nu, E_\mu^{\rm min}) = N_{\rm A}\,\sigma_{\rm CC}(E_\nu)\langle R(E_\nu; E_\mu^{\rm min})\rangle, \tag{14}$$

where $\langle R(E_\nu; E_\mu^{\rm min})\rangle$ is the average range of a muon in rock.

Using the formulae above, event rates for a variety of sitiuations can be calculated. We give a set of sample rates below, from Ref [7].

Let us consider for illustration a detector with effective area $A = 0.1$ km^2. This choice of size is intermediate to existing detectors and the planned future facilities. We show in Tables 1 and 2 the annual event rates for upward-going muons with observed energies exceeding 1 TeV and 10 TeV, respectively. We tabulate rates for the full upward-going solid angle of 2π, as well as for the detection of "nearly horizontal" muons with nadir angle θ between 60° and 90°. The predicted event rates, shown here for the CTEQ4–DIS parton distributions, are very similar for other modern parton distributions.

We note that the atmospheric background overwhelms the signal when the threshold is 1 TeV. However, signals should emerge above background for 10 TeV and 100 TeV thresholds. Above 100 TeV, we have the rare case of "all signal and no background", as the atmospheric muon flux disappears. Also evident in Table 2 is the effect of shadowing, with a majority of signal events in the "nearly horizontal" direction, where the neutrino traverses less rock prior to detection.

In addition to the (upward) partially contained events and the (downward) fully contained ones, it may be possible to detect neutrino interactions at extremely high energies ($\geq 10^{17}$ eV) in the proposed Pierre Auger Cosmic Ray Observatory [10], which will consist of both a ground array with detectors distributed over a very large area ($\approx 3000 km^2$) and nitrogen fluorescence detectors. At these energies, the probability of a horizontally incoming neutrino interacting with the atmosphere is non-negligible. The events are thus rendered distinguishable from CR showers initiated by proton or other primaries by their incoming direction and their tendency to shower late into the atmosphere. Since the Auger observatory is expected to observe 50-100 CR events even above 10^{20} eV, one can hope that UHE neutrino detection above its threshold of 10^{17} eV is not a far-fetched possibility. The predictions for these are given in [7].

7 Conclusions

We have made an attempt to review the essential motivations and phenomenological issues related to the interactions and detection of UHE neutrinos and provided expected event rates for water/ice Cerenkov detectors.

Although present detector sizes do not allow them to see above backgrounds, with upgradations and several new large scale experiments underway, the detection of UHE neutrinos from astrophysical sources may be very likely in the near future. Besides the potential for serendipitous discovery, the unmatched (terrestially) energies may lead to new particle physics discoveries. In addition, these experiments may help clarify the astrophysical mechanism responsible for the ultra-relativistic acceleration of matter in these sources and answer important questions in CR physics. For those working in UHE neutrino physics and astronomy, the anticipation is palpable.

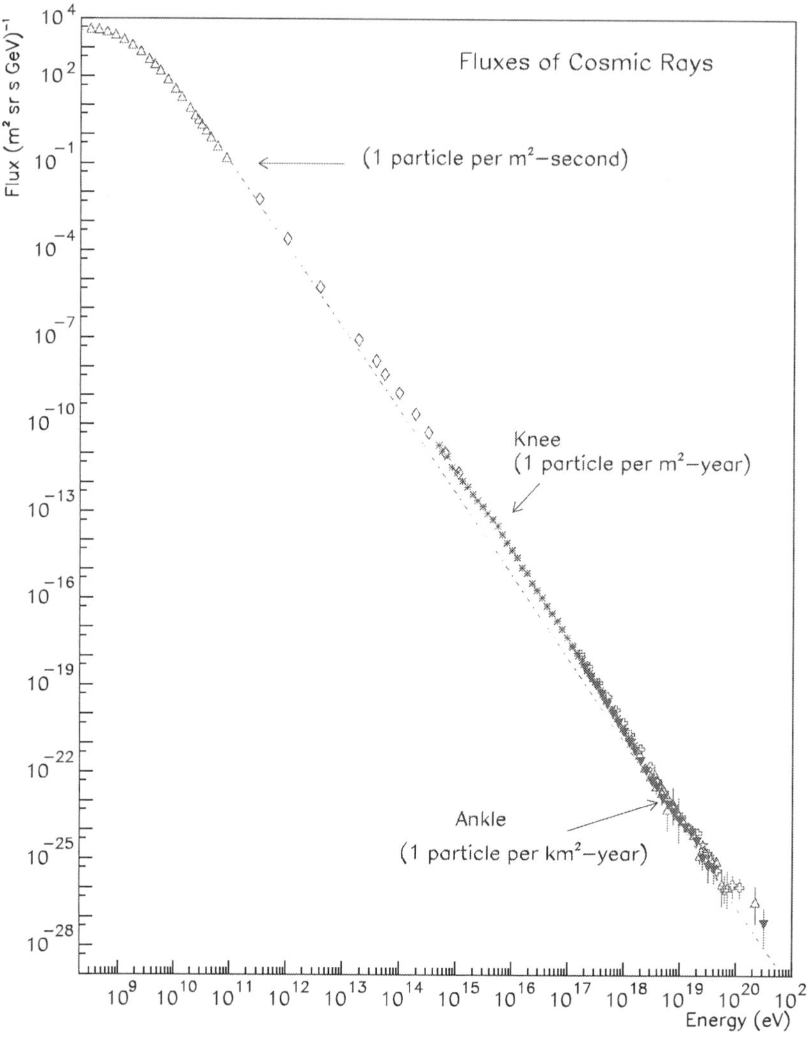

Figure 1. The CR all particle spectrum. Approximate integral fluxes are also shown.

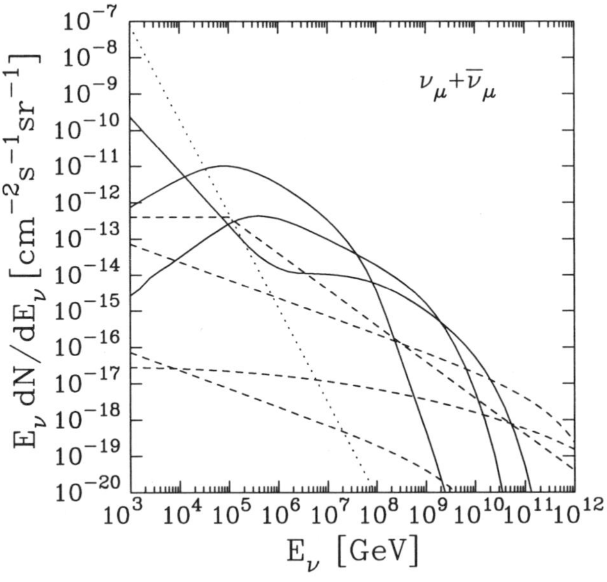

Figure 2. Muon neutrino plus antineutrino fluxes scaled by neutrino energy at the Earth's surface. Solid lines represent AGN fluxes. In decreasing magnitude at $E_\nu = 10^3$ GeV, they are AGN-M95, AGN-SS91 scaled by 0.3, and AGN-P96 ($p\gamma$). The dashed lines, in the same order, represent the GRB-WB, TD-WMB12, TD-WMB16, and TD-SLSC fluxes. The dotted line is the angle-averaged atmospheric (ATM) neutrino flux.

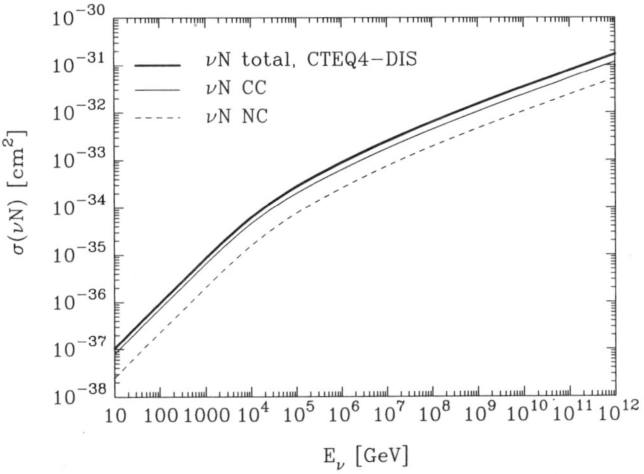

Figure 3. Cross sections for $\nu_\ell N$ interactions at high energies, according to the CTEQ4–DIS parton distributions: dashed line, $\sigma(\nu_\ell N \to \nu_\ell + \text{anything})$; thin line, $\sigma(\nu_\ell N \to \ell^- + \text{anything})$; thick line, total (charged-current plus neutral-current) cross section.

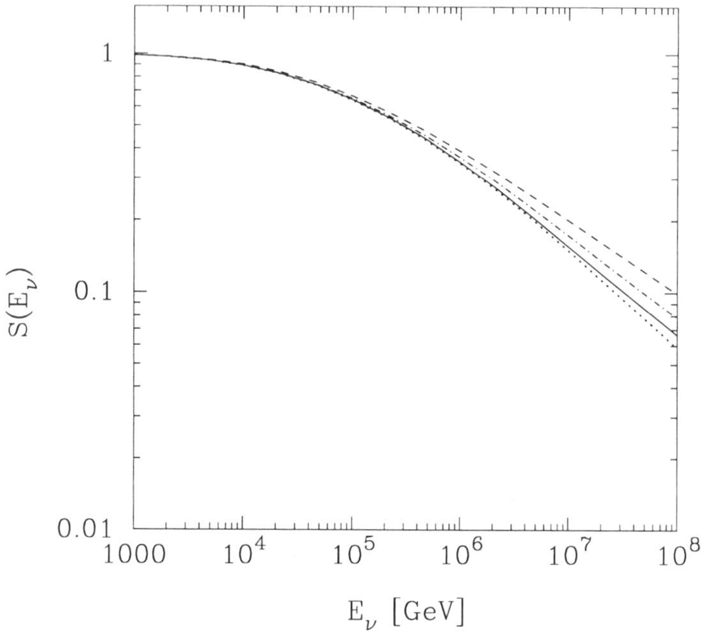

Figure 4. The shadow factor $S(E_\nu)$ for upward-going neutrinos assuming that $\sigma = \sigma_{\rm tot}$ in (1) for CTEQ-DIS (solid line), CTEQ-DLA (dot-dashed) and D_- (dotted) parton distribution functions. Also shown is the shadow factor using the EHLQ cross sections (dashed line).

Table 1. Upward $\mu^+ + \mu^-$ event rates per year arising from $\nu_\mu N$ and $\bar{\nu}_\mu N$ interactions in rock, for a detector with effective area $A = 0.1$ km^2 and muon energy threshol d $E_\mu^{\rm min} =$ 1 TeV. The rates are shown integrated over all angles below the horizon and restricted to "nearly horizontal" nadir angles $60° < \theta < 90°$.

	nadir angular acceptance	
Flux	$0° < \theta < 90°$	$60° < \theta < 90°$
ATM [61]	1100.	570.
ATM [61] + charm [63]	1100.	570.
AGN-SS91 [53]	500.	380.
AGN-M95 $(p\gamma)$ [52]	31.	18.
AGN-P96 $(p\gamma)$ [54]	45.	39.
GRB-WB [50]	12.	8.1
TD-SLSC [62]	0.005	0.0046
TD-WMB12 [56]	0.50	0.39
TD-WMB16 [56]	0.00050	0.00039

Table 2. Upward $\mu^+ + \mu^-$ event rates per year arising from $\nu_\mu N$ and $\bar{\nu}_\mu N$ interactions in rock, for a detector with effective area $A = 0.1$ km^2 and muon energy threshold $E_\mu^{\rm min} =$ 10 TeV. The rates are shown integrated over all angles below the horizon and restricted to "nearly horizontal" nadir angles $60° < \theta < 90°$.

	nadir angular acceptance	
Flux	$0° < \theta < 90°$	$60° < \theta < 90°$
ATM [61]	17.	10.
ATM [61] + charm [63]	19.	11.
AGN-SS91 [53]	270.	210.
AGN-M95 $(p\gamma)$ [52]	5.7	4.3
AGN-P96 $(p\gamma)$ [54]	28.	25.
GRB-WB [50]	5.4	4.0

References

1. Y. Fukuda *et. al*, Phys. Lett. **B 433**, 9 (1998); Phys. Rev. Lett. **81**, 1562 (1998).
2. T. Kajita, hep-ex/9810001, to appear in the proceedings of XVIII Int. Conf. on Neutrino Physics and Astrophysics (Neutrino '98) , Takayama, Japan, 4-9 June 1998.
3. For a discussion and references see, for instance, J. Conrad, "Recent Results on Neutrino Oscillations", hep-ex 9811009, 5 Nov. 1998.
4. LSND Collaboration; C. Athanassopoulos et. al; Phys. Rev. Lett. 81, 1774 (1998).
5. R. Gandhi, C. Quigg, M. H. Reno and I. Sarcevic, Astroparticle Physics, **5**, 81, (1996);
6. F. Halzen, Lectures at the TASI School, July 1998, astro-ph 9810368.
7. R. Gandhi, C. Quigg, M. H. Reno and I. Sarcevic, Phys. Rev. **D58** (1998) 093009.
8. See http://amanda.berkeley.edu/; see also F. Halzen, New Astron. Rev 42 (1999) 289; E. Andres et al., e-print astro-ph/9906203, submitted to Astropart. Phys.; AMANDA collaboration, e-print astro-ph/9906205, talk presented at the 8th Int. Workshop on Neutrino Telescopes, Venice, February, 1999.
9. For general information see http://www.ifh.de/baikal/baikalhome.html; also see Baikal Collaboration, e-print astro-ph/9906255, talk given at the 8th Int. Workshop on Neutrino Telescopes, Venice, Feb 1999.
10. J. W. Cronin, Nucl. Phys. B (Proc. Suppl.) 28B (1992) 213; The Pierre Auger Observatory Design Report (2nd edition), March 1997; see also http://http://www.auger.org/ and http://www-lpnhep.in2p3.fr/auger/welcome.html.
11. See http://www.roma1.infn.it/nestor/nestor.html.
12. For general information see http://www.ps.uci.edu/ ice-cube/workshop.html; F. Halzen, Am. Astron. Soc. Meeting 192 (1998) # 62 28; AMANDA collaboration, e-print astro-ph/9906205, talk presented at the 8th Int. Workshop on Neutrino Telescopes, Venice, Feb 1999.
13. For general information see http://antares.in2p3.fr/antares/antares.html; see also S. Basa, e-print astro-ph/9904213; ANTARES Collaboration, e-print astro-ph/9907432.
14. For general information see http://kuhep4.phsx.ukans.edu/ ice-man/index.html; also see D. Seckel, these proceedings.
15. T. Montaruli, preprint hep-ex 9905019.

16. V. S. Berezinsky, S. V. Bulanov, V. A. Dogiel, V. L. Ginzburg, and V. S. Ptuskin, *Astrophysics of Cosmic Rays* (North-Holland, Amsterdam, 1990).
17. T. K. Gaisser, *Cosmic Rays and Particle Physics*, Cambridge University Press (Cambridge, England, 1990)
18. Proc. 24th *International Cosmic Ray Conference* (Istituto Nazionale Fisica Nucleare, Rome, Italy, 1995)
19. Proc. 25th *International Cosmic Ray Conference*, eds.: M. S. Potgieter et al. (Durban, 1997).
20. Proc. 26th *International Cosmic Ray Conference*, (Utah, 1999).
21. T. H. Burnett et al., Phys. Rev. Lett. 51 (1983) 1010; M. Cherry et al., in [19], vol. 4, p. 1.
22. See, e.g., V. A. Dogiel, Nuovo Cimento 19C (1996) 671, and rapporteur talk in [18].
23. Proc. International Symposium on *Astrophysical Aspects of the Most Energetic Cosmic Rays*, eds. M. Nagano and F. Takahara (World Scientific, Singapore, 1991)
24. N. N. Efimov et al., Ref. [23], p. 20.
25. N. Hayashida et al., Phys. Rev. Lett. 73 (1994) 3491; S. Yoshida et al., Astropart. Phys. 3 (1995) 105; M. Takeda et al., Phys. Rev. Lett. 81 (1998) 1163; see also http://icrsun.icrr.u-tokyo.ac.jp/as/project/agasa.html.
26. D. J. Bird et al., Phys. Rev. Lett. 71 (1993) 3401; Astrophys. J. 424 (1994) 491; ibid. 441 (1995) 144.
27. See, e.g., K. Boothby et al., Astrophys. J. 491 (1997) L35.
28. For general information see http://ik1au1.fzk.de/KASCADE_home.html.
29. F. Arqueros et al. (HEGRA collaboration), e-print astro-ph/9908202, to appear in Astron. Astrophys.
30. P. Bhattacharjee and G. Sigl, preprint astro-ph 9811011, to appear in Physics Reports.
31. J. Linsley, Phys. Rev. Lett. 10 (1963) 146; Proc. 8th *International Cosmic Ray Conference* 4 (1963) 295.
32. R. G. Brownlee et al., Can. J. Phys. 46 (1968) S259; M. M. Winn et al., J. Phys. G 12 (1986) 653; see also http://www.physics.usyd.edu.au/hienergy/sugar.html.
33. See, e.g., M. A. Lawrence, R. J. O. Reid, and A. A. Watson, J. Phys. G Nucl. Part. Phys. 17 (1991) 733, and references therein; see also http://ast.leeds.ac.uk/haverah/hav-home.html.
34. K. Greisen, Phys. Rev. Lett. 16 (1966) 748.
35. G. T. Zatsepin and V. A. Kuzmin, Pis'ma Zh. Eksp. Teor. Fiz. 4 (1966) 114 [JETP. Lett. 4 (1966) 78].

36. M. Derrick, *et al.* (ZEUS Collaboration), *Phys. Lett.* **B316** (1993) 412.
37. I. Abt, *et al.* (H1 Collaboration), *Nucl. Phys.* **B407** (1993) 515.
38. G. Wolf, "HERA Physics," in *High Energy Phenomenology,* Proceedings of the 42nd Scottish Universities Summer School in Physics, St. Andrews, 1–21 August 1993, edited by K. J. Peach and L. L. J. Vick (SUSSP Publications, Edinburgh, and Institute of Physics Publishing, London, 1994), p. 135.
39. M. Derrick, *et al.* (ZEUS Collaboration), *Z. Phys. C***65** (1995) 379.
40. T. Ahmed, *et al.* (H1 Collaboration), *Nucl. Phys.* **B439** (1995) 471.
41. H. Lai, *et al.* (CTEQ Collaboration), *Phys. Rev. D***51** (1995) 4763.
42. G. Altarelli and G. Parisi, *Nucl. Phys.* **B126** (1977) 298. See also V. N. Gribov and L. N. Lipatov, *Yad. Fiz.* **15** (1972) 781 [English transl.: *Sov. J. Nucl. Phys.* **15** (1972) 438]; L. N. Lipatov, *Yad. Fiz.* **20** (1974) 181 [English transl.: *Sov. J. Nucl. Phys.* **20** (1974) 94]; Yu. L. Dokshitser, *Zh. Eksp. Teor. Fiz.* **73** (1977) 1216 [English translation: *Sov. Phys.-JETP* **46** (1977) 641].
43. E. A. Kuraev, L. N. Lipatov, and V. S. Fadin, *Zh. Eksp. Teor. Fiz.* **71** (1976) 840 [English translation: *Sov. Phys.-JETP* **44** (1976) 443]; *Zh. Eksp. Teor. Fiz.* **72** (1977) 377 [English translation: *Sov. Phys.-JETP* **45** (1977) 199]; Ya. Ya. Balitskiĭ, and L. N. Lipatov, *Yad. Fiz.* **28** (1978) 1597 [English transl.: *Sov. J. Nucl. Phys.* **28** (1978) 822]; L. N. Lipatov, *Zh. Eksp. Teor. Fiz.* **90** (1986) 1536 [English translation: *Sov. Phys.-JETP* **63** (1986) 904]; L. N. Lipatov, "Pomeron in Quantum Chromodynamics," in *Perturbative Quantum Chromodynamics,* edited by A. H. Mueller (World Scientific, Singapore, 1989), p. 411.
44. J. Kwiecinski, A. Martin and A. Stasto, hep-ph 9905307.
45. G. Burdman, F. Halzen, and R. Gandhi, Phys. Lett. **B417,** 107 (1998).
46. M. A. Doncheski and R. W. Robinett, Phys. Rev. D. **56,** 7412 (1997); R. W. Robinett, Phys. Rev. D. **37,** 84 (1988).
47. M. Carena, D. Choudhury, S. Lola, and C. Quigg, Phys. Rev. **D58** (1998) 095003
48. J. Bordes, *et al., Astropart. Phys.* **8** (1998) 135.
49. T. Piran, Nucl. Phys. (Proc. Suppl.) 70 (1999) 431; Phys. Rept. 314 (1999) 575.
50. E. Waxman and J. N. Bahcall, Phys. Rev. Lett. 78 (1997) 2292.
51. M. Vietri, Phys. Rev. Lett. 80 (1998) 3690.
52. K. Mannheim, *Astropart. Phys.* **3** (1995) 295.
53. F. W. Stecker, C. Done, M. H. Salamon, and P. Sommers, Phys. Rev. Lett. **66,** 2697 (1991); *ibid.* **69** (1992) 2738E. Revised estimates of the neutrino flux appear in F. W. Stecker and M. H. Salamon, "High

Energy Neutrinos from Quasars," Space Science Reviews **75,** 341 (1996) (electronic archive: astro–ph/9501064), where it is assumed that all the observed AGN x-ray luminosity arises from hadronic interactions. We have scaled this flux by a factor of 0.3.
54. R. J. Protheroe, "High-Energy Neutrinos from Blazars" (electronic archive: astro–ph/9607165) in *Accretion Phenomena and Related Outflows, IAU Colloquium 163,* Volume 121 of the ASP Conference Series, ed. D. T. Wickramasinghe, G. V. Bicknell, and L. Ferrario (Astronomical Society of the Pacific, San Francisco, 1997).
55. E. Waxman and J. Bahcall, Phys. Rev. Lett. **78,** 2292 (1997).
56. U. F. Wichoski, J. H. MacGibbon, and R. H. Brandenberger, "High Energy Neutrinos, Photons and Cosmic Rays from Non-Scaling Cosmic Strings," BROWN–HET–1115 (electronic archive: hep–ph/9805419).
57. G. Sigl, S. Lee, D. N. Schramm and P. Coppi, Phys. Lett. **B392,** 129 (1997).
58. R. Protheroe, Talk at Neutrino 98, astro-ph 9809144.
59. E. Waxman and J. Bahcall, Phys. Rev. **D59**, (1999) 023002.
60. J Rachen *et al*; asto-ph 9908031.
61. L. V. Volkova, *Yad. Fiz.* **31** (1980) 1510 [English transl.: *Sov. J. Nucl. Phys.* **31** (1980) 784].
62. G. Sigl, S. Lee, D. N. Schramm and P. Coppi, Phys. Lett. **B392,** 129 (1997).
63. L. Pasquali, M. H. Reno, and I. Sarcevic, "Lepton fluxes from atmospheric charm" (electronic archive: hep–ph/9806428).
64. N. Arkani-Hamed, S. Dimopoulos, and G. Dvali, Phys. Lett. B 429 (1998) 263; I. Antoniadis, N. Arkani-Hamed, S. Dimopoulos, and G. Dvali, Phys. Lett. B 436 (1998) 257; N. Arkani-Hamed, S. Dimopoulos, and G. Dvali, Phys. Rev. D 59 (1999) 086004.
65. G. Domokos and S. Kovesi-Domokos, Phys. Rev. Lett. 82 (1999) 1366.
66. S. Nussinov and R. Shrock, Phys. Rev. D 59 (1999) 105002.

FROM AMANDA TO ICECUBE: CURRENT AND FUTURE HIGH ENERGY NEUTRINO TELESCOPES AT THE SOUTH POLE

T. MILLER, G. SPICZAK

Bartol Research Institute, University of Delaware, Newark, DE, USA

A. BIRON, A. BOUCHTA, P. DESIATI, M. GAUG, S. HUNDERTMARK, M. KOWALSKI, H. LEICH, M. LEUTHOLD, P. NIESSEN, T. SCHMIDT, A. SILVESTRI, C. SPIERING, P. STEFFEN, O. STREICHER, T. THON, C. WIEBUSCH, R. WISCHNEWSKI

DESY-Zeuthen, Zeuthen, Germany

S. CARIUS, P. LINDAHL

Dept. of Physics, Kalmar University, Sweden

L. BERGSTRÖM, E. DALBERG, J. EDSJÖ, P. EKSTRÖM, A. GOOBAR, P. O. HULTH, J. RODRIGUEZ MARTINO, H. RUBINSTEIN, C. WALCK

Dept. of Physics, Stockholm University, Stockholm, Sweden

S. W. BARWICK, J. BOOTH, J. KIM, P. MOCK, R. PORRATA, D. ROSS, E. SCHNEIDER, W. WU, G. YODH, S. YOUNG

Dept. of Physics, UC Irvine, Irvine, CA, USA

D. F. COWEN, F. M. NEWCOMER, I. TABOADA

Dept. of Physics, University of Pennsylvania, Philadelphia, PA, USA

G. BAROUCH, M. CARLSON, C. COSTA, T. DEYOUNG, P. DOKSUS, F. HALZEN, R. HARDTKE, G. HILL, A. KARLE, B. KOCI, I. LIUBARSKY, R. MORSE, K. RAWLINS, P. ROMENESKO, S. TILAV

Dept. of Physics, University of Wisconsin, Madison, WI, USA

K. BECKER

Dept. of Physics, University of Wuppertal, Wuppertal, Germany

W. CHINOWSKY, A. GOLDSCHMIDT, J. JACOBSEN, J. LAMOUREUX, J. LUDVIG, H. MATIS, D. NYGREN, G. PRZYBYLSKI, G. SMOOT, R. STOKSTAD

Lawrence Berkeley Laboratory, Berkeley, CA, USA

R. BAY, D. CHIRKIN, D. M. LOWDER, P. MIOCINOVIC, P. B. PRICE,

W. RHODE, M. SOLARZ, K. WOSCHNAGG
Dept. of Physics, UC Berkeley, Berkeley, CA, USA

T. FESER, M. HELLWIG, L. KOEPKE, T. NEUNHOEFFER, H. SANDER, C. WEINHEIMER
Mainz University, Mainz, Germany

E. ANDRES, X. BAI, S. RICHTER, R. SCHWARZ, N. STARINSKI
South Pole Station, Antarctica

D. BERTRAND, J. DEWULF, M. VANDER DONCKT
ULB - IIHE - CP230, Boulevard du Triomphe, B-1050 Bruxelles, Belgium

D. BESSON, G. FRICHTER, I. KRAVCHENKO
University of Kansas, Lawrence, KS, USA

O. BOTNER, J. CONRAD, A. HALLGREN, P. LOAIZA, P. MARCINIEWSKI, C. PÉREZ DE LOS HEROS
University of Uppsala, Uppsala, Sweden

The AMANDA Collaboration has successfully deployed a 10-string array (AMANDA-B10) at a depth of 1.5 to 1.9 km in ice at the South Pole, and subsequently used it to identify upward moving neutrino-induced muons. I will explain the detector's principles of operation and current performance. Simulation studies indicate that AMANDA-B10 has an effective area of $\sim 10,000 m^2$ for through-going muons and angular resolution of $\sim 4°$. I will then describe future plans to expand AMANDA. First, the AMANDA-II array, which will be completed in January, 2000 and have an effective area of $\sim 0.1 km^2$ and pointing resolution of $\sim 1°$. Finally, a proposal is in progress to build IceCube during the next decade, which will instrument a cubic kilometer of ice for high energy neutrino detection. Discussion of the scientific motivation and capabilities of these detectors is given in the paper by F. Halzen in this volume.

1 Introduction

Techniques now being developed will soon allow astronomers to observe the universe in a new way: with high energy ($\sim$ 1 TeV) neutrinos. The unique properties of neutrinos allow them to travel to Earth through large amounts of intervening matter without deflection, thus complementing observations made with high energy photons and cosmic rays. Several authors have recently discussed expected fluxes from possible sources of high energy neutrinos [9,12,14,20,10]. Well motivated potential sources include some of the most energetic phenomena observed in the universe: active galactic nuclei (AGN) and gamma ray bursts (GRB). Arguments presented in the above papers conclude that a detector of order 1 km^3 will be necessary to observe the expected fluxes.

The Antarctic Muon and Neutrino Detector Array (AMANDA) has been operating and under continual expansion at the South Pole for the past several years. The current AMANDA detector has an effective area of $\sim 10^4\ m^2$ and will soon be expanded to $\sim 10^5\ m^2$. A proposal is in progress to construct the IceCube array, which will realize the desired 1 km^2 size over the next decade. I will first describe the operating principles of AMANDA, then the current status of the project. Finally, I will describe the plans for the IceCube observatory.

2 Principles of Operation

The only proven technology for detection of high energy neutrinos that is reliable and expandable in an affordable way is to observe Cherenkov emission from muons in deep water or ice. The muons are produced by upward moving ν_μ in charged current interactions in or below the detector. The Cherenkov light is observed by a large array of photomultiplier tubes (PMTs), and the direction of the traversing muon, which is within 1° of the initial neutrino detection at such high energies, is reconstructed from the timing and pulse height information recorded by the PMTs.

High energy neutrino detectors must be placed as deep under water or ice as possible in order to reduce the background from downward moving muons created by cosmic ray interactions high in the atmosphere above the detector. After rejection of the downward moving muon background one is still left with an irreducible background of upward moving muons created by atmospheric neutrinos produced in cosmic ray interactions on the far side of the Earth. Even at a depth of 1.5 km the raito of downward moving to upward moving muons is of order 10^6.

There are several advantages to locating such an array in ice as opposed to deep ocean water. The central electronics is directly above the array, making extremely long site-to-shore cables unnecessary. Each PMT signal is sent directly to the surface electronics, with no need for extensive data processing within the ice, making the design very parallel and reliable. In addition, the low temperatures and low natural radioactive backgrounds in the ice result in very low counting rates for each PMT. Finally, the proximity of the SPASE-2 air shower array provides the opportunity to study cosmic ray composition and to calibrate AMANDA or IceCube with a beam of muons from a known direction (see Section 3).

3 Status

The current AMANDA detector is shown in Figure 1. The upper detector, known as AMANDA-A, was installed at a depth of 810-1000 m during the winter of 1993-94. It consists of four vertical strings of PMTs spaced 30 m apart. Each of these strings contains 20 PMTs spaced at 10 m intervals, making the entire array about 200 m tall by 30 m in diameter. The ice at 1 km depth was found to contain air bubbles that scatter light with a small effective scattering length that ranged from 40 cm at the top of the array to 80 cm at the bottom [2]. This short scattering length made use of AMANDA-A for neutrino astronomy problematic, so subsequent AMANDA strings were placed at much greater depths in order to reach bubble-free ice.

The first four strings of the deeper array, called AMANDA-B, were installed in January, 1996, at a depth of 1520 m to 2000 m. Each of these strings containes 20 PMTs and uses coaxial cable to transmit signals to the surface. One string also contains an additional 6 PMTs below the rest of the array. The next six strings of AMANDA-B were then installed in December, 1996 and January, 1997, at a depth of 1500 to 1950 m, completing the 10 string detector AMANDA-B10. Each of these strings contains 36 PMTs and uses twisted pair electrical cables to send signals to the surface. After transmission through the signal cable, the rise time of single photoelectron pulses was typically 180(60) ns and the FWHM 600(280) ns for coaxial (twisted pair) cables. After appropriate setting of discriminator thresholds and pulse height corrections, the timing accuracy was $\approx$ 4 ns. Altogether, AMANDA-B10 is approximately 400 m tall by 120 m in diameter, and contains 302 PMTs. The ice at the depth of AMANDA-B is much more clear than at AMANDA-A, with an effective scattering length of $\approx$ 30 m and a maximum absorption length of over 100 m [3].

3.1 Simulations and Analysis

The primary goal of the current AMANDA detector is to demonstrate that the optical properties of deep ice are good enough to perform the required angular resolution and up-down discrimination. Detailed descriptions of the performance of AMANDA-B10 are given in [17] and [15]. A brief summary is given below.

Up-Down Rejection

As of summer, 1999, an initial analysis of 85 live days of data from 1997 has been done. Atmospheric muons were recorded at a rate of 70 Hz, whereas upward going atmospheric muons are expected to trigger AMANDA-B10 at

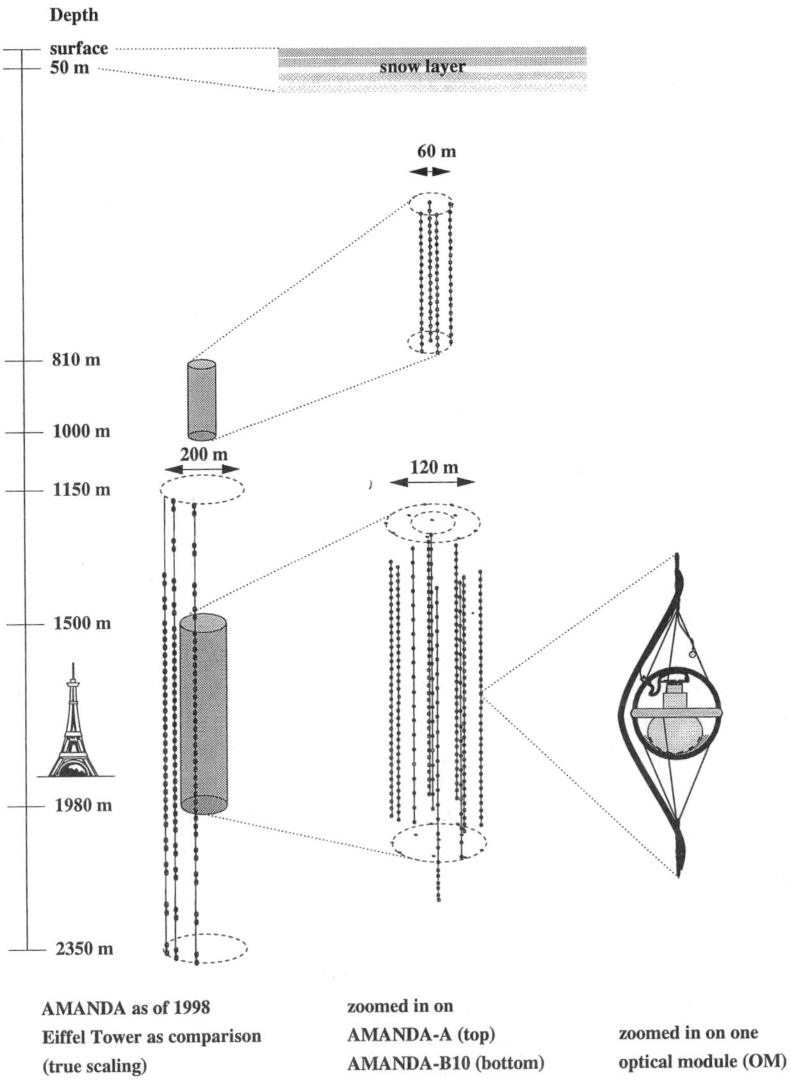

Figure 1. The AMANDA detector as of 1999.

a rate of only 25 events per day. About 90% of the cosmic ray muon triggers were eliminated by a simple filter method using the correlation of arrival photon arrival times with depth. The remaining events were reconstructed by fitting the Cherenkov light cone of a relativistic particle to the observed arrival times using a maximum likelihood procedure [21]. After reconstruction a set of quality cuts were applied to suppress the background of cosmic ray muons that were mistakenly reconstructed as upward moving. The most important cut was the number of "direct hits", defined as the number of photons that are detected within [-5,+25] ns of the fitted Cherenkov cone in an event. Another important cut required that the hits are distributed over a track length of at least 100 m. Finally, an "edge cut" required that the event not be confined to the top or bottom edge of the array. The same set of cuts were applied to three data sets:

1. Experimental data (4.9×10^8 events, corresponding to 85 live days)
2. Monte Carlo simulation of downgoing atmospheric muons (3.4×10^6 events, corresponding to 14 live hours)
3. Monte Carlo simulation of upgoing atmospheric neutrinos (2000 events, corresponding to 85 live days)

After all cuts (described in detail in [17]) were applied, all Monte Carlo background muon events were eliminated, 21 Monte Carlo atmospheric neutrino events remained, and 17 experimental data events remained.

The reconstructed zenith angle distribution at different cut levels is shown in Figure 2, and a display of one very good neutrino candidate is shown in Figure 3. The angular distribution of the observed upward moving tracks agrees well with the expectation from atmospheric neutrinos. It should be noted that the current analysis does not use all event information, such as spatial hit topology and hit amplitudes. Inclusion of additional event information, such as fits to energy, are in progress and should improve both the rejection power and the sensitivity.

Pointing Resolution

Two methods are used to determine the angular resolution of AMANDA. First, simulated signal events are reconstructed and compared to their true direction. Second, events triggering both AMANDA and the nearby surface air shower array SPASE-2 are reconstructed by both and the directions of the two analyses compared.

The space angle error distribution determined from fitting Monte Carlo events generated with an E^{-2} spectrum, for events passing Level 3 cuts [17], is shown in Figure 4. The median resolution is 3.4° [16]. The median angular resolution does depend upon cut level, and further improvements are in progress.

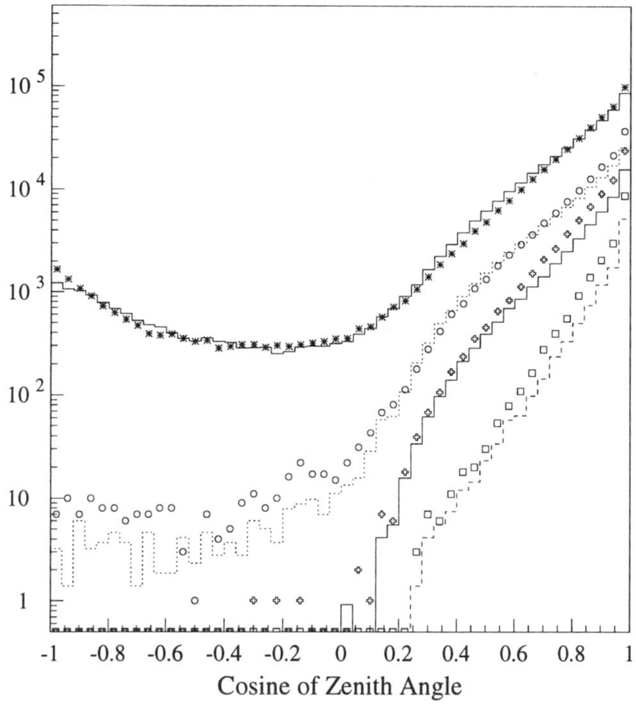

Figure 2. Reconstructed zenith angle distribution at different cut levels, for data (dots) and MC (line). Applying progressively stricter cuts eliminates the misreconstructed events that appear to be upward going.

This result can be compared to that obtained from SPASE-AMANDA coincidence events. The ability to test simulated reconstruction results with data from a surface array is a unique advantage for AMANDA, as no other existing or planned large high energy neutrino telescope is located near an air shower array. The primary goal of the coincidence analysis is determination of cosmic ray composition in the knee region [7]. The coincidence data is also useful for investigating other aspects of AMANDA, including optical module location and ice transparency[18].

The 1997 coincidence data set has been reconstructed by both arrays, and the mismatch in zenith angle for different cut levels are shown in Figure 6. We expect cosmic ray protons at the threshold of SPASE-2 ($\approx$ 100 TeV) to generate an average of 1.2 muons that penetrate to the depth of AMANDA-

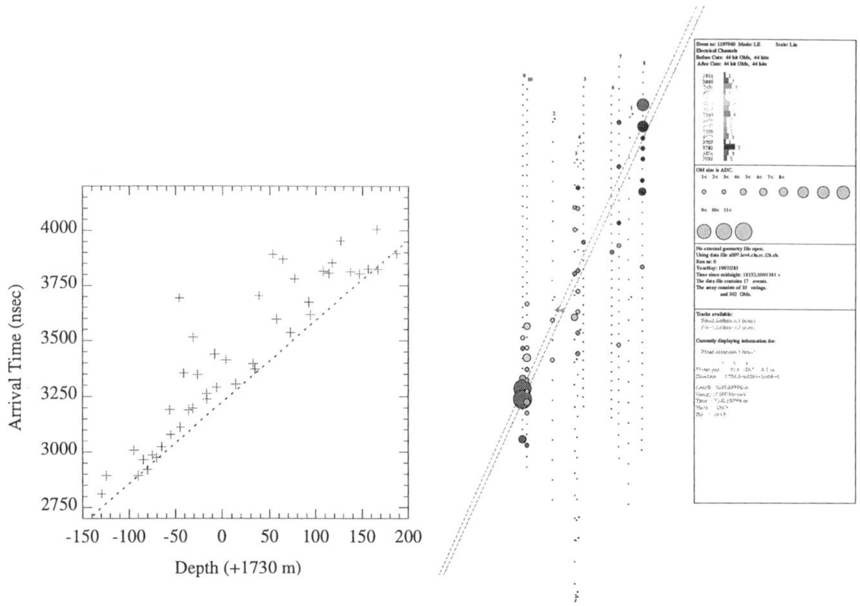

Figure 3. Event 1197960. Left: Arrival times of photons vs depth of the ovserving PMT. The slope of the dashed line represents a zenith angle of 155° . Right: A display of the event. The shading (size) of the points represents the time (amplitude) of the triggered PMTs.

B, following very nearly a Poisson distribution [8]. The pointing resolution of SPASE-2 is known from both Monte Carlo and subarray comparisons [6] to be betweend 1 and 2° for the data shown in the figure. If the SPASE and AMANDA pointing errors are uncorrelated, this implies a zenith angle resolution of $\sim$ 2.5 to 3° for AMANDA-B10. This is not inconsistent with the space angle error obtained above from the Monte Carlo. However, the muon multiplicites and energies are not identical for the two event samples, so they cannot be compared directly.

3.2 Other Results

In addition to observing atmospheric neutrinos and demonstrating both up-down discrimination and angular resolution, several other results have been obtained with AMANDA-B10. These are described in detail elsewhere, and

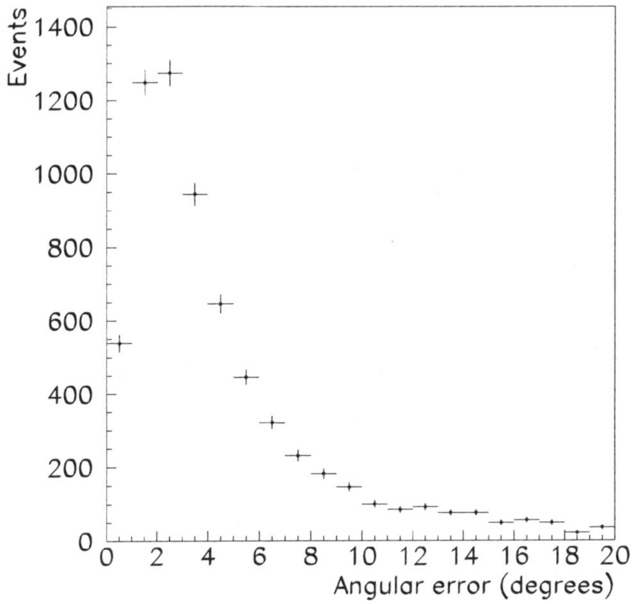

Figure 4. Space angle error of neutrino events from a source with an E^{-2} spectrum.

include limits on high energy neutrinos from gamma ray bursts [4], limits on relativistic monopoles [19], limits on point sources of high energy neutrinos [16], searches for WIMP dark matter trapped within the Earth [5], and searches for galactic supernovae [22].

4 Future Plans

4.1 AMANDA-II and IceCube

A potential layout for IceCube, including AMANDA-II, is shown in Figure 7. The first three strings of AMANDA-II have already been deployed in January, 1998, and the remaining six strings will be deployed during the 1999-2000 austral summer. The completed AMANDA-II will be approximately 500 m tall by 200 m in diameter, and contain about 700 PMT's, including those already comprising AMANDA-B10.

Construction of IceCube will occur over the next decade, with a possible timeline shown in Table 1. A straw-man design calls for 80 strings spaced at $\approx$ 100 m intervals, with ~60 PMTs per string at depths of 1400 to 2300 m,

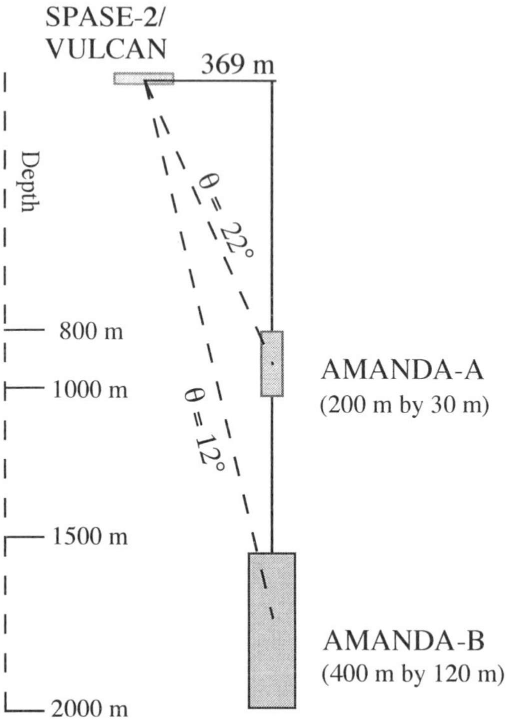

Figure 5. Schematic side view of SPASE and AMANDA.

for a total of ~5000 optical modules.

Extrapolating from the current performance of AMANDA, achieving 1° angular resolution for long muon tracks in IceCube should be straightforward. However, IceCube will also have several advantages over AMANDA-B and AMANDA-II besides its larger size. A detector of cubic kilometer size will allow us to fully exploit the properties of deep Antarctic ice. As shown earlier, AMANDA has demonstrated that scattering due to residual dust in bubble-free ice is not a significant problem, and that absorption lengths in deep ice, at over 100 m, are much greater than in deep ocean water. The ability to detect the prompt unscattered photons allows one to reconstruct incident neutrino direction, while the additional ability to collect the scattered photons over large distances and time scales improves energy measurement. Energy determination is very important, especially for very large detectors, because at

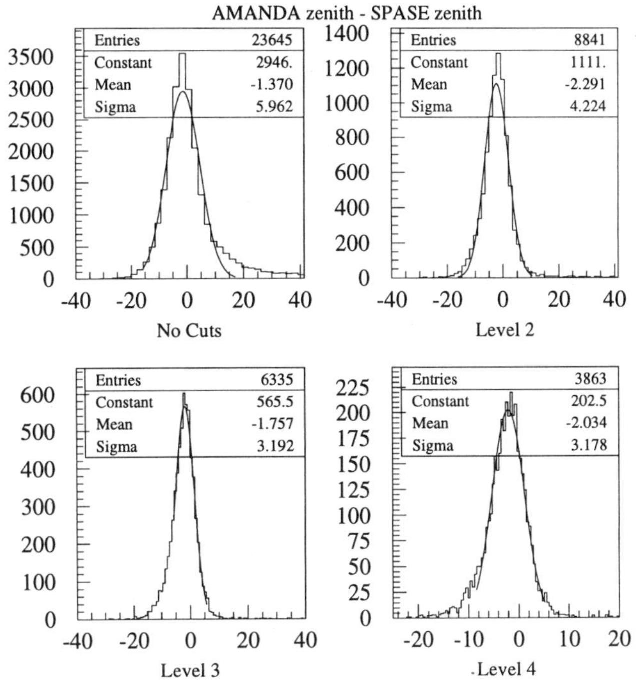

Figure 6. Mismatch in zenith angle reconstructed by SPASE-2 and AMANDA-B10, for different AMANDA cut levels.

above 100 TeV there should be essentially no background from atmospheric neutrinos. Simulations run so far indicate that IceCube will have a much higher efficiency to reconstruct tracks, map showers from electron and tau neutrinos, and measure neutrino energy than AMANDA. Initial simulations indicate that the direction of showers can be reconstructed to better than 10° in both θ and ϕ above 10 TeV. Energy resolution is predicted to be about 25%, much better than the logarithmic energy resolution currently possible in AMANDA.

4.2 Technology

So far, all AMANDA data analysis has been done using signals sent over coaxial or twisted pair electrical cables. The 14 stage Hamamatsu R5912-0 PMT produces a pulse of approximately 1 V amplitude for a single detected

Table 1. Major IceCube milestones.

Milestone	Date
Submit IceCube pre-proposal	1999 June
Submit IceCube proposal	2000 January
Proposal Reviews	2000/2001
Funding Start	2001 April
Final Design Review	2001 June
Stage 1 Deployment (8 strings)	2002/2003 season
Stage 2 Deployment (16 strings)	2003/2004 season
Stage 3 Deployment (16 strings)	2004/2005 season
Stage 4 Deployment (16 strings)	2005/2006 season
Stage 5 Deployment (16 strings)	2006/2007 season
Stage 6 Deployment (8 strings)	2007/2008 season
Routine Operation Continues	2008-2013

photoelectron, large enough to be transmitted over 2 km of electrical cable. Despite attenuation of the signal to only a few mV at the surface, a timing resolution of 4 ns has been achieved. However, the high gain at which the PMT must be run limits the dynamic range to only a few photoelectrons per channel.

The collaboration is currently investigating technologies to greatly improve signal transmission. Optical modules using fiber optics have already been installed on AMANDA-B and the first three strings of AMADNA-II. The PMT pulse is converted by an LED to a 1300 nm light pulse and transmitted over multi-mode optical fiber to the surface. The attenuation is only 1.5 dB/km, and the original PMT pulse shape is preserved with almost no distortion. The dynamic range should be larger than 50 photoelectrons over 7 ns. Note that since the photons typically arrive over time scales longer than the PMT pulse width, the true dynamic range will be much larger if the charge is integrated or waveforms are digitized at the surface.

If the results from the final AMANDA-II deployment are satisfactory, optical fiber may become the default technology for IceCube. However, another method, waveform digitization within the OM, is also being investigated. The digitized waveform is transmitted over a serial link to the surface. One string of AMANDA-II will be instrumented with digital optical modules in order to evaluate the effectiveness and robustness of this method compared to optical fiber.

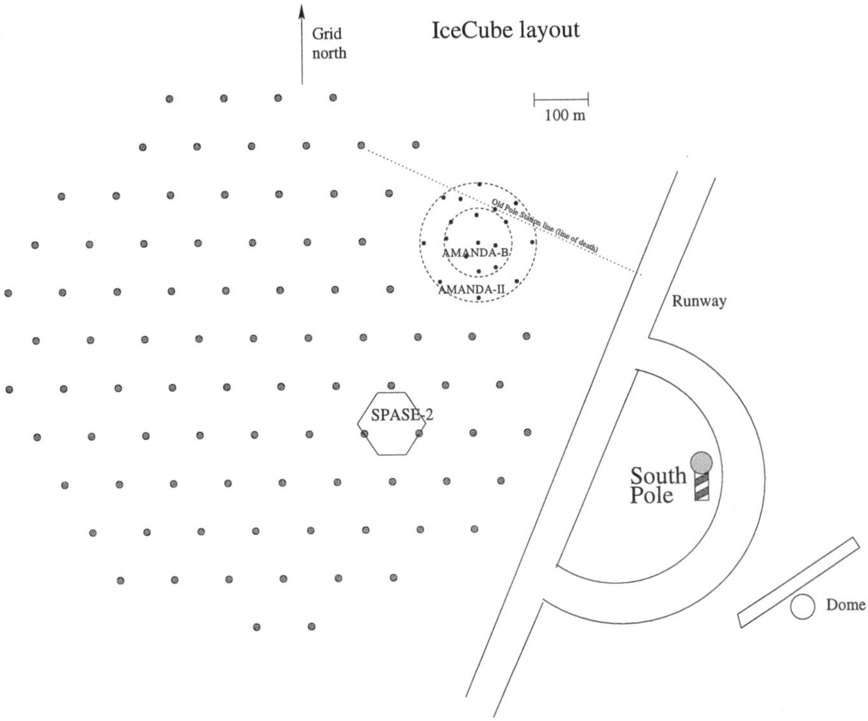

Figure 7. Top view of one possible IceCube layout, showing it in relation to the existing AMANDA-B and AMANDA-II arrays.

Acknowledgments

This research was supported by the following agencies:

- U.S. National Science Foundation, Office of Polar Programs
- U.S. National Science Foundation, Physics Division
- University of Wisconsin Alumni Research Foundation
- U.S. Department of Energy

- U.S. National Energy Research Scientific Computing Center (supported by the Office of Energy Research of the U.S. Department of Energy)
- Swedish Natural Science Research Council
- Swedish Polar Research Secretariat
- Knut and Alice Wallenberg Foundation, Sweden
- Deutsches Elektronen-Synchrotron (DESY)

C.P.H. acknowledges the support of the European Commission through TMR contract No. ERBFMBICT91551.

D.Z.B. acknowledges support from the Cottrell Research Corporation.

References

1. E. Andres, *et al.*, accepted for publication by Astroparticle Physics.
2. P. Askebjer, it et al., Science **267**, 1147 (1995).
3. P. Askebjer, it et al., Applied Optics **36**, 4168 (1995).
4. R. Bay, *et al.*, *Proc. 26th Intl. Cosmic Ray Conference*, Salt Lake City, Utah, HE 4.2.06 (1999).
5. E. Dalberg, *et al.*, *Proc. 26th Intl. Cosmic Ray Conference*, Salt Lake City, Utah, HE 5.3.06 (1999).
6. J. Dickinson, *et al.*, 1999a and 1999b, accepted for publication by NIM A.
7. J. Dickinson, *et al.* 1998c, *Proc. 26th Intl. Cosmic Ray Conference*, Salt Lake City, Utah, OG 1.2.05.
8. T.K. Gaisser, *Cosmic Rays and Particle Physics*, (Cambridge University Press, 1990).
9. T.K. Gaisser, F. Halzen, T. Stanev, Phys. Rep. **258**, 173 (1995).
10. R. Gandhi, C. Quigg, M.H. Reno, I. Sarcevic, Astrop. Phys. **5**, 81 (1996).
11. F. Halzen, "Large Natural Cherenkov Detectors: Water and Ice", talk presented at the *5th International Workshop on Topics in Astroparticle and Underground Physics (TAUP97)*, Gran Sasso, Italy, Sept. 1997.
12. F. Halzen, these proceedings.
13. Y.D. He, & P.B. Price, J. Geophys. Res. **103**, 17041 (1998).
14. G. Hill, Astrop. Phys. **6**, 215 (1997).
15. G. Hill, *et al.*, *Proc. 26th Intl. Cosmic Ray Conference*, Salt Lake City, Utah, HE 6.3.02 (1999).
16. J.H. Kim, *et al.*, *Proc. 26th Intl. Cosmic Ray Conference*, Salt Lake City, Utah, HE4.1.14 (1999).

17. A. Karle, *et al.*, *Proc. 26th Intl. Cosmic Ray Conference*, Salt Lake City, Utah, HE 4.2.05 (1999).
18. T.C. Miller, *et al.*, *Proc. 26th Intl. Cosmic Ray Conference*, Salt Lake City, Utah, HE 6.3.11 (1999).
19. P. Niessen, *et al.*, *Proc. 26th Intl. Cosmic Ray Conference*, Salt Lake City, Utah, HE 5.3.05 (1999).
20. R.J. Protheroe, High Energy Neutrino Astrophysics, invited talk at Neutrino 98, Takayama, Japan.
21. C. Wiebusch, *et al.*, *Proc. Intern. Workshop on Simulation and Analysis Methods for Large Neutrino Detectors*, Zeuthen, Germany, DESY-PROC-1999-01 (in press) (1998).
22. R. Wischnewski, *et al.*, *Proc. 26th Intl. Cosmic Ray Conference*, Salt Lake City, Utah,HE 4.2.07 (1999).
23. K. Woschnagg, *et al.*, *Proc. 26th Intl. Cosmic Ray Conference*, Salt Lake City, Utah, HE 4.1.15 (1999).

DETECTING ULTRA HIGH ENERGY NEUTRINOS ($E_\nu > 10^{18}$EV) WITH A LARGE RADIO ARRAY

D. SECKEL

Bartol Research Institute, University of Delaware, Newark DE 19716, USA
E-mail: seckel@bartol.udel.edu

It is widely anticipated that high energy neutrinos are produced in nature[1], either as a result of the same processes that make high energy cosmic-rays, or as a by product of the interactions of those cosmic rays with intervening target matter. This contribution discusses some reasonable expectations for the neutrino fluxes and the broad issues concerning their detection. An overview of proposed experimental techniques is given, with a focus on three versions of RICE (Radio Ice Cherenkov Experiment).

1 Introduction

There are two broad classes of source models for ultra high energy cosmic rays (UHECR's, $E > 1$ EeV). In what may be called the astrophysical, or bottom up, models ordinary matter is accelerated to high energies in a source and then propagates across some distance to interact at Earth. The sources may be compact (AGN, GRB's) or diffuse (termination shocks of the jets from AGN in the intergalactic medium). The presence of TeV-PeV gamma rays eminating from compact sources suggests acceleration of protons and production of π^0's in either proton-nucleon or proton-photon interactions. If so, it seems reasonable to speculate that charged π's may be produced as well, which implies a flux of neutrinos when the π's decay. In such models, cosmic rays are accelerated as protons, but escape as neutrons after a charge changing collision. Since such interactions imply production of a charged pion, one may roughly estimate the flux of neutrinos as one per escaped nucleon, but at an average energy about a factor of 10 smaller, i.e. for each proton above 10 EeV of energy there will be roughly one neutrino with energy above 1 EeV. If UHECR protons are accelerated in diffuse termination shocks, then leakage dominates loss from the acceleration region, and production of gamma-rays and neutrinos may not occur efficiently.

The other type of model is the top down models, in which UHECR's result from energy released in the decay or annihlation of supermassive particles or via the evaporation of topological defects left over from the early universe. In these models, neutrinos may be produced directly in the energy release process; for example as direct decay products of X bosons of a grand unified theory. Quark and gluon jets may arise at the same level, but in this case nucleon production would be only a few percent of meson production. Thus, in such models for each nucleon observed, one expects a larger number of neutrinos and they may be higher in energy as well.

No matter what the source of UHECR's, once produced they will propagate through the intergalactic medium. Specifically, they pass through the cosmic microwave background, and for energies much greater than 10 EeV this implies interactions with CMBR photons via the Griessen, Zatsepin, Kuzmin (GZK) process. This causes energy loss for the nucleons in a way that necessarily produces π's and neutrinos of energy approximately one tenth that of the GZK cutoff energy.

To distinguish between source models of UHECR's it is then interesting to form the ratio

$$R = 1 + \alpha = \frac{\int_{\text{EeV}} \phi_\nu}{\int_{10\text{EeV}} \phi_p}. \tag{1}$$

The "1" represents the neutrinos produced in the GZK process. That contribution comes from propagation of the UHECR's after production, and is always there. The quantity α represents production of neutrinos in the source. For compact astrophysical sources where escape is dominated by charge changing collisions, $\alpha \simeq 1$. In diffuse astrophysical models, $\alpha \ll 1$. In topdown models, $\alpha \gg 1$. By observing the flux of ultra high energy neutrinos (UHEN) and comparing to the UHECR flux, one may distinguish between the different source models.

As given in Eq. (1), the ratio R is determined at the source, however, if one desires the ratio at Earth one must account for propagation effects. Below 10 EeV, neutrinos and protons both propagate free of interactions; however, for super GZK UHECR's one can only see sources that are closer than a few Mpc, approximately 1% of the column depth to the horizon. Thus, for these super high energies, the flux of UHECRs is suppressed by about a factor of 100.

2 Detecting GZK neutrinos

Since $R > 1$, but may be only of order one, it is useful to plan UHEN detectors on the basis of observing the GZK produced neutrinos. Since these fluxes may be estimated in a fairly reliable way by comparing directly to the UHECR data, it is possible to estimate the necessary detector mass in a nearly model independent way. Detector sizes are large, but since particle energies are also high, instrumentation of the detector may be fairly sparse and based only on the ability to detect the hadronic and electromagnetic cascades caused by

UHEN interactions.

2.1 Detector mass

Detecting neutrinos is a challenge due to the small cross-sections with matter. At high energies though, the cross-sections increase; enough so that for energies above 1 EeV the interaction path length for neutrinos is between 100 and 1000 km of water. If one imagines a detector with thickness equivalent to a kilometer of water, then the efficiency for interactions is somewhat less than a percent. This is comparable to the 1% column depth suppression of the super GZK cosmic rays. It follows that in order to measure the ratio R, one needs comparable area detectors for both the UHECR's and for the UHEN's. Indeed, taking a conservative cosmological model for source evolution from Yoshida and Teshima[2], and utilizing the cross-sections of Ghandi, et al.[3], one estimates an event rate of approx 10^{-2} km^{-3} yr^{-1}. To do go beyond discovery level science, one may desire 100's of events per year, suggesting a detector area of approximately 10^4 km^2, i.e. a detector comparable in size to the Pierre Auger cosmic ray observatories.

2.2 Detector geometry

Most astrophysicists may be familiar with the idea that at low energies, 1-100 GeV, the most efficient strategy for neutrino detection is to optimize the detector geometry to observe upward going muons, typically through their optical cherenkov radiation, taking advantage of the long stopping range of the μ. By selecting for upward μ's one may avoid a large background of μ's produced in cosmic ray air showers. At high energies ($>$ PeV), these considerations are altered. The Earth becomes opaque to neutrinos, and so only a few upward neutrinos are observable near the horizon. Secondly, the background from cosmic ray produced μ's becomes relatively unimportant due to a loss of efficieny in producing such muons, and also due to the increasing neutrino cross-section which makes neutrino detection more efficient. Thus, the overall detector design for UHEN's should be one optimized to detect neutrinos incident from above, not below.

The obvious possiblity is to design a detector to take advantage of the optical properties of ice or deep ocean water, and indeed to detect neutrinos with energies up to a PeV, this may be reasonable. To produce interesting numbers of events on an annual basis would require a detector with size of order 1 km^3. Efforts are underway at South Pole (AMANDA/ICECUBE) and in the Mediterranean (ANTARES, NESTOR). However, for EeV neutrino astronomy, instrumenting 10^4 km^3 of volume with phototubes seems like a dubious proposition. An alternative is to look for cherenkov radiation at radio frequencies, 100 MHz - 1 GHz, and this is the basis for the RICE (Radio Ice Cherenkov Experiment) projects described below. Another proposal is to look for "horizontal" air showers. The term horizontal implies that the air shower is initiated deep in the atmosphere, at a depth far deeper than a nucleon could penetrate. Large areas can be monitored efficiently by atmospheric flourescence techniques or by large area ground arrays. For neutrino astronomy, however, the difficulty is that the column depth of the atmosphere is only equivalent to 10 m of water. Thus, instead of a 10^4 km^2 array one needs a detector 1000 km on a side, i.e 100 to 1000 times the effective area of the Pierre Auger Observatories. To gain larger area, the OWL/AIRWATCH team proposes to utilize the air-flourescent technique from space. Whether or not this proves efficient will depend on the threshold for detection and the cost of a space based platform. Lastly, an effort has been made to utilize the radio cherenkov technique by searching for pulses of radio emission from the limb of the moon[4].

2.3 Neutrino interactions and flavor

Neutrino's may interact with matter via a number of processes, the most important of which is deep inelastic scattering from nucleons. Such processes produce an outgoing lepton of reduced energy, and a spray of hadronic particles in the recoil jet. In a neutral current reaction the outgoing lepton will be a neutrino, which is unlikely to reinteract in the detector. With an effectively invisible lepton, neutral current interactions must be detected via the hadronic side of the event. Charged current events produce a charged lepton, either e, μ, or τ, and detection strategies depend on the flavor and energy of the lepton.

Whether detecting the leptonic or the hadronic side of the event, the key will be detection of either electromagnetic or hadronic cascades, produced either at the neutrino interaction vertex or as part of the energy loss process for the energetic charged leptons. The main factor determining the evolution of electromagnetic cascades is the radiation length in the medium, typically ~ 50 cm. At energies above a few PeV (depending on the medium) the Landau-Pomeranchuk-Migdal (LPM) effect causes the radiation length to increase as $\sqrt{E}$ which may affect the observability of electromagnetic cascades. At very high energies, the radiation length may become so long that photons interact hadronically before they pair produce[5].

Similarly, hadronic cascades depend on the hadronic interaction length. In a dense medium, charged mesons typically interact without decaying, but below a few PeV, π^0's decay and so hadronic cascades evolve into electromagnetic ones, albeit with larger fluctuations. Above a few PeV, however, even π^0's interact, and most of the energy evolves in a purely hadronic fashion.

It is now possible to characterize the events expected from different flavor UHEN's. Independent of flavor, neutral current events make hadronic cascades. Average particle energies are initially so high that the cascades remain hadronic. When energies drop to a PeV, the energy starts to leak into the electromagnetic component. Since this is below the LPM threshold, the electromagnetic and hadronic components have similar evolution lengths. By the time particle densities reach a maximum, almost all the energy is electromagnetic.

For charged current interactions of UHE electron neutrinos, the charged lepton is an electron and an electromagnetic cascade is formed at the neutrino interaction vertex. This cascade will be affected by the LPM effect (if in water/ice), and so may be distinguished from the hadronic cascade by its length, providing a flavor tag for ν_e's.

For ν_μ's, the charged lepton loses most of its energy over about a km. Losses to pair production, bremstrahlung and photoproduction on hadrons are all important. Detection of this dE/dx trail would tag charged current ν_μ interactions.

In the case of tau neutrinos, one must deal with the lower dE/dx of the τ ($\sim$ 0.1 that for a muon[6]) and also with the decay of the τ. At an EeV a τ has a decay length of some 50 km.

3 RICE

It has been proposed for many years that neutrino induced electromagnetic cascades in dense media may be detected via radio cherenkov emission. The basic mechanism is quite easy to understand. Two processes lead to a significant charge asymmetry in the evolution of an electromagnetic cascade in matter: electrons in the medium can be upscattered into the cascade, and positrons in the cascade can annihilate on electrons in the medium. The result is net negative charge flowing relativistically with the cascade, and a static positive charge left behind in the medium. The charge asymmetry is of order 20%[7]. At shower max there is roughly one particle per 80 MeV of energy, so the net charge of the cascade is $Q \sim .25E/\mathrm{GeV}$, where E is the energy in the cascade. For wavelengths large compared to its transverse dimensions, the cascade will radiate as a point particle moving relativistically through the medium. If the index of refraction is greater than 1, that implies cherenkov radiation with a very large source strength (e.g. 10^9 for an EeV cascade). These conditions are all satisfied for radio frequencies in ice. Peak power occurs for frequencies of a few GHz.

Other considerations may lead to an experiment operating in the range of several hundred MHz. For the technique to be effective, the medium must be transparent. Salt water doesn't work, nor will rock with significant water content. Cold fresh water ice, however, is promising. The absorption lengths depend on frequency and temperature. Generally the colder the better. The best ice is the top 1500 m. Below that, proximity to the geothermal heat source warms the ice and radio transparency is poor. Above a GHz, opacity increases as well, but below a GHz the absorption length is greater than a km.

Since the radiating cherenkov path is finite in length, and the individual particles are not exactly parallel, the radiation pattern has some width around the cherenkov angle. These effects have been modeled extensively by Zas, Halzen and Stanev[7] (ZHS) for electromagnetic cascades with energy of 1 PeV. Those calculations were used by Frichter, Ralston and Mckay[8] to show that an extensive array of radio antennas could be used effectively to detect high energy astrophysical neutrinos. More recently, Zas and Muniz-Alvarez[9] have extended the calculations of ZHS to higher energy including LPM effects and also modeling hadronic cascades. Seckel and Frichter[10] have used these improvements to study the array parameters appropriate to detection of EeV neutrinos using the radio technique.

The result of these studies is that three versions of RICE are being considered. Basic RICE is an exploratory effort underway at South Pole, taking advantage of the infrastructure developed for AMANDA. Progress with this modest size array is described below. $RICE^3$ would extend the RICE effort to the km^3 scale, possibly in conjunction with a successful ICECUBE optical cherenkov neutrino experiment. On the largest scale, X-RICE would deploy antennas on a spacing of 1 per km over 10^4 km^2 such as is necessary to carry out the EeV neutrino astronomy program described above.

3.1 RICE

The RICE collaboration has been active since 1995. The goal has been to deploy an array of antennas with a coincident trigger and a data acquisition system that allows one to distinguish neutrino induced radio pulses from other noise and background events which may occur near the array. As of 1999, the array consists of 12 in ice receivers, 3 surface horns and 4 in ice transmitters that can be used for calibration and ice transparency studies. The DAQ requires at least four amplitude discriminated hits from 12 channels in order to form a trigger. With each trigger, time delays are calculated, and with four hits an event vertex can be generated. By doing a vertex reconstruction cut, a real time veto can eliminate pulses generated at the surface in the building which houses the RICE and AMANDA electronics. The number of hits also helps lower the signal to noise threshold required on each channel. These systems are now in place and tested. A description of the veto system may be found in the 1999 ICRC proceedings[11]. The array sensitive volume is roughly .01 km^3 at 10 PeV. It is not likely that an array this small will detect sources of astrophysical neutrinos: a more pragmatic goal is technology demonstration and upper limits on some of the more optimistic models for PeV neutrinos.

3.2 RICE^3

The second level of development for the radio technique would be an array occupying roughly a km^3 of instrumented volume. This is still not large enough to do EeV astronomy, but should be capable of PeV neutrino astronomy. The goal is to colocate with ICECUBE, a km^3 expansion of the AMANDA experiment. With similar senstive volumes, the challenge is to instrument the radio array densely enough to push the threshold down to a PeV, so that a reasonable number of events would be detected both by ICECUBE and RICE^3. A straw design includes 330 antennas, much fewer and individually cheaper than the phototubes needed for the optical array. A dual deployment would have many advantages. Foremost would be confirmation of the neutrino events seen by each technique. The dual observations would also allow for enhanced flavor identification, source location, etc, thus improving the science return at a modest increase in cost. From the point of view of proceeding to detection of EeV neutrinos, the most important benefit would be an *in situ* calibration of the radio technique. This is an essential task before commiting resources to a much larger detector.

3.3 X-RICE

The largest experiment envisioned is X-RICE, consisting of roughly 1 antenna per km^2 with a sensitive volume of 10^4 km^3. Compared to RICE^3 the low density of antennas is possible since the design threshold is significantly higher - an EeV as opposed to a PeV. The higher threshold translates into an effective charge 1000 times higher, or equivalently 10^6 times the signal power. With so much energy in the radio pulse, antennas several km from the event may still register hits. As a test of the idea, Seckel and Frichter[10] simulated various array specifications, varying the spacing of the ice holes in which antennas would be deployed and the number of antennas per hole. They found that an array spacing of one antenna per hole at 1/2 km spacing or four antennas per hole at one km spacing would both give roughly 1 km effective depth for the array, i.e. an array of 10^4 km^2 would have an effective volume of 10^4 km^3.

It appears that the design concept is valid in terms of sensitivity. More difficult issues concern logistics, data acquisition over such an extended area, trigger formation, operation in the polar winter, etc. Such an experiment would have the combined headaches of AMANDA/ICECUBE and AUGER. Optimistically, AMANDA/ICECUBE is demonstrating the feasibility of performing neutrino astronomy in Antarctic ice. AUGER is demonstrating the ability to operate an experiment distributed on the larger physical scale needed for X-RICE, form triggers in real time, and do data acquisition. Although daunting, there are precedents for dealing with the technical challenges.

4 Summary

Detection of EeV astrophysical neutrinos could be a discriminating factor between models of ultra high energy cosmic rays above the GZK cutoff. Although it hasn't been mentioned so far, such observations would also have implications for particle physics. At a minimum, it would provide a very long baseline neutrino oscillation experiment. Confirmation of top down models would be a window into GUT scale physics. Observation of UHEN's coincident with a Gamma Ray Burst would provide some of the most sensitive tests of special and general relativity.

Assuming conservative source models, a detector with mass of 10^{13} tons is required. Such a detector, utilizing the radio cherenkov technique, could consist of a 10^4 km^2 sparse radio array planted in the Antarctic ice. A three step development project is outlined. The first step, RICE, is nearing its goals as a demonstration project. The second step, RICE^3 would overlap ICECUBE, enhancing the science gain of that project while serving to calibrate the radio technique. The third step would be an independent radio array on the scale necessary to do EeV neutrino astronomy.

Acknowledgments

This work was partially supported by DOE grant DE-FG02-91ER40626.

References

1. Gaisser, T.K., F. Halzen, and T. Stanev, *Phys. Rep.* **258**, 173 (1995).
2. Yoshida, S. and M. Teshiima 1993, Prog. Theor. Phys. 89, 833.

3. Gandhi, R., C. Quigg, H. Reno, I. Sarkevic 1998, Phys. Rev. D58 (093009).
4. Gorham, P.W., K.M. Liewer and C.J. Naudet 1999, Proceedings 826th Int. Cos. Ray Conf., Salt Lake City (astro-ph/9906504).
5. Spencer Klein, Private Communication, 1999.
6. Reno, M.H., I. Sarkevic, and D. Seckel 1999, In preparation.
7. Zas, E., F. Halzen, and T. Stanev 1992, Phys. Rev. D 45, 362.
8. Frichter, G., J. Ralston and D. McKay 1996, Phys. Rev. D 53, 1684.
9. Alarez-Muniz, J. and E. Zas 1997, Phys. Lett. B 411, 218, Alarez-Muniz, J. and E. Zas 1998, Phys. Lett. B 434, 396.
10. Seckel, D. and G. Frichter 1999, HE 6.3.12, Proc. 26th Int. Cos. Ray Conf., Salt Lake City.
11. Frichter, G. et al. 1999, Proceedings 26th Int. Cos. Ray Conf., Salt Lake City.

ULTRA-HIGH ENERGY NEUTRINOS WITH AUGER

STÉPHANE COUTU
104 Davey Laboratory, The Pennsylvania State University, University Park, PA 16802, USA
E-mail: coutu@phys.psu.edu

XAVIER BERTOU, PIERRE BILLOIR
L.P.N.H.E., Universities Paris VI and VII, 4 Place Jussieu, Paris, France
E-mail: Xavier.Bertou@ens.fr, billoir@in2p3.fr

(FOR THE AUGER COLLABORATION)

The advent of the next generation of cosmic-ray observatories will open a new energy frontier in the study of neutrinos. Ultra-high energy neutrinos are predicted to exist due to interactions of the highest energy cosmic rays with the radiation fields permeating the Universe, but have never been observed. Moreover, it is conceivable that such neutrinos are produced by exotic mechanisms, such as the decay of topological defects. Naturally occurring neutrinos at energies beyond about 10^{18} eV will be detected and studied with the Pierre Auger Observatory, revealed as nearly horizontal air showers initiating very deep in the atmosphere. In this paper, the triggering and sensitivity of the Auger Observatory to ultra-high energy neutrinos are examined.

1 Ultra-High Energy Cosmic Rays

1.1 Introduction

The origin of the highest energy cosmic rays, the most energetic particles in the Universe, remains one of the foremost problems in particle astrophysics today. Since 1962, when the first cosmic-ray event with an energy approaching 10^{20} eV was observed [1], only a handful of similarly energetic events have been detected world-wide, the maximum to date reported at 3×10^{20} eV [2–5]. Substantial progress has been made in understanding the nature of cosmic rays of relatively modest energy, beyond about 10^{15} eV, leading to the possibility that the highest energy primary particles, beyond about 10^{19} eV, have an entirely different physical origin. The nature of these highest energy particles, and the mechanisms by which they have acquired such tremendous energy, remain as enigmatic today as when they were first observed.

The flux of cosmic rays above 10^{19} eV is extraordinarily low: on the order of one particle per square kilometer per year. This is illustrated in Figure 1, showing a compilation of experimental measurements of the cosmic-ray dif-

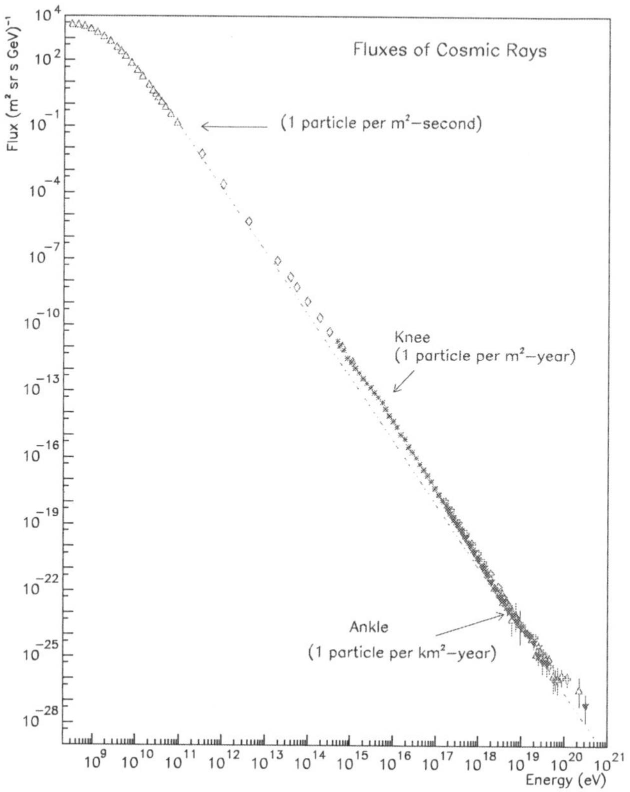

Figure 1. Differential energy spectrum of cosmic rays. The dotted line illustrates an E^{-3} spectrum for comparison. Approximate integral fluxes are indicated at places where changes occur in the behavior of the distribution.

ferential energy spectrum [6]. Only detectors of exceptional size, thousands of square kilometers, can hope to acquire a significant number of events at the highest energies. The nature of the primary particles must be inferred from properties of the air showers they initiate. Two techniques are typically used: 1) surface arrays of detectors sample the lateral density profile of the muon and electromagnetic components of the air shower front; 2) atmospheric fluorescence detectors observe the evolution of air showers – their growth and subsequent attenuation – as they develop. Theoretical understanding of the production and propagation of 10^{20} eV primaries is problematic as well. At such energies, cosmic rays typically have a mean free path less than 50 Mpc

($\sim$ 150 million light-years) – a short distance on a cosmological scale. Cosmic rays lose energy due to interactions with the 2.7K cosmic microwave background radiation: protons photo-produce pions, nuclei photodisintegrate via the giant dipole resonance; photons interact with the radio background to produce e^+e^- pairs; only neutrinos propagate freely. Thus, if the highest energy cosmic rays are cosmological in origin, their observed energy spectrum is expected to exhibit a cut-off near 5×10^{19} eV, an effect known as the Greisen-Zatsepin-Kuz'min (GZK) cut-off [7,8]. Consequently, the source or sources of any observed 10^{20} eV cosmic rays (other than neutrinos) must be relatively nearby, within 50 - 100 Mpc. The interactions that these cosmic rays undergo with the universal background radiation must almost certainly result in the production of neutrinos of exceptional energy. Some theoretical scenarios proposed to explain the origin of the highest-energy particles also invoke the creation of ultra-high energy (UHE) neutrinos. Therefore, projects currently under construction to study the highest-energy cosmic rays, such as the Pierre Auger Observatory, carry the exciting promise of a new frontier in the study of neutrinos.

1.2 Theoretical Understanding

Models of the acceleration of protons or nuclei to the extreme energies considered here have proven very difficult to construct. Few sufficiently energetic astrophysical environments lie within the requisite distance. If particle acceleration proceeds by the Fermi shock-acceleration mechanism [9], *i.e.*, high energies are achieved through repeated encounters with moving magnetized plasmas, then a maximum attainable energy is given by:

$$E_{max} \approx \beta c \times Z e \times B \times L, \tag{1}$$

where βc is the velocity of the shock associated with the moving plasma, Ze is the charge of the accelerated particle, B is the average magnetic field strength in the plasma, and L is the characteristic size of the accelerating region. Only a few known astrophysical sites appear able to generate particle energies above 10^{20} eV [10]. Generally, large structures associated with galaxies or groups of galaxies seem to have sufficient size and field strength to merit consideration as acceleration sites. Moreover, when energy losses within the accelerating region are properly taken into account, the ability of the conventional acceleration scenarios to generate the requisite energies is even more limited.

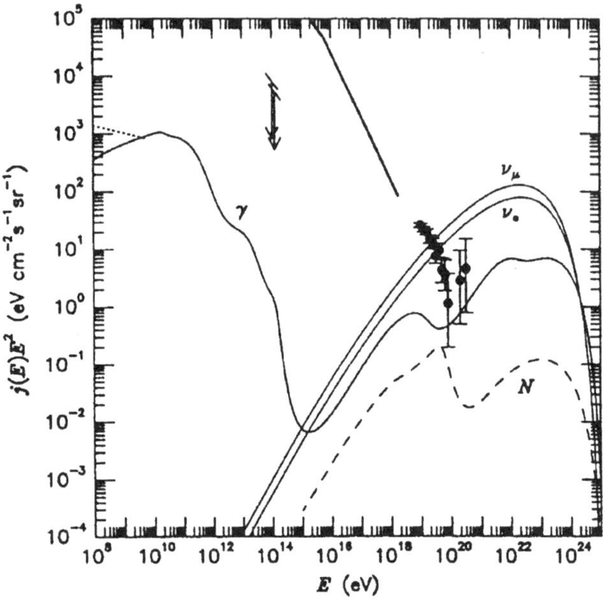

Figure 2. Expected spectrum and composition of the highest energy cosmic rays in a topological defect decay model.

1.3 Ultra-High Energy Neutrinos

Other possibilities for the origin of the highest energy cosmic rays include exotic sources such as topological defects formed during phase transitions as the early Universe cooled, a product of spontaneous symmetry breaking implicit in some Grand Unified Theories (GUTs) [11,12]. Relic topological defects, such as ordinary and superconducting cosmic strings, domain walls, or magnetic monopoles, are relatively stable topologically, but can release part of their energy in the form of X particles, if they collapse or annihilate. These X particles, with typical GUT scale masses on the order of 10^{24} eV, subsequently decay into leptons and quarks, including neutrinos. The spectrum of cosmic rays generated by this mechanism could extend to extraordinarily high energies, perhaps as much as 10^{24} eV. Figure 2 illustrates the spectrum and composition of the highest energy cosmic rays obtained in one topological defect decay model [13]. Note that the UHE spectrum is dominated by neutrinos, so that here again is an interesting prediction of significant numbers of neutrinos being present at the upper end of the spectrum. Alternatively, cosmologically distant sources of neutrinos, such as active galactic nuclei,

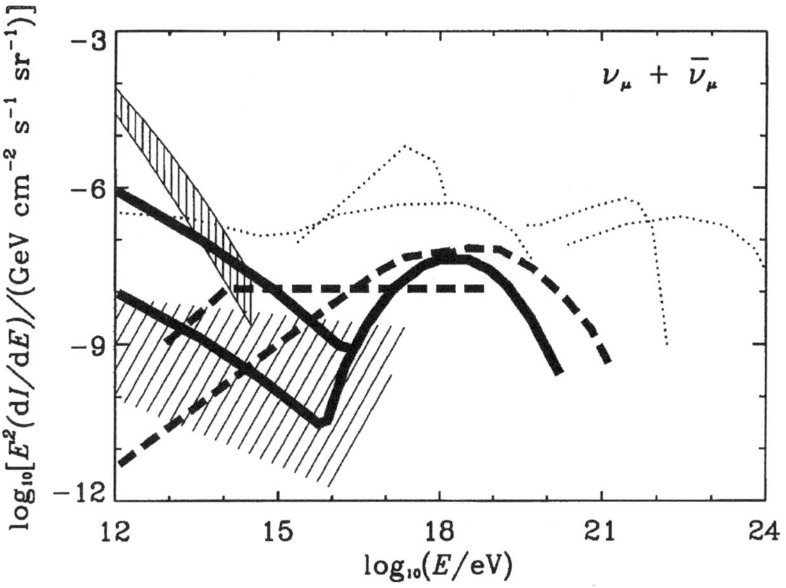

Figure 3. Summary of predicted ν_μ fluxes from different models. The various curves are explained in the text.

have been proposed as the origin [14]. The UHE neutrinos would resonantly annihilate with dark matter neutrinos from the cosmological neighborhood, producing 10^{20} eV protons pointing back to the source. It has been pointed out that the existing few 10^{20} eV events are each consistent with a radio-loud quasar as the source [15]. Once more, neutrinos play a key role.

A summary [16] of theoretical predictions for neutrino fluxes (in this case ν_μ) at energies beyond 10^{12} eV from various models is shown in Figure 3. On it, the closed hatched region indicates atmospheric neutrinos over a range of zenith angles, the open hatched region neutrinos resulting from cosmic-ray interactions within the intergalactic medium in galactic clusters, and thick solid lines represent neutrinos from cosmic-ray interactions with the interstellar medium in our own Galaxy (the upper and lower curves are for extrema of galactic latitude). These predictions are deemed well founded, and neutrinos from these sources are certain. The thick dashed curves and the dotted curves below about 10^{20} eV represent diffuse neutrinos produced in association with gamma ray bursts, in proton blazar models. These sources are thought to be somewhat more speculative, but still fairly certain. Finally, the dotted curves beyond about 10^{20} eV are for neutrinos from topological defect decay models, quite speculative. Figure 4 depicts a few more model predictions [17] for the UHE neutrino flux, including the same atmospheric neutrino contribution as in Figure 3 (leftmost cross-hatched region). Also shown are predictions for neutrinos from optically thin pion photoproduction sources: proton blazars (dotted curve) and gamma ray bursters (dashed curve). The curve labelled "Frejus" is an experimental upper limit. Some UHE neutrino production models may have already been ruled out by constraints from measurements of the diffuse low energy gamma ray flux (below about 10^{14} eV) and of the UHE cosmic ray spectrum itself [18]. This is illustrated in Figure 4 as the solid curve labelled "W&B." This argument is under some debate, and the exact limits imposed on the neutrino flux may not be so stringent. For instance, the rightmost cross-hatched region of Figure 4 represents potentially allowed neutrino fluxes under modified sets of assumptions regarding the input spectrum of extragalactic UHE cosmic rays [17]. The curves labelled "$\tau_{n\gamma} \gg 1$" and "$\tau_{n\gamma} < 1$" are for optically thick and thin pion photoproduction sources, respectively. (The thick dashed curve illustrates the limit of sensitivity of Auger and is described in Section 3.)

Uncertainties in all of the model calculations are appreciable. At any rate, the possibility of detecting and studying neutrinos of such extreme energies, which are certain to exist at some level, is a fascinating one that may yield insight into fundamental astrophysical and particle physics processes. The Pierre Auger Observatory, recently approved and currently under construction, will bring about the opportunity to open this new window on the Universe.

2 The Pierre Auger Observatory

2.1 Scientific Goals

The Auger Observatory [19] is a large-scale, international experiment whose primary goal is to solve the mystery of the highest energy cosmic rays. Two separate, essentially identical detector systems, one in the Northern Hemisphere (in Millard County, Utah) and one in the Southern Hemisphere (in the Province of Mendoza, Argentina), will measure the arrival direction and energy of the UHE primary particles, the muon content of the extensive air showers they induce, and for a subset of events, the longitudinal development of the showers in the atmosphere. Analysis of these shower parameters, guided by extensive and detailed Monte Carlo simulations, can be used to select event samples enriched in either light ($A \leq 16$) or heavy ($A \geq 16$) primaries. Correlations among the various parameters – energy, arrival direction, and mass

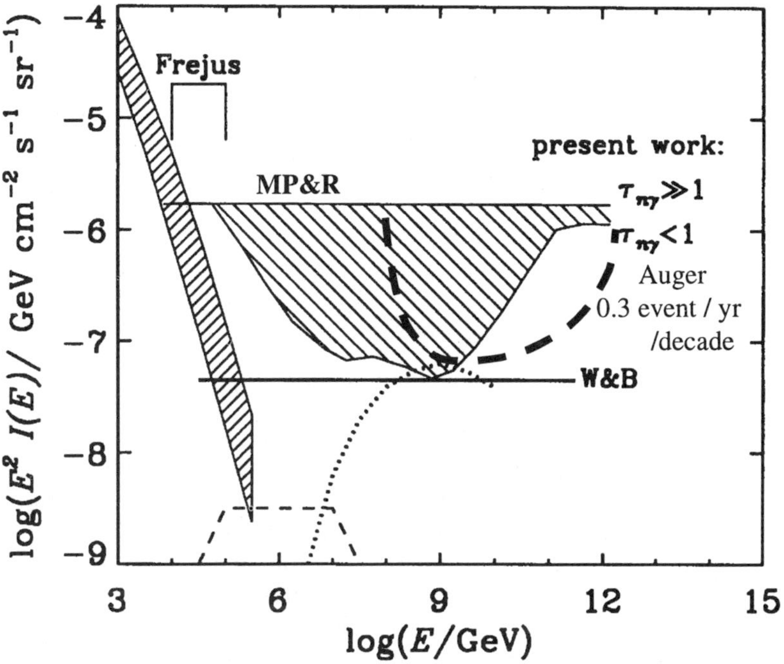

Figure 4. Flux limits for UHE neutrinos from pion photoproduction sources, and limit of the sensitivity of Auger. The various curves are explained in the text.

– may reveal that a number of different components contribute to the total observed cosmic-ray flux. For example, it is possible that a significant fraction of the primaries around 10^{19} eV are heavy nuclei produced in our Galaxy. If so, their arrival directions ought to be correlated with the Galactic mass distribution, *i.e.*, they should appear to come from the direction of the plane, as opposed to the poles, of the Galaxy. This component, nearly unaffected by the cosmic microwave background radiation, would exhibit an energy spectrum directly reflecting the production mechanism within the Galactic accelerators. If the light primaries, including protons, are isotropic and exhibit a cut-off near 10^{20} eV, they would represent a universal extragalactic component which has suffered attenuation due to its interaction with the cosmic microwave background.

It is the origin of the very highest energy ($E \geq 10^{20}$ eV) cosmic rays that is most perplexing, and there is virtually no information as to their fundamental nature. It would be crucial to determine whether they are protons, nuclei, neutrinos, or perhaps exotic particles such as magnetic monopoles, and to determine how and where they acquire such enormous energies. The study of their arrival directions will determine whether they cluster about any known, particularly energetic astrophysical objects, or about any extended mass distribution such as the super-galactic plane. If they do not exhibit any directional anisotropy, then they could be among the products created in the decay of topological defects, in which case their energy distribution should be characteristic of a "top down" source: the product of a cascade from still higher energies.

2.2 Detector Design

The Auger Observatory is a hybrid detector, employing two complementary and well established techniques to observe extensive air showers. A giant array of particle counters will measure the lateral and temporal distribution of shower particles at ground level. An optical air fluorescence detector will measure the air shower development in the atmosphere above the surface array. Operating together, the surface array and fluorescence detector characterize showers to a greater degree than either technique alone. The surface array resembles that successfully employed by the Haverah Park group for over twenty years [5], although the Auger array is of much increased dimensions. The optical system uses the fluorescence technique pioneered by the University of Utah's Fly's Eye detector [20]. Measurement of atmospheric fluorescence is possible only on clear, dark nights. About 10% (possibly more) of all Auger showers will be measured by both techniques simultaneously.

Each of the two surface arrays of the Auger Observatory consists of about 1700 detectors spaced on a triangular grid with about 1.5 km separation between individual detectors. Each array encompasses an area of 3000 km^2. The detector configuration for the Southern half of the Observatory is illustrated in Figure 5. The angular and energy resolutions of a ground array alone (without coincident fluorescence data) are typically better than 1.5° and 20%, respectively. If an event trigger is assumed to require five detectors above threshold, the array is fully efficient at 10^{19} eV. New technologies are employed, making it practical to operate thousands of detectors spread over such an area. Each detector will be solar powered and will communicate via modern wireless techniques. Inter-detector relative timing is accomplished by individual Global Positioning Satellite receivers.

The surface stations consist of Water Čerenkov Detectors (WCD), cylindrical water tanks of 10 m^2 area and 1.2 m depth with an efficient, diffusely

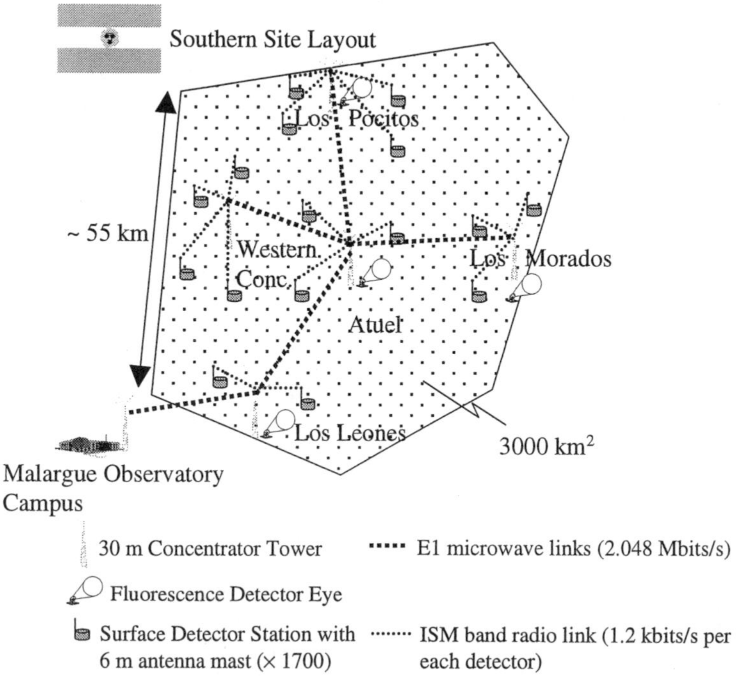

Figure 5. Layout of the Southern half of the Auger Observatory in Argentina.

reflective lining. The Čerenkov light produced by the shower particles is viewed by three downward-facing photomultiplier tubes (PMTs). A WCD, in comparison to a scintillation detector or a gas chamber, is simple, stable, inexpensive, and has a significantly better sensitivity to showers at large zenith angles, which is important in the study of neutrino-induced air showers. Air showers at 10^{19} eV may contain 10^{10} charged particles at the ground and extend over an effective area of twenty square kilometers. Due to the 1.5 km detector separation in Auger, the properties of showers at a distance of a kilometer or more from the core are the most relevant to the detector design. At a distance of 1 km from the core, the charged particle density has decreased to a few per square meter, with gamma rays about 10 times more abundant. Although the electromagnetic particles (electrons, positrons, and photons) are 100 times more numerous than the muons at the core, and 10 times at 1 km, their mean energy is only 10 MeV while that of the muons is about 1 GeV. The thickness of the shower front, measured as the time it takes to pass through the detector, is several microseconds. The electrons in the shower front produce a large number of relatively small Čerenkov pulses whereas the muons produce a small number of large pulses. Given the spread of the arrival times, it is feasible, using only moderately fast flash-ADCs, to estimate the muonic and electromagnetic components of the shower independently. Similarly, the rise-time of the shower front signal is related to the relative proportion of muons in the shower (muons are typically "leading" particles within the shower front). The ratio of muons to the electromagnetic component is sensitive to the nuclear type of the primary cosmic-ray particle.

The Auger fluorescence detectors consist of many meter-sized mirrors, each of which is equipped with a cluster of a hundred or more photomultiplier tubes. As shown in Figure 5, for the Southern Auger site the fluorescence detectors are arranged in four groups, corresponding to a total equivalent of three 360° "eyes." Each mirror, with its associated PMT cluster, views its own segment of the sky. Overall, the system of mirrors observes most of the sky above the surface array. The magnitude of the PMT signals yields the number of electromagnetic particles in the shower at a given point along the track across the sky, and the integral signal yields the energy. Fast timing of the sequence of signals yields the trajectory of air showers passing in the field of view of the detector. The measurement of the depth of shower maximum X_{max} yields further sensitivity to the primary particle mass.

Events which are observed by both the fluorescence and the surface detectors can be effectively reconstructed even if only two surface detectors have recorded signals. Thus, the event trigger for the hybrid arrangement can permit "sub-threshold" surface array triggers (*i.e.*, less than five stations). The hybrid threshold therefore extends down to about 10^{18} eV. Events in this lower energy range than that targeted by the experiment can provide a valuable comparison with the extensive data of previous (and current) cosmic-ray experiments. In the hybrid mode of operation, simulations indicate that the surface and fluorescence detectors together have a directional reconstruction resolution of about 0.36° for events near 10^{20} eV, and an energy resolution of 2.5%.

3 Neutrino Detection with Auger

Neutrino-induced air showers are identifiable as such only for nearly horizontal showers. Indeed, the Earth's atmosphere presents a target of thickness only about 1000 g/cm^2 to cosmic ray particles arriving vertically, but a target of thickness ~36,000 g/cm^2 to nearly horizontal particles. With a charged current interaction cross section of order 10^{-32} cm^2 at 10^{18} eV, a neutrino

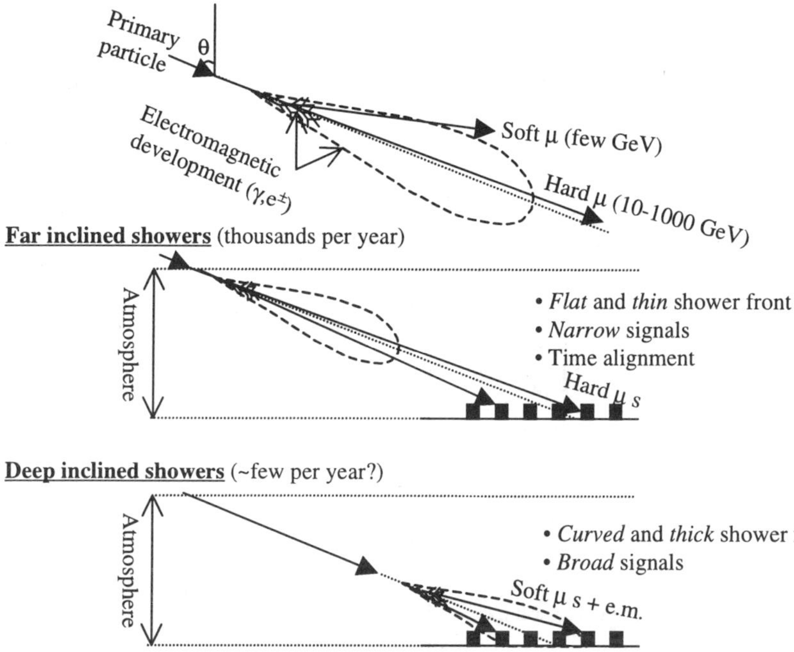

Figure 6. Schematic representation of a UHE air shower, and of its placement with respect to the ground and the Auger array. A "far inclined" shower is likely to be due to a hadronic cosmic ray, whereas a "deep inclined" shower can only be caused by a neutrino.

arriving at a large zenith angle has a signifiant probability of interacting in the atmosphere within a few kilometers of an Auger array covering 3,000 km^2. The hadrons produced in the neutrino interaction initiate an air shower which is in all respects similar to cosmic ray induced showers, except that the neutrino-induced shower can begin deep in the atmosphere. In the charged current interaction of a ν_e, a UHE electron is produced which also initiates a large electromagnetic cascade. In contrast, the charged current interaction of a ν_μ produces a muon which is not detectable by Auger.

For ordinary air showers occurring at large zenith angles, the electromagnetic component is attenuated long before the shower front reaches the ground, whereas the hard muon component (in the energy range 10 to 1000 GeV) can reach a ground based detector array, accompanied by a small amount of radiative products which arrive at the ground at the same time (within a few tens of nanoseconds). Therefore the shower front for such an event is rather

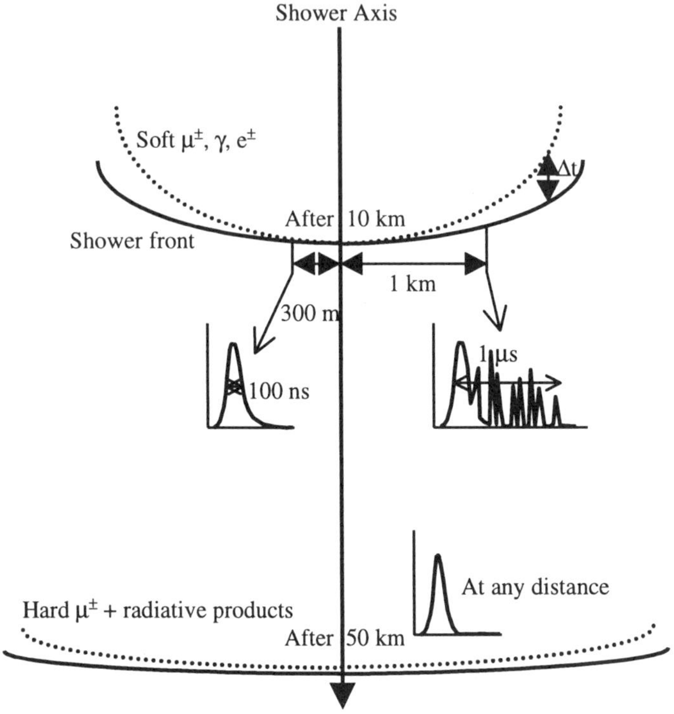

Figure 7. Time spread of the shower front as a function of propagation distance. Deep showers result in a narrow shower front at all distances from the shower core.

flat and thin, with all particles arriving within about 100 ns of a planar front. This represents a "far inclined" shower. In contrast, a neutrino induced event deep in the atmosphere (a "deep inclined" shower) yields a much larger electromagnetic contribution at the ground. This is illustrated in Figure 6 which shows the geometry of an inclined air shower in relation to a ground array. The shower front in the case of a deep shower departs measurably from a plane, and particle arrival times can be spread out over several microseconds. Figure 7 illustrates the spread in time of the UHE shower front as the shower develops in the atmosphere. In principle then, for a detector such as Auger capable of distinguishing between muons and electromagnetic activity, it is straightforward to distinguish between neutrino-induced events at large zeniths (from 70° to the horizon) from other types of cosmic rays.

With one Auger ground array alone, a neutrino acceptance of 18,000 km^{3}sr at 10^{18} eV is achieved. This can be estimated roughly by multiplying the surface area of 3,000 km^2 by an effective solid angle of 2 sr beyond 70° and by an effective production height above ground varying between 0 and 5 km, depending on zenith angle (the useful shower length being about 15 km at 10^{18} eV). At energies below 10^{18} eV, the acceptance is reduced because of the diminished lateral extent of the electromagnetic component, which then requires the shower front to strike the ground in a position with respect to the array that is favorable for triggering. At energies greater than 10^{18} eV, the acceptance increases smoothly with energy owing to an increased effective shower length. It is worth noting that 18,000 km^3sr of air corresponds to roughly 18 km^3sr of water, *i.e.*, a size comparable to the acceptance of proposed ice Čerenkov projects such as the ICECUBE detector. An important distinction however is due to the fact that the Earth is opaque to the 10^{18} eV neutrinos that will be detectable by Auger, for which the target is the atmosphere. Hence the ice Čerenkov projects will aim at a much lower energy regime, and Auger is truly complementary to these other efforts.

The considerations and simple estimates above are borne out by Monte Carlo calculations of the response of a simulated array to hadron-induced showers injected at various altitudes and zenith angles [21]. The different signals resulting from an ordinary, proton induced "far" shower and from a neutrino induced "deep" shower are illustrated in Figures 8 and 9, respectively. These represent events simulated using a custom Monte Carlo algorithm (the AIRES program, based in part on the MOCCA [22] and SIBYLL [23] algorithms). Figure 8 shows the muon particle densities at the ground (top two panels) for a 10^{20} eV proton induced shower injected at an altitude of 100 km and a zenith angle of 80°. Only muons penetrate the atmosphere through to the ground, so no distribution of electromagnetic energy density is shown. In the top panel, the shower is injected along a North-South direction, and in the middle panel along a West-East direction. The differences are due to deflections of the air shower secondaries in the Earth's magnetic field. On the figure, water Čerenkov tanks that have a sufficient signal to satisfy the standard Auger trigger are shown as white circles with a T inside. It is clear that such events can trigger the array and constitute a potential background to neutrino studies via horizontal shower detection. Tanks below threshold that still carry information usable in shower reconstruction are shown as small white circles. The bottom panel of Figure 8 shows the time lag of the muons behind a fictitious planar shower front. It is apparent that all muons arrive within about 100-200 ns of a planar front. In sharp contrast, Figure 9 shows the simulated response to a neutrino induced shower deep within the atmosphere, approximated here by a 10^{19} eV proton induced shower injected at an altitude of 3 km and a zenith angle of 80°. In this case, a signifiant electromagnetic component exists, shown in the middle panel. Given the ability to distinguish the muons from the electromagnetic signals in the Auger water tanks, as described in Section 2.2, the presence of the electromagnetic activity in this event would be a signature for a deep shower. Moreover, the distribution of muon time delays from a planar front, shown in the lower panel, illustrates the curvature and thickness of the front, with particle arrival times spread over many hundreds of nanosecond or more.

Expected event rates due to horizontal UHE neutrinos depend on a number of factors, such as the source of the neutrinos, the behavior of the neutrino interaction cross-section with energy, the details of the array triggering efficiency, etc. Nevertheless, approximate rates can be calculated and are summarized in Table 1. Shown are expected numbers of events per year for two different, bracketing expectations for the neutrino interaction cross-section (labeled MSR [24] and GRV [25], respectively). Event rates are calculated for neutrinos from cosmic ray interactions with the cosmic microwave background (CMBR) and from topological defect decays (TD). Two different water tank triggering conditions are considered, a threshold of 0.1 or 1 electron/m^2. The exact instrumental triggering scheme is still being optimized, but is likely to be close to 1 electron/m^2. With this sensitivity and additional criteria (time spread of the signals, rough compatibility with a planar front), background studies with prototype tanks show that the rate of accidental coincidences between four stations or more may be reduced to well below 1 event/yr; coincidences between three stations could be significant. Backgrounds from the muonic tails of distant showers are still under study, and should be rejectable based on the electromagnetic activity of the shower. From Table 1, it appears that event rates of a few per year from essentially certain sources will be just sufficient to claim detection, and that a larger rate would point to new and fundamental physics. An estimate of the maximum sensitivity of the Auger Observatory to UHE neutrinos is indicated by the thick dashed curve of Figure 4, representing 0.3 event detected per year per energy decade. Thus, Auger will be sensitive not only to secondary UHE neutrinos from cosmic ray interactions with the CMBR and to exotic TD sources, but also to much of the region of interest for pion photoproduction sources.

4 Conclusions

The Pierre Auger Observatory now under construction has been optimized for the study of UHE cosmic rays beyond 10^{18} eV. The study of nearly horizontal air showers will permit the identification of air showers induced by UHE neutrinos deep in the atmosphere. Because of Auger's very large ac-

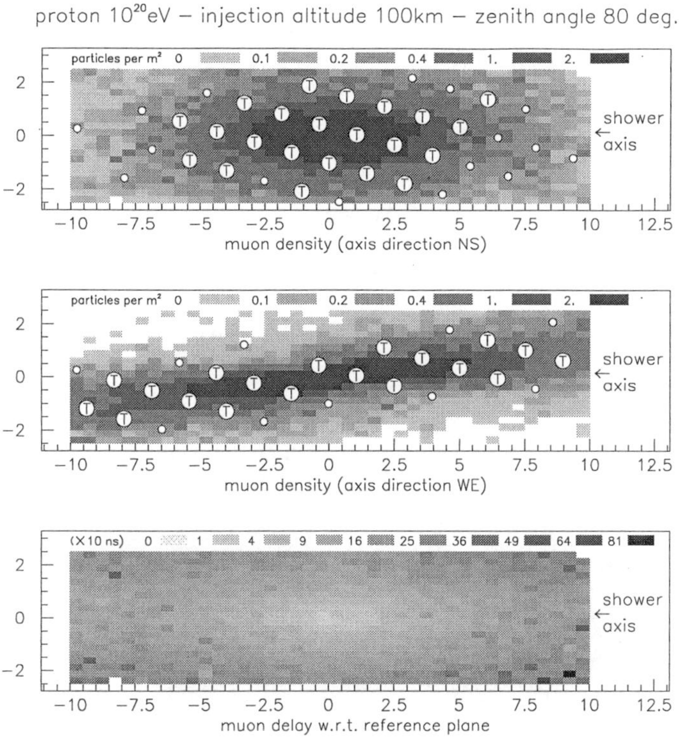

Figure 8. Simulated muon density and muon time delay distributions at the ground for a 10^{20} eV proton induced shower, with an injection height of 100 km at a zenith angle of 80°. The x and y axes are coordinates on the ground (in km), and the gray scales represent particle densities (top two panels) and shift in time from a planar shower front (bottom panel). For the top panel, the shower axis is injected in a North-South direction, whereas a West-East direction is used for the middle panel. Circles with a T represent triggered water Čerenkov tanks, and small white circles are tanks below threshold with information usable in shower reconstruction.

ceptance, rates of UHE neutrino induced events appear to be sufficient to detect neutrinos from cosmic ray interactions with the CMBR, and possibly from primary UHE cosmic ray pion photoproduction sources. Larger than expected neutrino rates could indicate a more exotic source such as topological defect decays. With the projected operation of the Auger Observatory of some twenty years or more, sufficient statistical samples will be acquired to

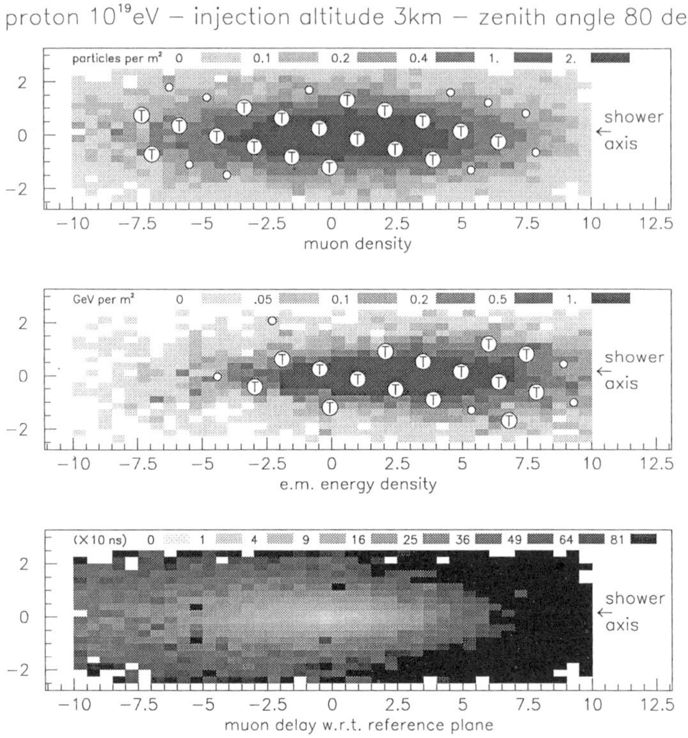

Figure 9. Simulated muon density, electromagnetic energy density and muon time delay distributions at the ground for a 10^{19} eV proton induced shower, with an injection height of 3 km at a zenith angle of 80°. The x and y axes are coordinates on the ground (in km), and the gray scales represent particle densities (top two panels) and shift in time from a planar shower front (bottom panel). Circles with a T represent triggered water Čerenkov tanks, and small white circles are tanks below threshold with information usable in shower reconstruction.

make unambiguous statements on the origin of the UHE neutrinos.

Acknowledgments

This work has been supported in part by the National Science Foundation.

ρ_e^{th} (m^{-2})	MSR cross-section	GRV cross-section
	CMBR	CMBR
1	1 ev/yr	0.5 ev/yr
0.1	3 ev/yr	1 ev/yr
	TD	TD
1	27 ev/yr	8 ev/yr
0.1	51 ev/yr	17 ev/yr

Table 1. Expected horizontal neutrino-induced event rates per year under two bracketing expectations for the neutrino interaction cross-section, labeled MSR and GRV, for neutrinos from cosmic ray interactions with the cosmic microwave background and from topological defect decays. Two possible tank trigger threshold conditions are considered, where the minimum required electromagnetic particle density ρ_e^{th} is either 1 or 0.1 particle per m^2. The final instrumental trigger is likely to use a threshold close to 1 m^{-2}.

References

1. J. Linsley, Phys. Rev. Lett. **10**, 146 (1963).
2. D. J. Bird *et al.*, Astrophys. J. **441**, 144 (1995).
3. N. N. Efimov *et al.*, Proc. ICRR International Symposium on the Astrophysical Aspects of the Most Energetic Cosmic Rays (eds. M. Nagano, F. Takahara), p. 20 (1991).
4. N. Hayashida *et al.*, Phys. Rev. Lett. **73**, 3491 (1994).
5. M. A. Lawrence, R. J. O. Reid and A. A. Watson, J. Phys. G **17**, 733 (1991).
6. S. Swordy, private communication. The points are from published results of the LEAP, Proton, Akeno, AGASA, Fly's Eye, Haverah Park, and Yakutsk experiments.
7. K. Greisen, Phys. Rev. Letters **16**, 748 (1966).
8. G. T. Zatsepin and V. A. Kuz'min, JETP Letters **4**, 78 (1966).
9. E. Fermi, Phys. Rev. **75**, 1169 (1949).
10. A.M. Hillas, Ann. Rev. Astron. Astrophys. **22**, 425 (1984).
11. C. T. Hill, Nucl. Phys. B **224**, 469 (1983); C. T. Hill, D. N. Schramm, and T. P. Walker, Phys. Rev. D **36**, 1007 (1987); P. Bhattacharjee, Phys. Rev. D **40**, 3968 (1989); P. Bhattacharjee and N. C. Rana, Phys. Lett. **B246**, 365 (1990); G. Sigl, Space Sc. Rev. **75**, 375 (1996).
12. P. Bhattacharjee, C. T. Hill, and D. N. Schramm, Phys. Rev. Lett. **69**, 567 (1992).
13. G. Sigl, S. Lee, D. N. Schramm, and P. Coppi, Phys. Lett. **B392**, 129 (1997).

14. T. Weiler, Proc. of the Workshop on Observing Giant Air Showers from Space, AIP Conf. Proc. **433**, 246 (1998).
15. G. Farrar & P. Biermann, Phys. Rev. Lett. **81**, 3579 (1998).
16. R. J. Protheroe, astro-ph/9809144, Proc. of the Neutrino 98 Conf., Takayama, Japan (1998).
17. K. Mannheim, R.J. Protheroe & J.P. Rachen, astro-ph/9812398, submitted to Phys. Rev. D (1999).
18. E. Waxman & J. Bahcall, Phys. Rev. D **59**, 023002 (1999).
19. Auger Collaboration, The Pierre Auger Observatory Design Report, 2nd Edition (1997), available on the World-Wide Web from http://www.auger.org/admin/DesignReport/index.html
20. R. M. Baltrusaitis *et al.*, Nucl. Inst. Methods A **240**, 410 (1985).
21. K. S. Capelle *et al.*, Astropart. Phys. **8**, 321 (1998); P. Billoir, Auger internal memo 97-049 (1997); X. Bertou, P. Billoir, and T. Pradier, Auger internal memo 97-058 (1997); Auger internal memos are available on the World-Wide Web at http://www.auger.org/admin/GAP_Notes/index.html
22. A. M. Hillas, Proc. of the 19th Int. Cosmic Ray Conf. (La Jolla) **1**, 155 (1985).
23. R. S. Fletcher *et al.*, Phys. Rev. D **50**, 5710 (1994).
24. A. D. Martin, R. G. Roberts, and W. J. Stirling, Phys. Lett. B **387**, 419 (1996).
25. M. Glück, E. Reya, and A. Vogt, Z. Phys. C **67**, 433 (1995).

SOLAR NEUTRINOS

J. N. BAHCALL
Institute for Advanced Study, Princeton, NJ 08540, USA
E-mail: jnb@sns.ias.edu

I summarize the current state of solar neutrino research.

1 Introduction

The most important result from solar neutrino research is, in my view, that solar neutrinos have been detected experimentally with fluxes and energies that are qualitatively consistent with solar models that are constructed assuming that the sun shines by nuclear fusion reactions. The first experimental result, obtained by Ray Davis and his collaborators in 1968, [1,2] has now been confirmed by four other beautiful experiments, Kamiokande, [3] SAGE, [4] GALLEX, [5] and SuperKamiokande. [6]

The observation of solar neutrinos with approximately the predicted energies and fluxes establishes empirically the theory [7] that main sequence stars derive their energy from nuclear fusion reactions in their interiors and has inaugurated what we all hope will be a flourishing field of observational neutrino astronomy. The detections of solar neutrinos settle experimentally the debate over the age and energy source of the sun that raged for many decades, beginning in the middle of the 19th century. The leading theoretical physicists of the 19th century argued convincingly that the sun could not be more than 10^7 years old because that was the maximum lifetime that could be fueled by gravitational energy ("No other natural explanation, except chemical action, can be conceived.", Lord Kelvin). [8] On the other hand, geologists and evolutionary biologists argued that the sun must be $> 10^9$ years old in order to account for observed geological features and for evolutionary processes [9a]. Today we know that the biologists and geologists were right and the theoretical physicists were wrong, which may be a historical lesson to which we physicists should pay attention.

[a]The arguments of Lord Kelvin and his theoretical physics associates were so persuasive that in later editions Darwin removed all mention of time scales from "The Origin of the Species."

I will discuss predictions of the combined standard model in the main part of this review. By 'combined' standard model, I mean the predictions of the standard solar model and the predictions of the minimal electroweak theory. We need a solar model to tell us how many neutrinos of what energy are produced in the sun and we need electroweak theory to tell us how the number and flavor content of the neutrinos are changed as they make their way from the center of the sun to detectors on earth. For all practical purposes, standard electroweak theory states that nothing happens to solar neutrinos after they are created in the deep interior of the sun.

Using standard electroweak theory and fluxes from the standard solar model, one can calculate the rates of neutrino interactions in different terrestrial detectors with a variety of energy sensitivities. The combined standard model also predicts that the energy spectrum from a given neutrino source should be the same for neutrinos produced in terrestrial laboratories and in the sun and that there should not be measurable time-dependences (other than the seasonal dependence caused by the earth's orbit around the sun). The spectral and temporal departures from standard model expectations are expected to be small in all currently operating experiments [10] and have not yet yielded definitive results. Therefore, I will concentrate here on inferences that can be drawn by comparing the total rates observed in solar neutrino experiments with the combined standard model predictions.

I will begin by reviewing in Section 2 the quantitative predictions of the combined standard solar model and then describe in Section 3 the three solar neutrino problems that are established by the chlorine, Kamiokande, SAGE, GALLEX, and SuperKamiokande experiments. In Section 4, I detail the uncertainties in the standard model predictions and then show in Section 5 that helioseismological measurements indicate that the standard solar model predictions are accurate for our purposes. In Section 5, I discuss the implications for solar neutrino research of the precise agreement between helioseismological measurements and the predictions of standard solar models. Next, ignoring all knowledge of the sun, I cite analyses in Section 6 that show that one cannot fit the existing experimental data with neutrino fluxes that are arbitrary parameters, unless one invokes new physics to change the shape or flavor content of the neutrino energy spectrum. I summarize in Section 7 the characteristics of the best-fitting neutrino oscillation descriptions of the experimental data. Finally, I will discuss and summarize the results in Section 8.

If you want to obtain numerical data or subroutines that are discussed in this talk, or to see relevant background information, you can copy them from my Web site: http://www.sns.ias.edu/~jnb .

Before we begin the detailed discussion, I want to make just a brief historical diversion. Nearly all of the current interest in solar neutrinos centers around the opportunity to use the sun as a neutrino source in a very long baseline oscillation experiment. In preparing for a talk in honor of Fred Reines a few weeks ago, I ran across a long forgotten 1972 letter from Bruno Pontecorvo, the originator of the hypothesis that oscillations may be observed in solar neutrino experiments. For your interest, I enclose a reproduction of this letter in Fig. 1.

For the benefit of neutrino pioneers of today, it is perhaps worth remarking that Ray Davis and I never considered the possibility that solar neutrinos could be used to learn more about neutrinos when, in the early 1960's, we were first analyzing the potentialities of a practical chlorine experiment. We sold the experiment as a fundamental test of the hypothesis that the sun shines by nuclear fusion reactions in its interior. Only after the first results of the chlorine experiment showed in 1968 a conflict with the solar model calculations and Gribov and Pontecorvo published their epochal 1969 paper on vacuum oscillations of solar neutrinos did we begin to consider the possibility that solar neutrinos might tell us something new about particle physics. Maybe there are previously unimagined physics treasures to be discovered in future neutrino experiments.

2 Standard Model Predictions

Table 1 gives the neutrino fluxes and their uncertainties for our best standard solar model, hereafter BP98. [11] Figure 2 shows the predicted neutrino fluxes from the dominant *p-p* fusion chain.

The BP98 solar model includes diffusion of heavy elements and helium, makes use of the nuclear reaction rates recommended by the expert workshop held at the Institute of Nuclear Theory, [12] recent (1996) Livermore OPAL radiative opacities, [13] the OPAL equation of state, [14] and electron and ion screening as determined by the recent density matrix calculation. [15,16] The neutrino absorption cross sections that are used in constructing Table 1 are the most accurate values available [17,18] and include, where appropriate, the thermal energy of fusing solar ions and improved nuclear and atomic data. The validity of the absorption cross sections has recently been confirmed experimentally using intense radioactive sources of ^{51}Cr. The ratio, R, of the capture rate measured (in GALLEX and SAGE) to the calculated ^{51}Cr capture rate is $R = 0.95 \pm 0.07$ (exp) $+$ $^{+0.04}_{-0.03}$ (theory) and was discussed extensively at Neutrino 98 by Gavrin and by Kirsten. The neutrino-electron scattering cross sections, used in interpreting the Kamiokande and SuperKamiokande experiments, now include electroweak radiative corrections. [19]

ОБЪЕДИНЕННЫЙ ИНСТИТУТ ЯДЕРНЫХ ИССЛЕДОВАНИЙ

JOINT INSTITUTE FOR NUCLEAR RESEARCH

Москва, Главный почтамт п/я 79. Head Post Office, P.O. Box 79, Moscow, USSR

№ 994/31 April 6/10 19 72

Prof. J.N.Bahcall
The Institute for Advanced Study
School of Natural Science
Princeton, New Jersey 08540, USA

Dear Prof. Bahcall,

Thank you very much for your letter and the abstract of the new Davis investigation the numerical results of which I did not know. It starts to be really interesting! It would be nice if all this will end with something unexpected from the point of view of particle physics. Unfortunately, it will not be easy to demonstrate this, even if nature works that way.

I will attend the Balaton meeting on neutrinos and looking forward to see you there.

Yours sincerely,

B.Pontecorvo

BMP/nn

Figure 1. Letter from Bruno Pontecorvo in 1972

Figure 3 shows for the chlorine experiment all the predicted rates and the estimated uncertainties (1σ) published by my colleagues and myself since the first measurement by Ray Davis and his colleagues in 1968. This figure should give you some feeling for the robustness of the solar model calculations. Many hundreds and probably thousands of researchers have, over three decades, made great improvements in the input data for the solar models, including nuclear cross sections, neutrino cross sections, measured element abundances on the surface of the sun, the solar luminosity, the stellar radiative opacity, and the stellar equation of state. Nevertheless, the most accurate predictions of today are essentially the same as they were in 1968 (although now they can be made with much greater confidence). For the gallium experiments, the neutrino fluxes predicted by standard solar models, corrected for diffusion, have been in the range 120 SNU to 141 SNU since 1968. [17] A SNU is a convenient unit with which to describe the measured rates of solar neutrino experiments: 10^{-36} interactions per target atom per second.

There are three reasons that the theoretical calculations of neutrino fluxes are robust: 1) the availability of precision measurements and precision calculations of input data; 2) the connection between neutrino fluxes and the measured solar luminosity; and 3) the measurement of the helioseismological frequencies of the solar pressure-mode (p-mode) eigenfrequencies. I have discussed these reasons in detail in another talk. [21]

Figure 4 displays the calculated ^{7}Be and ^{8}B neutrino fluxes for all 19 standard solar models which have been published in the last 10 years in refereed science journals. The fluxes are normalized by dividing each published value

Table 1. Standard Model Predictions (BP98): solar neutrino fluxes and neutrino capture rates, with 1σ uncertainties from all sources (combined quadratically).

Source	Flux (10^{10} cm^{-2}s^{-1})	Cl (SNU)	Ga (SNU)
pp	$5.94\,(1.00^{+0.01}_{-0.01})$	0.0	69.6
pep	$1.39 \times 10^{-2}\,(1.00^{+0.01}_{-0.01})$	0.2	2.8
hep	2.10×10^{-7}	0.0	0.0
^{7}Be	$4.80 \times 10^{-1}\,(1.00^{+0.09}_{-0.09})$	1.15	34.4
^{8}B	$5.15 \times 10^{-4}\,(1.00^{+0.19}_{-0.14})$	5.9	12.4
^{13}N	$6.05 \times 10^{-2}\,(1.00^{+0.19}_{-0.13})$	0.1	3.7
^{15}O	$5.32 \times 10^{-2}\,(1.00^{+0.22}_{-0.15})$	0.4	6.0
^{17}F	$6.33 \times 10^{-4}\,(1.00^{+0.12}_{-0.11})$	0.0	0.1
Total		$7.7^{+1.2}_{-1.0}$	129^{+8}_{-6}

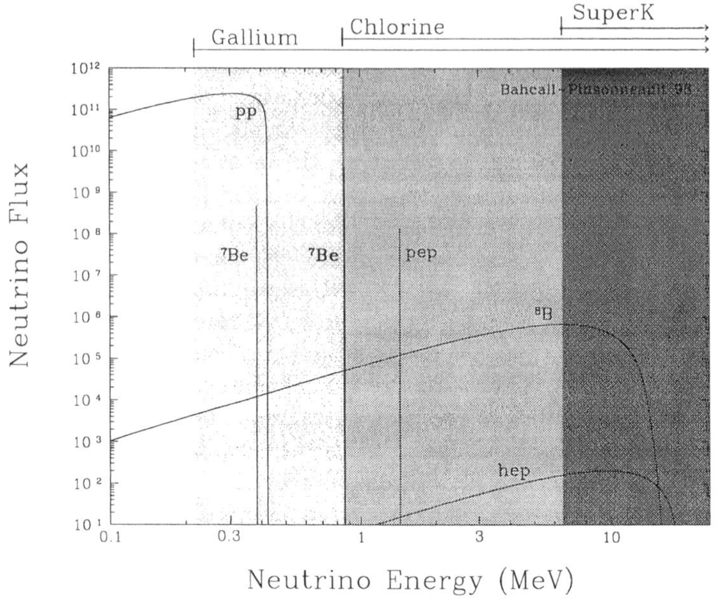

Figure 2. The energy Spectrum of neutrinos from the pp chain of interactions in the Sun, as predicted by the standard solar model. Neutrino fluxes from continuum sources (such as pp and 8B) are given in the units of counts per cm2 per second. The pp chain is responsible for more than 98% of the energy generation in the standard solar model. Neutrinos produced in the carbon-nitrogen-oxygen CNO chain are not important energetically and are difficult to detect experimentally. The arrows at the top of the figure indicate the energy thresholds for the ongoing neutrino experiments.

by the flux from the BP98 solar model;[11] the abscissa is the normalized ^{8}B flux and the ordinate is the normalized ^{7}Be neutrino flux. The rectangular box shows the estimated 3σ uncertainties in the predictions of the BP98 solar model.

All of the solar model results from different groups fall within the estimated 3σ uncertainties in the BP98 analysis (with the exception of the Dar-Shaviv model whose results have not been reproduced by other groups). This agreement demonstrates the robustness of the predictions since the calculations use different computer codes (which achieve varying degrees of precision) and involve a variety of choices for the nuclear parameters, the equation of state, the stellar radiative opacity, the initial heavy element abundances, and the physical processes that are included.

The largest contributions to the dispersion in values in Figure 4 are due to

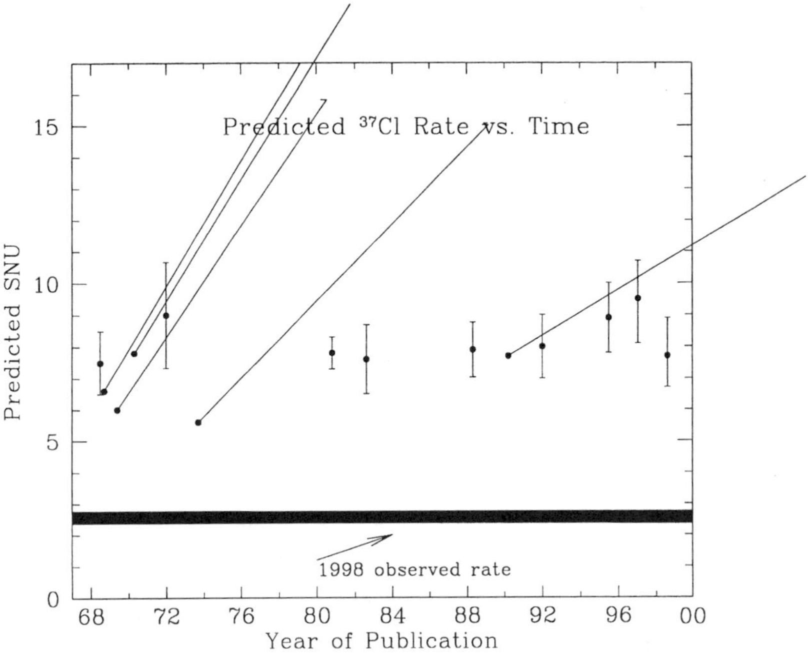

Figure 3. The predictions of John Bahcall and his collaborators of neutrino capture rates in the ^{37}Cl experiment are shown as a function of the date of publication (since the first experimental report in 1968. [1]) The event rate SNU is a convenient product of neutrino flux times interaction cross section, 10^{-36} interactions per target atom per sec. The format is from Figure 1.2 of the book Neutrino Astrophysics. [20] The predictions have been updated through 1998.

the choice of the normalization for S_{17} (the production cross-section factor for ^{8}B neutrinos) and the inclusion, or non-inclusion, of element diffusion in the stellar evolution codes. The effect in the plane of Fig. 4 of the normalization of S_{17} is shown by the difference between the point for BP98 (1.0,1.0), which was computed using the most recent recommended normalization, [12] and the point at (1.18,1.0) which corresponds to the BP98 result with the earlier (CalTech) normalization. [23]

Helioseismological observations have shown [11,24] that element diffusion is occurring and must be included in solar models, so that the most recent models shown in Fig. 4 now all include helium and heavy element diffusion. By comparing a large number of earlier models, it was shown that all published standard solar models give the same results for solar neutrino fluxes to an accuracy of better than 10% if the same input parameters and physical processes are included. [25,26]

Bahcall, Krastev, and Smirnov [10] have compared the observed rates with the calculated, standard model values, combining quadratically the theoretical solar model and experimental uncertainties, as well as the uncertainties in the neutrino cross sections. Since the GALLEX and SAGE experiments measure the same quantity, we treat the weighted average rate in gallium as one experimental number. We adopt the SuperKamiokande measurement as the most precise direct determination of the higher-energy ^8B neutrino flux.

Using the predicted fluxes from the BP98 model, the χ^2 for the fit to the three experimental rates (chlorine, gallium, and SuperKamiokande, see Fig. 5) is

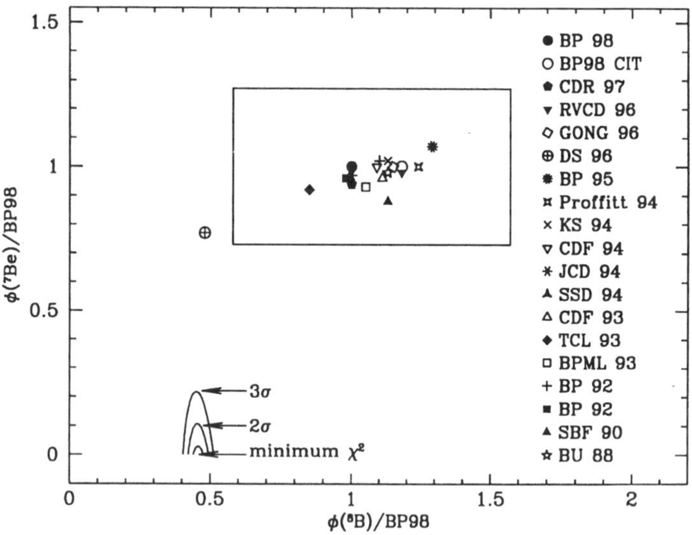

Figure 4. Predictions of standard solar models since 1988. This figure, which is Fig. 1 of Bahcall, Krastev and Smirnov (1998)[10], shows the predictions of 19 standard solar models in the plane defined by the ^{7}Be and ^{8}B neutrino fluxes. The abbreviations that are used in the figure to identify different solar models are defined in the bibliographical item, Ref. 22. The figure includes all standard solar models with which I am familiar that were published in refereed journals in the decade 1988-1998. All of the fluxes are normalized to the predictions of the Bahcall-Pinsonneault 1998 solar model, BP98. [11] The rectangular error box defines the 3σ error range of the BP98 fluxes. The best-fit ^{7}Be neutrino flux is negative. At the 99% C.L., there is no solution [10] with all positive neutrino fluxes (see discussion in Section 6). All of the standard model solutions lie far from the best-fit solution, even far from the 3σ contour.

$$\chi^2_{\rm SSM}(3 \text{ experimental rates}) = 61 . \tag{1}$$

The result given in Eq. (1), which is approximately equivalent to a 20σ discrepancy, is a quantitative expression of the fact that the standard model predictions do not fit the observed solar neutrino measurements.

3 Three Solar Neutrino Problems

I will now compare the predictions of the combined standard model with the results of the operating solar neutrino experiments.

We will see that this comparison leads to three different discrepancies between the calculations and the observations, which I will refer to as the three solar neutrino problems.

Figure 5 shows the measured and the calculated event rates in the five ongoing solar neutrino experiments. This figure reveals three discrepancies between the experimental results and the expectations based upon the combined standard model. As we shall see, only the first of these discrepancies depends in an important way upon the predictions of the standard solar model.

3.1 Calculated versus Observed Absolute Rate

The first solar neutrino experiment to be performed was the chlorine radiochemical experiment, [2] which detects electron-type neutrinos that are more energetic than 0.81 MeV. After more than a quarter of a century of operation of this experiment, the measured event rate is 2.56 ± 0.23 SNU, which is a factor of three less than is predicted by the most detailed theoretical calculations, $7.7^{+1.2}_{-1.0}$ SNU. [11] Most of the predicted rate in the chlorine experiment is from the rare, high-energy ^{8}B neutrinos, although the ^{7}Be neutrinos are also expected to contribute significantly. According to standard model calculations, the *pep* neutrinos and the CNO neutrinos (for simplicity not discussed here) are expected to contribute less than 1 SNU to the total event rate.

This discrepancy between the calculations and the observations for the chlorine experiment was, for more than two decades, the only solar neutrino problem. I shall refer to the chlorine disagreement as the "first" solar neutrino problem.

3.2 Incompatibility of Chlorine and Water Experiments

The second solar neutrino problem results from a comparison of the measured event rates in the chlorine experiment and in the Japanese pure-water

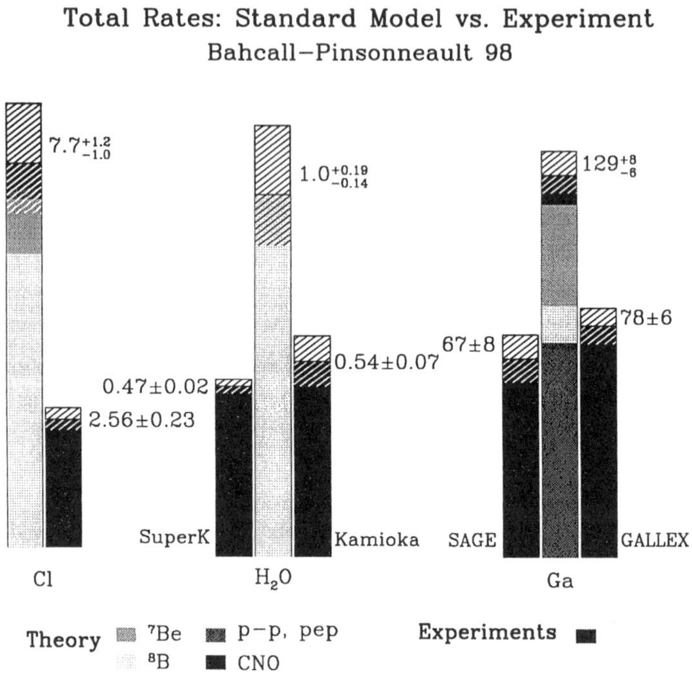

Figure 5. Comparison of measured rates and standard-model predictions for five solar neutrino experiments. [2,3,4,5,6] The unit for the radiochemical experiments (chlorine and gallium) is SNU (see Fig. 3 for a definition); the unit for the water-Cerenkov experiments (Kamiokande and SuperKamiokande) is the rate predicted by the standard solar model plus standard electroweak theory. [11] .

experiments, Kamiokande [3] and SuperKamiokande. [6] The water experiments detect higher-energy neutrinos, most easily above 7 MeV, by by observing the Cerenkov radiation from neutrino-electron scattering: $\nu + e \longrightarrow \nu' + e'$. According to the standard solar model, ^{8}B beta decay, and possibly the *hep* reaction, [27] are the only important source of these higher-energy neutrinos.

The Kamiokande and SuperKamiokande experiments show that the observed neutrinos come from the sun. The electrons that are scattered by the incoming neutrinos recoil predominantly in the direction of the sun-earth vector; the relativistic electrons are observed by the Cerenkov radiation they produce in the water detector. In addition, the water Cerenkov experiments measure the energies of individual scattered electrons and therefore provide information about the energy spectrum of the incident solar neutrinos.

The total event rate in the water experiments, about 0.5 the standard model value (see Fig. 5), is determined by the same high-energy ^8B neutrinos that are expected, on the basis of the combined standard model, to dominate the event rate in the chlorine experiment. I have shown elsewhere [28] that solar physics changes the shape of the ^8B neutrino spectrum by less than 1 part in 10^5 . Therefore, we can calculate the rate in the chlorine experiment (threshold 0.8 MeV) that is produced by the ^8B neutrinos observed in the Kamiokande and SuperKamiokande experiments at an order of magnitude higher energy threshold.

If no new physics changes the shape of the ^8B neutrino energy spectrum, the chlorine rate from ^8B alone is 2.8 ± 0.1 SNU for the SuperKamiokande normalization (3.2 ± 0.4 SNU for the Kamiokande normalization), which exceeds the total observed chlorine rate of 2.56 ± 0.23 SNU.

Comparing the rates of the SuperKamiokande and the chlorine experiments, one finds–assuming that the shape of the energy spectrum of ^8B ν_e's is not changed by new physics–that the net contribution to the chlorine experiment from the *pep*, ^7Be, and CNO neutrino sources is negative: -0.2 ± 0.3 SNU. The contributions from the *pep*, ^7Be, and CNO neutrinos would appear to be completely missing; the standard model prediction for the combined contribution of *pep*, ^7Be, and CNO neutrinos is a relatively large 1.8 SNU (see Table 1). On the other hand, we know that the ^7Be neutrinos must be created in the sun since they are produced by electron capture on the same isotope (^7Be) which gives rise to the ^8B neutrinos by proton capture.

Hans Bethe and I pointed out [29] that this apparent incompatibility of the chlorine and water-Cerenkov experiments constitutes a "second" solar neutrino problem that is almost independent of the absolute rates predicted by solar models. The inference that is usually made from this comparison is that the energy spectrum of ^8B neutrinos is changed from the standard shape by physics not included in the simplest version of the standard electroweak model.

3.3 Gallium Experiments: No Room for ^7Be Neutrinos

The results of the gallium experiments, GALLEX and SAGE, constitute the third solar neutrino problem. The average observed rate in these two experiments is 73 ± 5 SNU, which is accounted for in the standard model by the theoretical rate of 72.4 SNU that is calculated to come from the basic *p-p* and *pep* neutrinos (with only a 1% uncertainty in the standard solar model *p-p* flux). The ^8B neutrinos, which are observed above 6.5 MeV in the Kamiokande experiment, must also contribute to the gallium event rate. Using the standard shape for the spectrum of ^{8}B neutrinos and normalizing to the rate observed in Kamiokande, ^{8}B contributes another 6 SNU. (The contribution predicted by the standard model is 12 SNU, see Table 1.) Given the measured rates in the gallium experiments, there is no room for the additional 34 ± 3 SNU that is expected from ^{7}Be neutrinos on the basis of standard solar models (see Table 1).

The seeming exclusion of everything but *p-p* neutrinos in the gallium experiments is the "third" solar neutrino problem. This problem is essentially independent of the previously-discussed solar neutrino problems, since it depends strongly upon the *p-p* neutrinos that are not observed in the other experiments and whose theoretical flux can be calculated accurately.

The missing ^{7}Be neutrinos cannot be explained away by a change in solar physics. The ^{8}B neutrinos that are observed in the Kamiokande experiment are produced in competition with the missing ^{7}Be neutrinos; the competition is between electron capture on ^{7}Be versus proton capture on ^{7}Be. Solar model explanations that reduce the predicted ^{7}Be flux generically reduce much more (too much) the predictions for the observed ^{8}B flux.

The flux of ^{7}Be neutrinos, $\phi(^7\mathrm{Be})$, is independent of measurement uncertainties in the cross section for the nuclear reaction $^7\mathrm{Be}(p,\gamma)^8\mathrm{B}$; the cross section for this proton-capture reaction is the most uncertain quantity that enters in an important way in the solar model calculations. The flux of ^{7}Be neutrinos depends upon the proton-capture reaction only through the ratio

$$\phi(^7\mathrm{Be}) \;\propto\; \frac{R(e)}{R(e)+R(p)}, \tag{2}$$

where $R(e)$ is the rate of electron capture by ^{7}Be nuclei and $R(p)$ is the rate of proton capture by ^{7}Be. With standard parameters, solar models yield $R(p) \approx 10^{-3}R(e)$. Therefore, one would have to increase the value of the $^7\mathrm{Be}(p,\gamma)^8\mathrm{B}$ cross section by more than two orders of magnitude over the current best-estimate (which has an estimated uncertainty of $\sim 10\%$) in order to affect significantly the calculated ^{7}Be solar neutrino flux. The required change in the nuclear physics cross section would also increase the predicted neutrino event rate by more than 100 in the Kamiokande experiment, making that prediction completely inconsistent with what is observed.

I conclude that either: 1) at least three of the five operating solar neutrino experiments (the two gallium experiments plus either chlorine or the two water Cerenkov experiments, Kamiokande and SuperKamiokande) have yielded misleading results, or 2) physics beyond the standard electroweak model is required to change the energy spectrum of ν_e after the neutrinos are produced

Table 2. Average uncertainties in neutrino fluxes and event rates due to different input data. The flux uncertainties are expressed in fractions of the total flux and the event rate uncertainties are expressed in SNU. The ^{7}Be electron capture rate causes an uncertainty of $\pm 2\%$ [30] that affects only the ^{7}Be neutrino flux. The average fractional uncertainties for individual parameters are shown.

<Fractional uncertainty>	pp	^{3}He^{3}He	^{3}He^{4}He	^{7}Be + p	Z/X	opac	lum	age	diffuse
	0.017	0.060	0.094	0.106	0.033		0.004	0.004	
Flux									
pp	0.002	0.002	0.005	0.000	0.002	0.003	0.003	0.0	0.003
^{7}Be	0.0155	0.023	0.080	0.000	0.019	0.028	0.014	0.003	0.018
^{8}B	0.040	0.021	0.075	0.105	0.042	0.052	0.028	0.006	0.040
SNUs									
Cl	0.3	0.2	0.5	0.6	0.3	0.4	0.2	0.04	0.3
Ga	1.3	0.9	3.3	1.3	1.6	1.8	1.3	0.20	1.5

in the center of the sun.

4 Uncertainties in the Flux Calculations

I will now discuss uncertainties in the solar model flux calculations.

Table 2 summarizes the uncertainties in the most important solar neutrino fluxes and in the Cl and Ga event rates due to different nuclear fusion reactions (the first four entries), the heavy element to hydrogen mass ratio (Z/X), the radiative opacity, the solar luminosity, the assumed solar age, and the helium and heavy element diffusion coefficients. The ^{14}N + p reaction causes a 0.2% uncertainty in the predicted pp flux and a 0.1 SNU uncertainty in the Cl (Ga) event rates.

The predicted event rates for the chlorine and gallium experiments use recent improved calculations of neutrino absorption cross sections. [17,18] The uncertainty in the prediction for the gallium rate is dominated by uncertainties in the neutrino absorption cross sections, +6.7 SNU (7% of the predicted rate) and −3.8 SNU (3% of the predicted rate). The uncertainties in the chlorine absorption cross sections cause an error, ±0.2 SNU (3% of the predicted rate), that is relatively small compared to other uncertainties in predicting the rate for this experiment. For non-standard neutrino energy spectra that result from new neutrino physics, the uncertainties in the predictions for currently favored solutions (which reduce the contributions from the least well-determined ^{8}B neutrinos) will in general be less than the values quoted here for standard spectra and must be calculated using the appropriate cross section uncertainty for each neutrino energy. [17,18]

The nuclear fusion uncertainties in Table 2 were taken from Adelberger *et al.*, [12] the neutrino cross section uncertainties from Bahcall (1997)[17] and Bahcall *et al.* (1996),[18] the heavy element uncertainty was taken from helioseismological measurements, [31] the luminosity and age uncertainties were adopted from BP95, [26] the 1σ fractional uncertainty in the diffusion rate was taken to be 15%, [32] which is supported by helioseismological evidence, [24] and the opacity uncertainty was determined by comparing the results of fluxes computed using the older Los Alamos opacities with fluxes computed using the modern Livermore opacities. [25] To include the effects of asymmetric errors, the now publicly-available code for calculating rates and uncertainties (see discussion in previous section) was run with different input uncertainties and the results averaged. The software contains a description of how each of the uncertainties listed in Table 2 were determined and used.

The low energy cross section of the ^{7}Be$+p$ reaction is the most important quantity that must be determined more accurately in order to decrease the error in the predicted event rates in solar neutrino experiments. The ^{8}B neutrino flux that is measured by the Kamiokande, [3] Super-Kamiokande, [6] and SNO [33] experiments is, in all standard solar model calculations, directly proportional to the ^{7}Be $+ p$ cross section. If the 1σ uncertainty in this cross section can be reduced by a factor of two to 5%, then it will no longer be the limiting uncertainty in predicting the crucial ^{8}B neutrino flux (cf. Table 2).

5 How Large an Uncertainty Does Helioseismology Suggest?

Could the solar model calculations be wrong by enough to explain the discrepancies between predictions and measurements for solar neutrino experiments? Helioseismology, which confirms predictions of the standard solar model to high precision, suggests that the answer is probably "No."

Figure 6 shows the fractional differences between the most accurate available sound speeds measured by helioseismology [34] and sound speeds calculated with our best solar model (with no free parameters). The horizontal line corresponds to the hypothetical case in which the model predictions exactly match the observed values. The rms fractional difference between the calculated and the measured sound speeds is 1.1×10^{-3} for the entire region over which the sound speeds are measured, $0.05R_\odot < R < 0.95R_\odot$. In the solar core, $0.05R_\odot < R < 0.25R_\odot$ (in which about 95% of the solar energy and neutrino flux is produced in a standard model), the rms fractional difference between measured and calculated sound speeds is 0.7×10^{-3}.

Helioseismological measurements also determine two other parameters

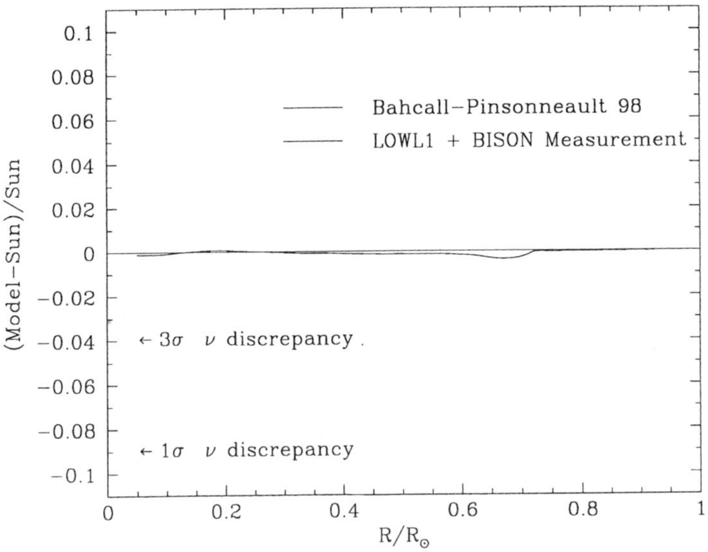

Figure 6. Predicted versus Measured Sound Speeds. This figure shows the excellent agreement between the calculated (solar model BP98, Model) and the measured (Sun) sound speeds, a fractional difference of 0.001 rms for all speeds measured between $0.05R_\odot$ and $0.95R_\odot$. The vertical scale is chosen so as to emphasize that the fractional error is much smaller than generic changes in the model, 0.04 to 0.09, that might significantly affect the solar neutrino predictions.

that help characterize the outer part of the sun (far from the inner region in which neutrinos are produced): the depth of the solar convective zone (CZ), the region in the outer part of the sun that is fully convective, and the present-day surface abundance by mass of helium ($Y_{\rm surf}$). The measured values, $R_{\rm CZ} = (0.713 \pm 0.001)R_\odot$, [35] and $Y_{\rm surf} = 0.249 \pm 0.003$, [31] are in satisfactory agreement with the values predicted by the solar model BP98, namely, $R_{\rm CZ} = 0.714R_\odot$, and $Y_{\rm surf} = 0.243$. However, we shall see below that precision measurements of the sound speed near the transition between the radiative interior (in which energy is transported by radiation) and the outer convective zone (in which energy is transported by convection) reveal small discrepancies between the model predictions and the observations in this region.

If solar physics were responsible for the solar neutrino problems, how large would one expect the discrepancies to be between solar model predictions and helioseismological observations? The characteristic size of the discrepancies can be estimated using the results of the neutrino experiments and scaling laws for neutrino fluxes and sound speeds.

All recently published solar models predict essentially the same fluxes from the fundamental pp and pep reactions (amounting to 72.4 SNU in gallium experiments, cf. Table 1), which are closely related to the solar luminosity. Comparing the measured gallium rates and the standard predicted rate for the gallium experiments, the ^{7}Be flux must be reduced by a factor N if the disagreement is not to exceed n standard deviations, where N and n satisfy $72.4 + (34.4)/N = 72.2 + n\sigma$. For a 1σ (3σ) disagreement, $N = 6.1(2.05)$. Sound speeds scale like the square root of the local temperature divided by the mean molecular weight and the ^{7}Be neutrino flux scales approximately as the 10th power of the temperature. [36] Assuming that the temperature changes are dominant, agreement to within 1σ would require fractional changes of order 0.09 in sound speeds (3σ could be reached with 0.04 changes), if all model changes were in the temperature[b]. This argument is conservative because it ignores the ^{8}B and CNO neutrinos which contribute to the observed counting rate (cf. Table 1) and which, if included, would require an even larger reduction of the ^{7}Be flux.

I have chosen the vertical scale in Fig. 6 to be appropriate for fractional differences between measured and predicted sound speeds that are of order 0.04 to 0.09 and that might therefore affect solar neutrino calculations. Fig. 6 shows that the characteristic agreement between solar model predictions and helioseismological measurements is more than a factor of 40 better than would be expected if there were a solar model explanation of the solar neutrino problems.

6 Fits Without Solar Models

Suppose (following the precepts of Hata *et al.*, [37] Parke, [38] and Heeger and Robertson [39]) we now ignore everything we have learned about solar models over the last 35 years and allow the important *pp*, ^{7}Be, and ^{8}B fluxes to take on any non-negative values. What is the best fit that one can obtain to the solar neutrino measurements assuming only that the luminosity of the sun is supplied by nuclear fusion reactions among light elements (the so-called

[b]I have used in this calculation the GALLEX and SAGE measured rates reported by Kirsten and Gavrin at Neutrino 98. The experimental rates used in BP98 were not as precise and therefore resulted in slightly less stringent constraints than those imposed here. In BP98, we found that agreement to within 1σ with the then available experimental numbers would require fractional changes of order 0.08 in sound speeds (3σ could be reached with 0.03 changes.)

'luminosity constraint')? [40]

The answer is that the fits are bad, even if we completely ignore what we know about the sun. I quote the results from Bahcall, Krastev and Smirnov (1998).

If the CNO neutrino fluxes are set equal to zero, there are no acceptable solutions at the 99% C. L. ($\sim 3\sigma$ result). The best-fit is worse if the CNO fluxes are not set equal to zero. All so-called 'solutions' of the solar neutrino problems in which the astrophysical model is changed arbitrarily (ignoring helioseismology and other constraints) are inconsistent with the observations at much more than a 3σ level of significance. No fiddling of the physical conditions in the model can yield the minimum value, quoted above, that was found by varying the fluxes independently and arbitrarily.

Figure 4 shows, in the lower left-hand corner, the best-fit solution and the 1σ -3σ contours. The 1σ and 3σ limits were obtained by requiring that $\chi^2 = \chi^2_{\rm min} + \delta\chi^2$, where for 1σ $\delta\chi^2 = 1$ and for 3σ $\delta\chi^2 = 9$. All of the standard model solutions lie far from the best-fit solution and even lie far from the 3σ contour.

Since standard model descriptions do not fit the solar neutrino data, we will now consider models in which neutrino oscillations change the shape of the neutrino energy spectra.

7 Neutrino Oscillations

The experimental results from all five of the operating solar experiments (chlorine, Kamiokande, SAGE, GALLEX, and SuperKamiokande) can be fit well by descriptions involving neutrino oscillations, either vacuum oscillations (as originally suggested by Gribov and Pontecorvo [41])or resonant matter oscillations (as originally discussed by Mikheyev, Smirnov, and Wolfenstein (MSW) [42]).

Table 3 summarizes the four best-fit solutions that are found in the two-neutrino approximation. [10,27] Only the SMA and vacuum oscillation solutions fit well the recoil electron energy spectrum measured in the SuperKamiokande experiment–if the standard value for the *hep* production reaction cross section (^{3}He + p $\rightarrow$ ^{4}He + e^+ + ν_e) is used. [10] However, for over a decade I have not given an estimated uncertainty for this cross section. [20] The transition matrix element is essentially forbidden and the actual quoted value for the production cross section depends upon a delicate cancellation between two comparably sized terms that arise from very different and hard to evaluate nuclear physics. I do not see anyway at present to determine from experiment or from first principles theoretical calculations a relevant, robust upper limit

Table 3. Neutrino Oscillation Solutions.

Solution	Δm^2	$\sin^2 2\theta$
SMA	5×10^{-6} eV^2	5×10^{-3}
LMA	2×10^{-5} eV^2	0.8
LOW	8×10^{-8} eV^2	0.96
VAC	8×10^{-11} eV^2	0.7

to the *hep* production cross section (and therefore the *hep* solar neutrino flux).

The possible role of *hep* neutrinos in solar neutrino experiments is discussed extensively in Bahcall and Krastev (1998) [27] The most important unsolved problem in theoretical nuclear physics related to solar neutrinos is the range of values allowed by fundamental physics for the *hep* production cross section.

8 Discussion and Conclusion

When the chlorine solar neutrino experiment was first proposed, [43] the only stated motivation was "...to see into the interior of a star and thus verify directly the hypothesis of nuclear energy generation in stars." This goal has now been achieved,

The focus has shifted to using solar neutrino experiments as a tool for learning more about the fundamental characteristics of neutrinos as particles. Experimental effort is now concentrated on answering the question: What are the probabilities for transforming a solar ν_e of a definite energy into the other possible neutrino states? Once this question is answered, we can calculate what happens to ν_e's that are created in the interior of the sun. Armed with this information from weak interaction physics, we can return again to the original motivation of using neutrinos to make detailed, quantitative tests of nuclear fusion rates in the solar interior. Measurements of the flavor content of the dominant low energy neutrino sources, *p-p* and ^{7}Be neutrinos, will be crucial in this endeavor and will require another generation of superb solar neutrino experiments.

Three decades of refining the input data and the solar model calculations has led to a predicted standard model event rate for the chlorine experiment, 7.7 SNU, which is very close to 7.5 SNU, the best-estimate value obtained in 1968. [44] The situation regarding solar neutrinos is, however, completely different now, thirty years later. Four experiments have confirmed the original chlorine detection of solar neutrinos. Helioseismological measurements are in excellent agreement with the standard solar model predictions and very strongly disfavor (by a factor of 40 or more) hypothetical deviations from the standard model that are require to fits the neutrino data (cf. Fig. 6). Just in the last two years, improvements in the helioseismological measurements have resulted in a five-fold improvement in the agreement between the calculated standard solar model sound speeds and the measured solar velocities (cf. Figure 2 of the Neutrino 96 talk [45] with Figure 6 of this talk).

Acknowledgments

I acknowledge support from NSF grant #PHY95-13835.

References

1. R. Davis, Jr., D.S. Harmer and K.C. Hoffman, *Phys. Rev. Lett.* **20**, 1205 (1968).
2. R. Davis, Jr., *Prog. Part. Nucl. Phys.* **32**, 13 (1994); B.T. Cleveland, T. Daily, R. Davis, Jr., J.R. Distel, K. Lande, C.K. Lee, P.S. Wildenhain, and J. Ullman, *Astrophys. J.* **496**, 505 (1998).
3. KAMIOKANDE Collaboration, Y. Fukuda *et al.*, *Phys. Rev. Lett.* **77**, 1683 (1996).
4. SAGE Collaboration, V. Gavrin *et al.*, in *Neutrino '96,* Proceedings of the 17th International Conference on Neutrino Physics and Astrophysics (Helsinki), eds. K. Enqvist, K. Huitu and J. Maalampi (World Scientific, Singapore, 1997), p. 14; SAGE collaboration, J.N. Abdurashitov *et al.*, *Phys. Rev. Lett.* **77**, 4708 (1996).
5. GALLEX Collaboration, P. Anselmann *et al.*, *Phys. Lett.* B **342**, 440 (1995); GALLEX Collaboration, W. Hampel *et al.*, *Phys. Lett.* B **388**, 364 (1996).
6. SuperKamiokande Collaboration, Y. Suzuki, in *Neutrino 98,* Proceedings of the XVIII International Conference on neutrino Physics and Astrophysics, Takayama, Japan, 4-9 June 1998, eds. Y. Suzuki and Y. Totsuka. To be published in *Nucl. Phys.* B (Proc. Suppl); the SuperKamiokande Collaboration, Y. Fukuda *et al.*, *Phys. Rev. Lett.* **81**, 1562 (1998); Y. Fukuda *et al.*, hep-ex/9812009; Y. Totsuka, in *Relativistic Astrophysics and Cosmology,* Proceedings of the 18th Texas Symposium,

December 15–20, 1996, Chicago, Illinois. eds. A. Olinto, J. Frieman and D. Schramm (World Scientific, Singapore, 1998) p. 114.
7. H.A. Bethe, *Phys. Rev.* **55**, 434 (1939).
8. W. Thompson, "On the Age of the Sun's Heat," *MacMilan's Magazine* **5**, 288 (1862); reprinted in *Treatise on Natural Philosphy*, W. Thompson and P. Gutherie, **Vol. 1**, appendix E (Cambridge Univ. Press, Cambridge, 1883), p. 485.
9. C. Darwin, *On the Origin of the Species*, 1859 (Reprinted: Harvard University Press, Cambridge, MA, 1964).
10. J.N. Bahcall, P.I. Krastev and A.Yu. Smirnov, *Phys. Rev.* D **58**, 096016-1 (1998); J.N. Bahcall, P.I. Krastev and A.Yu. Smirnov, *Phys. Rev.* D (November 1999), hep-ph/9905220.
11. J.N. Bahcall, S. Basu and M.H. Pinsonneault, *Phys. Lett.* B **433**, 1 (1998).
12. E. Adelberger *et al.*, *Rev. Mod. Phys.* **70**, 1265 (1998).
13. C.A. Iglesias and F.J. Rogers, *Astrophys. J.* **464**, 943 (1996); D.R. Alexander and J.W. Ferguson, *Astrophys. J.* **437**, 879 (1994). These references describe the different versions of the OPAL opacities.
14. F.J. Rogers, F.J. Swenson and C.A. Iglesias, *Astrophys. J.* **456**, 902 (1996).
15. A.V. Gruzinov and J.N. Bahcall, *Astrophys. J.* **504**, 996 (1998).
16. E.E. Salpeter, *Australian J. Phys.* **7**, 373 (1954).
17. J.N. Bahcall, *Phys. Rev.* C **56**, 3391 (1997).
18. J.N. Bahcall, E. Lisi, D.E. Alburger, L. De Braeckeleer, S.J. Freedman and J. Napolitano, *Phys. Rev.* C **54**, 411 (1996).
19. J.N. Bahcall, M. Kamionkowski and A. Sirlin, *Phys. Rev.* D **51**, 6146 (1995).
20. J.N. Bahcall, *Neutrino Astrophysics* (Cambridge University Press, Cambridge, 1989).
21. J.N. Bahcall, *Astrophys. J.* **467**, 475 (1996).
22. (GONG) J. Christensen-Dalsgaard *et al.*, GONG Collaboration, *Science* **272**, 1286 (1996); (BP95) J.N. Bahcall and M.H. Pinsonneault, *Rev. Mod. Phys.* **67**, 781 (1995); (KS94) A. Kovetz and G. Shaviv, *Astrophys. J.* **426**, 787 (1994); (CDF94) V. Castellani, S. Degl'Innocenti, G. Fiorentini, L.M. Lissia and B. Ricci, *Phys. Lett.* B **324**, 425 (1994); (JCD94) J. Christensen-Dalsgaard, *Europhys. News* **25**, 71 (1994); (SSD94) X. Shi, D.N. Schramm and D.S.P. Dearborn, *Phys. Rev.* D **50**, 2414 (1994); (DS96) A. Dar and G. Shaviv, *Astrophys. J.* **468**, 933 (1996); (CDF93) V. Castellani, S. Degl'Innocenti and G. Fiorentini, *Astron. Astrophys.* **271**, 601 (1993); (TCL93) S. Turck-Chièze and I. Lopes, *Astrophys. J.*

408, 347 (1993); (BPML93) G. Berthomieu, J. Provost, P. Morel and Y. Lebreton, *Astron. Astrophys.* **268**, 775 (1993); (BP92) J.N. Bahcall and M.H. Pinsonneault, *Rev. Mod. Phys.* **64**, 885 (1992); (SBF90) I.-J. Sackman, A.I. Boothroyd and W.A. Fowler, *Astrophys. J.* **360**, 727 (1990); (BU88) J.N. Bahcall and R.K. Ulrich, *Rev. Mod. Phys.* **60**, 297 (1988); (RVCD96) O. Richard, S. Vauclair, C. Charbonnel and W.A. Dziembowski, *Astron. Astrophys* **312**, 1000 (1996); (CDR97) F. Ciacio, S. Degl'Innocenti and B. Ricci, *Astron. Astrophys. Suppl. Ser.* **123**, 449 (1997).

23. C.W. Johnson, E. Kolbe, S.E. Koonin and K. Langanke, *Astrophys. J.* **392**, 320 (1992).
24. J.N. Bahcall, M.H. Pinsonneault, S. Basu and J. Christensen-Dalsgaard, *Phys. Rev. Lett.* **78**, 171 (1997).
25. J.N. Bahcall and M.H. Pinsonneault, *Rev. Mod. Phys.* **64**, 885 (1992).
26. J.N. Bahcall and M.H. Pinsonneault, *Rev. Mod. Phys.* **67**, 781 (1995).
27. J.N. Bahcall and P.I. Krastev, *Phys. Lett.* B **436**, 243 (1998).
28. J.N. Bahcall, *Phys. Rev.* D **44**, 1644 (1991).
29. J.N. Bahcall and H.A. Bethe, *Phys. Rev. Lett.* **65**, 2233 (1990).
30. A.V. Gruzinov and J.N. Bahcall, *Astrophys. J.* **490**, 437 (1997).
31. S. Basu and H.M. Antia, *Mon. Not. R. Astron. Soc.* **287**, 189 (1997).
32. A.A. Thoul, J.N. Bahcall, and A. Loeb, *Astrophys. J.* **421**, 828 (1994).
33. A.B. McDonald, in the *Proceedings of the 9th Lake Louise Winter Institute*, eds. A. Astbury *et al.* (World Scientific, Singapore, 1994), p. 1.
34. S. Basu *et al.*, *Mon. Not. R. Astron. Soc.* **292**, 234 (1997).
35. S. Basu and H.M. Antia, *Mon. Not. R. Astron. Soc.* **276**, 1402 (1995).
36. J.N. Bahcall and A. Ulmer, *Phys. Rev.* D **53**, 4202 (1996).
37. N. Hata, S. Bludman and P. Langacker, *Phys. Rev.* D **49**, 3622 (1994).
38. S. Parke, *Phys. Rev. Lett.* **74**, 839 (1995).
39. K.M. Heeger and R.G.H. Robertson, *Phys. Rev. Lett.* **77**, 3720 (1996).
40. J.N. Bahcall and P.I. Krastev, *Phys. Rev.* D **53**, 4211 (1996).
41. V.N. Gribov and B.M. Pontecorvo, *Phys. Lett.* B **28**, 493 (1969); B. Pontecorvo, *Sov. Phys. JETP* **26**, 984 (1968).
42. L. Wolfenstein, *Phys. Rev.* D **17**, 2369 (1978); S.P. Mikheyev and A.Yu. Smirnov, *Yad. Fiz.* **42**, 1441 (1985) [*Sov. J. Nucl. Phys.* **42**, 913 (1985)]; *Nuovo Cimento* C **9**, 17 (1986).
43. J.N. Bahcall, *Phys. Rev. Lett.* **12**, 300 (1964); R. Davis, Jr., *Phys. Rev. Lett.* **12**, 303 (1964); J.N. Bahcall and R. Davis, Jr., in *Stellar Evolution,* eds. R.F. Stein and A.G. Cameron (Plenum Press, New York, 1966), p. 241 [proposal first made in 1963 at this conference].

44. J.N. Bahcall, N.A. Bahcall and G. Shaviv, *Phys. Rev. Lett.* **20**, 1209 (1968).
45. J.N. Bahcall and M.H. Pinsonneault, in *Neutrino '96*, Proceedings of the 17th International Conference on Neutrino Physics and Astrophysics (Helsinki), eds. K. Enqvist, K. Huitu and J. Maalampi (World Scientific, Singapore, 1997), p. 56.

THE SUDBURY NEUTRINO OBSERVATORY AT TURN-ON

JOSHUA R. KLEIN, for the SNO Collaboration
Department of Physics and Astronomy, Universty of Pennsylvania, Philadelphia PA 19104-6396, USA
E-mail: jrk@upenn5.hep.upenn.edu

The Sudbury Neutrino Observatory (SNO) has been designed to study solar neutrinos using 1 kiloton of heavy water (D_2O) as both the interaction and detection medium. The D_2O gives SNO sensitivity to ν_e's exclusively through a charged current (CC) reaction as well as sensitivity to all other weakly interacting neutrino flavors through a neutral current (NC) reaction. SNO has begun taking initial data, and detector's performance has so far met or exceeded expectations.

1 Physics Motivation and Experimental Approach

The main goal of the Sudbury Neutrino Observatory (SNO) is to study the Solar Neutrino Problem. The measured deficit of neutrinos from the Sun relative to the predictions of the Standard Solar Model, combined with the apparent energy dependence of this deficit, argue strongly that ν_e's from the Sun are changing into another flavor before their arrival at the Earth. The recent atmospheric neutrino measurements from the SuperKamiokande Collaboration strongly support the idea that neutrino oscillations from one flavor to another do occur (in this case, probably $\nu_\mu \leftrightarrow \nu_\tau$). However, for the case of solar neutrinos, no single measurement as yet has demonstrated definitively that the flux of neutrinos from the Sun contains anything other than ν_e's.

SNO is designed to resolve this question through the direct comparison of the solar ν_e flux with the flux of all weakly interacting neutrino flavors. The comparison is done using heavy water (D_2O) as both the interaction and the detection medium. neutrinos interact in three ways in D_2O:

- $\nu_e + d \rightarrow p + p + e^-$ (Charged Current—CC)
- $\nu_x + d \rightarrow p + n + \nu_x$ (Neutral Current—NC)
- $\nu_{e,(\mu,\tau)} + e \rightarrow \nu_{e,(\mu,\tau)} + e^-$ (Elastic Scattering — ES)

The CC reaction is sensitive only to ν_e's while the NC reaction occurs for all weakly interacting neutrino flavors. Therefore comparison of the neutrino flux measured via the NC reaction and that measured via the CC reaction tells us directly whether the solar neutrino flux contains a significant fraction of non-electron neutrinos.

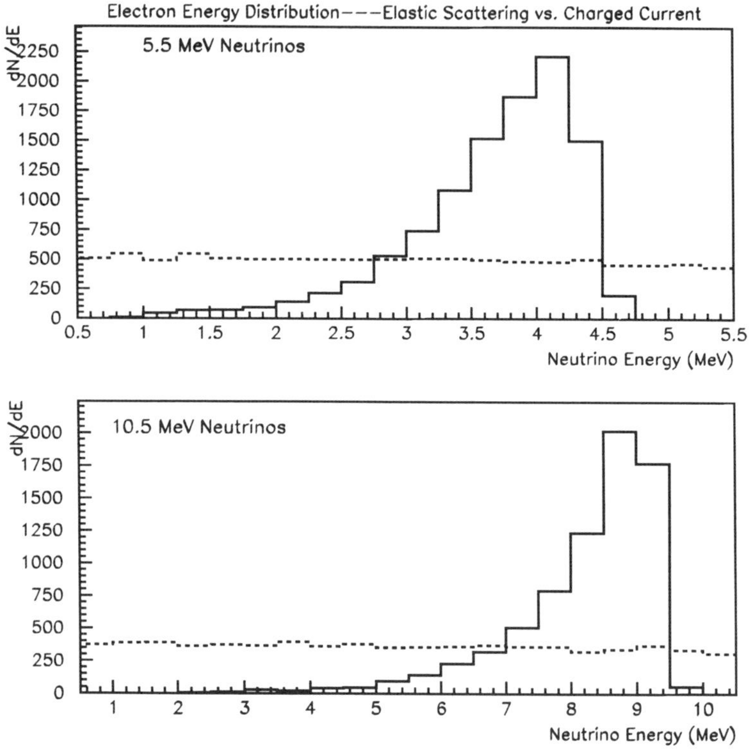

Figure 1: Comparison of the energy response of charge current and elastic scattering reactions. The ordinate is labelled in arbitrary units.

There are further advantages to this method, however. As is shown in Figure 1, the CC reaction is a sensitive measure of the incident neutrino energy—far more so than the elastic scattering reaction which produces a recoil electron energy spectrum that is nearly flat. SNO will therefore be able to produce a high sensitivity measurement of the solar ν_e spectrum (above about ~ 5 MeV). Optimistically, a sensitive measurement of this spectrum can tell us whether neutrino oscillations are occurring in an energy-dependent way, thereby distinguishing different oscillation scenarios.

Lastly, the ES reaction, which is sensitive predominantly to ν_e's but also to other neutrino flavors (with a much-reduced cross section) provides a nice consistency check of the total flux—both with other experiments and with SNO's own flux measurements from the CC and NC reactions.

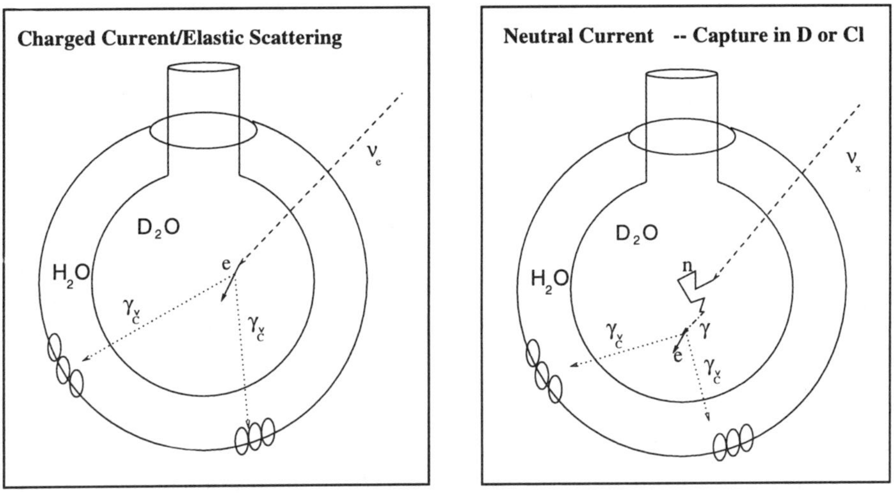

Figure 2: Neutrino detection mechanisms in SNO.

2 The SNO Detector

The SNO detector is located 2 km underground in the Creighton Mine near Sudbury, Ontario. One kiloton of D_2O is contained in a transparent acrylic vessel, and is surrounded by a geodesic sphere upon which are mounted roughly 9500 8" photomultiplier tubes (PMTs). The entire cavity volume outside the acrylic vessel is filled with $\sim$ 7 ktons of ultrapure H_2O, and the cavity is sealed with a radon impermeable material.

As is shown in Figure 2, SNO detects neutrinos in two ways. For both the CC and ES reactions, a relativistic electron is produced directly, and the resultant Cerenkov light is detected by the surrounding PMTs. Detection of the NC reaction, however, is a multistep process. With pure D_2O inside the acrylic vessel, the neutron liberated by the neutrino reaction is captuied by another deuteron, with the emission of a 6.25 MeV gamma. The γ then Compton scatters, producing one or more relativistic electrons which produce detectable Cerenkov light. There are other detection options for the NC reaction, however. Chlorine (in the form of $\sim$ 2 tons of MgCl) can be added to the D_2O, enhancing the capture probability of the neutrons by more than a factor of three as well as boosting the energy of the emitted γ to 8.6 MeV, making it easier to seperate from low energy backgrounds, and allowing a comparison of chlorine-in and chlorine-out rates. A second option is the addition of discrete, ^{3}He proportional counters, which will detect the neutrons directly.

3 Suppression of Backgrounds

SNO's success depends critically on the understanding and elimination of backgrounds. If we simply assume that the neutrino flux from the Sun is reduced by a factor of three (that is, no oscillations), then SNO would expect a NC event rate of three events per day, a CC rate of 9 events per day, and an ES rate of just 1 event per day above a threshold of 5 MeV. Each event typically has only 50 detected photons, and so any small amount of radioactive contamination which produces detectable light is a potential danger. Radioactivity which leads to the production of a free neutron (through photodisintegration of deuterium, for example), is particularly dangerous, as the capture of such a neutron cannot be distinguished from a neutron produced by a neutrino, regardless of the choice of neutral current detection method. The uranium and thorium decay chains are therefore the greatest concern, since they contain γ's of high enough energy to disintegrate deuterons in the detector.

SNO is designed to suppress the physics background from natural radioactivity as much as possible. The 5 km.w.e. depth reduces the rate of cosmic rays (and in turn their associated spallation products) to roughly one an hour through the D_2O. The ~ 4.5 m of light water between the cavity walls and the heavy water provides shielding from neutrons and high energy γ's from the surrounding rock, and the 2.5 m of light water between the photomultiplier tubes and the heavy water shields the radioactivity from the tubes themselves. In addition, the glass for the phototubes is custom made and has radioactivity levels five times lower than standard glass. The acrylic and the bonding material used to build the heavy water containment vessel are also very low radioactivity materials.

Ensuring and maintaining cleanliness of the water itself (both the light and heavy water) is one of the largest efforts undertaken by the SNO collaboration. The entire underground laboratory is maintained as a clean room ('class 2000') to prevent contamination from mine dust, which can have high levels of uranium and thorium. The water systems continually clean and purify the water, removing radioactivity through reverse osmosis and seeded ultrafiltration, killing off bacteria, and removing radon gas. The ultimate design goals for uranium levels in the H_2O and D_2O are $< 1 \times 10^{-12}\mathrm{g/cm^3}$ and $< 4.5\times 10^{-14}\mathrm{g/cm^3}$, respectively; for thorium the goals are $< 1.3\times 10^{-13}\mathrm{g/cm^3}$ for H_2O and $< 3 \times 10^{-15}\mathrm{g/cm^3}$ for the D_2O.

4 Rejection of Backgrounds

In addition to keeping radioactive contamination low, SNO has several handles on rejecting backgrounds, from both natural sources and from the detector itself. Most backgrounds associated with radioactivity (the exception is the photodisintegration of deuterium mentioned earlier) produce events whose energies lie well below the energies of ^8B neutrino events, and therefore measurement of energy alone (through the number of phototube hits) is a powerful discriminant itself.

The measurement of charge from each phototube in an event is also an important tool for eliminating backgrounds. For example, tube radioactivity may produce a large charge in a single tube, far more than would be anticipated from the number of total tube hits in that event. By the same token, detector backgrounds such as 'flasher events' can produce charge distributions which will look very different from the 1 photoelectron/phototube expected for a typical neutrino event.

By far the best handle on the rejection of background, however, is event reconstruction. As much of the backgrounds originate from outside the D_2O volume (and no signals except those from the ES reaction), rejection of events whose positions are inconsistent with an origin inside the acrylic vessel reduces the background substantially. For this to be effective, good phototube timing is critical. In addition, the ES reaction produces electrons which are forward peaked with respect to the Sun, while the CC reaction produces an angular distribution which falls as $1 - \frac{1}{3}\cos\theta$, peaked away from the solar direction. The reconstruction of direction is therefore a very powerful tool for identifying the solar signal statistically.

5 Detector Specifics

SNO's heavy water is surrounded by 9522 8" phototubes, each with a light concentrator, providing 55% photocathode coverage. The electronics for the detector provide three independent measurements of charge: two high gain measurements (one a $\sim$ 60 ns integration time and the other $\sim$ 400 ns) and a low gain measurement, for a full dynamic range of 0.1 to 3000 photoelectrons. Each channel also provides a relative tube time with a resolution $\sim$ 120 ps. The system as a whole provides a million event buffer with a burst capability slightly higher than 2 MHz, and each event is time-stamped with an absolute time from the Global Positioning System accurate to $\sim$ 300 ns.

The primary trigger for SNO is a based upon the number of tubes hit in a time window of programmable length, nominally set to 100 ns. The trigger is implemented as a fast analog sum of discrete signals from each tube originating with the firing of each channel's discriminator. In addition to this 'NHIT' trigger, there is a direct sum of shaped PMT pulses (thus giving greater weight to those tubes in a given event which see more light). Although the efficiency of such a trigger is naturally less than that of the NHIT trigger, it provides an excellent monitor of the status of all 9522 phototubes with a single oscilloscope trace. In addition, this 'ESUM' (for 'energy sum') trigger will aid SNO in looking at some of its more exotic backgrounds. Other triggers include a pre-scaled version of a low-threshold NHIT trigger and a zero-bias pulsed trigger for looking at very low energy backgrounds, and a reduced-volume trigger using a tighter coincidence (20 ns rather than 100 ns) for looking at low energy backgrounds near the center of the detector or at 'hot spots' in the detector through the use of programmable channel-to-channel delays.

The water systems continually purify and cool both the heavy and light water, and also provide the capability for periodic assays to determine the radioactive contamination which can be compared with the background levels determined through analysis.

6 Detector Calibrations

For understanding the detector's response to the neutrino signal, and for the efficient rejection of background, SNO needs a variety of calibration sources. These calibrations fall into a few categories: optical calibrations for the measurement of reflection coefficients, scattering lengths and attenuation; energy calibrations to obtain the energy scale (number of phototube hits/MeV) and resolution; and calibrations of the resolution of SNO's reconstruction algorithms.

A nitrogen/dye laser with a diffuser is the main source used for the optical calibrations. The laser's intensity can be varied over several orders of magnitude through the use of several neutral-density filters, and its wavelength is varied using several different dye units to study the wavelength-dependence of the detector's optical properties.

The primary source for energy calibrations is ^{16}N, which emits a 6.13 MeV γ. The subsequent Compton scattering of the γ provides Cerenkov electrons with a known energy spectrum which can then be used to determine both the energy scale and the energy resolution. For the higher energy scales, a 'pT' source— $^3\mathrm{H}(p,\gamma)^4\mathrm{He}$—yields a 19.8 MeV γ which can be used in the same way. In addition, a ^{8}Li source will be used to provide a spectrum of electron energies very similar to that of the electrons produced through the CC reaction of ^{8}B neutrinos in the detector.

For determination of the resolution of event reconstruction, all of the above sources will be used. In addition, an array of LED sources fixed to the phototube sphere will provide a check on the overall detector geometry. To measure the phototube timing to high accuracy, SNO employs a unique sonoluminescent source (in addition to the laser source). The sonoluminescent source uses resonant acoustic waves to create a sub-micron sized source of light with a time spread less than 50 ps. It is a high rate (23 kHz), isotropic source with very high intensity, and its combination of small size and short emission time make it particularly useful in measuring the phototube timing, as well as checking some of the optical properties measured by the laser.

7 Detector Turn-On

SNO's data history has had several stages. From April–September, 1998, during the completion of construction work, several sets of runs were made both in a completely dry detector ('Air Fill' data) and in a partially filled detector ('Partial Fill' data). The advantage of the Air Fill data is that it provided a 'beam off' view of backgrounds as well as experience with running the detector. From September, 1998–April 1999, the detector was mostly off as the water fill continued, and since April 30, 1999, SNO has been taking data with the full detector.

Most of the calibration sources mentioned in Section 6 were run in one or more of the Air Fill runs. The sonoluminescent source and the laser yielded measurements of the phototube timing as good as had been hoped for—transit jitter spreads of ~ 1.7 ns. In addition to running the calibration sources, muons were observed using the Cerenkov light produced by their passage through the walls of the (at that time) empty acrylic vessel. The empty detector also helped to identify a potential source of background—flasher events, in which a phototube emits light spontaneously. The rate of these events—up to a few/minute—was higher than had been anticipated, and they produced light in the same regime ($\sim$ 40-200 hit tubes) as SNO's expected neutrino signal. However, analysis efforts showed that these flasher events could be eliminated efficiently at better than the 99.8% level. In addition, once the water had been put into the detector, it was found that the rate of the flashers had become significantly lower, except during mine-related seismic events when typically a large burst of flasher events occured. This time-correlation makes the removal of the flashers even easier.

The data from the partially filled detector was taken mostly with the water level well below the halfway point of the cavity. From these runs the first water Cerenkov events were observed, allowing SNO to see both high energy γ's escaping the cavity walls and cosmic ray muons. As the water level went up, the rate of the cavity wall γ's subsided as expected, due to the shielding from the light water outside the detector's fiducial volume. In addition, it became apparent that there was a significant problem with high voltage breakdown

Table 1: Trigger Rates By Type, for May 1999 Data

Trigger	Threshold	Rate
Pulsed Trigger	Zero Bias	5 Hz
100 ns Coincidence	20 PMTs	3-5 Hz
20 ns Coincidence	18 PMTs	3-5 Hz
Energy Sum	$\sim$ 300 pe	$<$ 1 Hz
Prescaled (1:10000)	12 PMTs	$<$ 1 Hz

on the wet-end of the PMT cables. The cause of the breakdown was the marginal design of the wet-end high voltage connector, and the breakdown events presented a problem both for data acquistion (very high rates of giant events) and for the circuitry in the PMT bases. It was eventually realized that the problem was being exacerbated by the fact that SNO's water is de-gassed. The de-gassed water acted as a natural diffusion pump, evacuating the connector volume and creating a breakdown voltage much lower than when the volume was at atmospheric pressure. Dry nitrogen gas was added back into the water, and it was found that the number of breakdown events dropped to nearly zero.

From April 30, 1999, SNO has been running with the detector completely full. Only a handful of breakdown events have been identified. The livetime through the end of May was around 60%, dominated by a one-week shutdown for installation of the radon barrier. Both the laser and the ^{16}N source have been run to look at tube timing, detector and water optics, and SNO's energy response.

The most striking thing about the detector is that it is extremely quiet. At the current PMT thresholds (somewhat less than half of a photoelectron), the average phototube noise rates is below 800 Hz. During the May runs, the overall trigger rate was 10-15 Hz. Table 1 shows the rates for each of the triggers SNO was running during this time. The '100 ns Coincidence' trigger is the one which will be used for most of SNO's analysis.

In addition, over 98% of the channels were fully working. Most of those not working were damaged during the periods of breakdown before the re-gassing was initiated.

Beyond the quiet running conditions, the SNO detector has already shown us some beautiful events. Figure 3 is a picture of a partially contained, upward-going muon event. The muon can be seen exiting the top of the detector, with a secondary particle moving off toward the side. The low energy events are equally impressive—Figure 4 shows the Cerenkov ring produced by the Compton scatter of a 6.2 MeV gamma from the ^{16}N calibration source. Multiple scattering of the electron at these energies means that most of the events do not produce such nice rings, however it is very satisfying to see such clear events when they do occur. The histogram in the Figure shows the distribution of raw (uncalibrated) phototube times, and shows about 110 ns of the 400 ns full timing window. The in-time peak as well as a few reflection tubes can be seen, but otherwise the event is extremely clean. From the zero biased pulsed trigger data, we find that a typical event has just two noise tubes.

8 Summary

SNO is now working toward 100% livetime, including the times during detector calibrations. Production calibration data for both optical and energy calibrations are being taken, and analyses to study the levels of low energy backgrounds have begun.

SNO will soon give us:

- A measure of the solar ν_e flux.
- The first high-sensitivity measure of the solar ν_e spectrum.
- The first measurement of the faction of non-electron flavor neutrinos from the Sun.

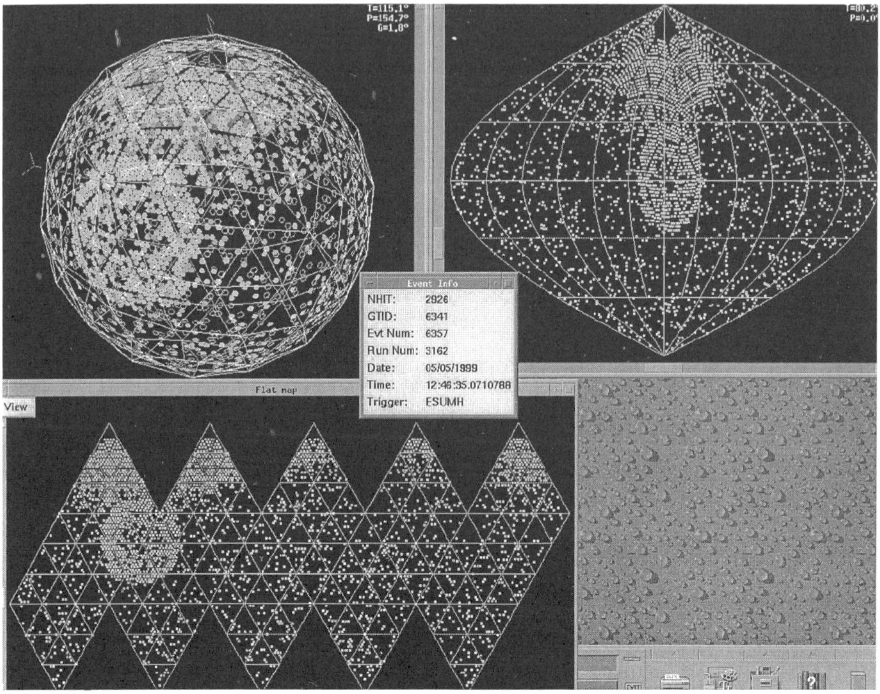

Figure 3: Partially contained upward-going muon event.

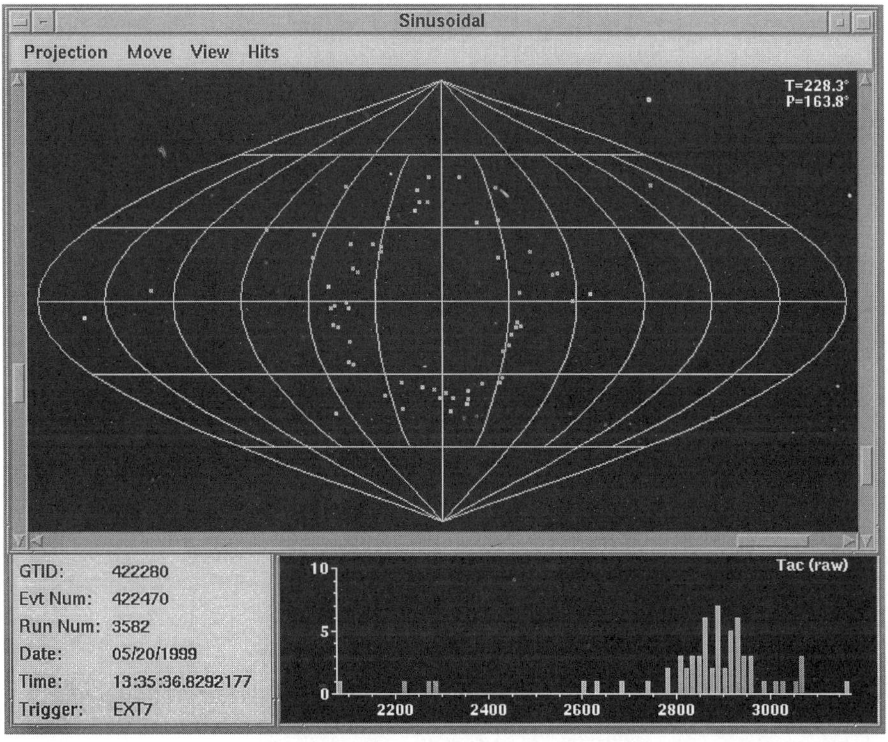

Figure 4: Cerenkov ring from Compton-scattered electron, using ^{16}N calibration source.

THE DISCOVERY OF NEUTRINO OSCILLATIONS AND MASS: THE DISAPPEARANCE OF MUON NEUTRINOS

JOHN G. LEARNED
Department of Physics and Astronomy, University of Hawaii
Honolulu, HI 96822, USA
E-mail: jgl@phys.hawaii.edu

With the announcement of strong new evidence for muon neutrino disappearance observed by the Super-Kamiokande experiment, the more than a decade old atmospheric neutrino anomaly moved from a possible indication for neutrino oscillations to an apparently inescapable fact. The evidence is reviewed, and new indications are presented that the oscillations are probably between muon and tau neutrinos. Implications and future directions are discussed.

1 Introduction

Herein we attempt to review the present understanding of the resolution of the atmospheric neutrino anomaly, put it in context with other results, and speculate upon future directions.

1.1 Product Warning

This paper documents the talk given by the author at the John's Hopkins Workshop on Neutrinos in the Next Millennium in June 1999. The subject to be covered was generally the situation with respect to the atmospheric neutrino oscillations. The understanding of this phenomenon is now dominated by the data announced by the Super-Kamiokande Collaboration in June 1998, of which group the present author is a member. Much of this report dwells upon those results and updates to them, and so credit for this work is due to the whole collaboration, listed in the Appendix, who have labored hard to bring this experiment to fruition and who have been ably lead by Prof. Yoji Totsuka of the University of Tokyo. That said, some of this report is the personal opinion of the author, particularly in matters of the previous history, the interpretation and future prospects for this line of research.

1.2 The Atmospheric Neutrino Anomaly

We will not dwell upon the past history, but note that the atmospheric neutrino anomaly has been around for some time, roughly fifteen years. Indeed the first notice of something peculiar going on was in the 1960's when the seminal underground experiments in South Africa and South India first detected the natural neutrinos and observed somewhat of an absolute rate deficit, but not convincingly as the flux predictions were rough and the statistics small.

The problem became serious after the activation of the IMB experiment and by 1983 their realization that the number of events containing muon decays was lower than expected[1]. Soon this was confirmed by the Kamioka experiment, which group extended the results with good particle identification giving a redundant measure of the relative muon deficit (as also did IMB). Some members of the IMB group and the Kamioka group began to proclaim that oscillations were the cause of the deficit, but the claim was not widely taken seriously[2,3]. This author acknowledges being one of the skeptics at that time.

The deficit in the ratio is characterized usually as an R value, the ratio of muon to electron neutrinos, observed to expected. This ratio of ratios is thus independent of the 20-25% uncertain absolute flux prediction, and itself systematically uncertain by less than 10%.

With the initial evidence, the oscillations could have been from muon neutrinos to others (eg. ν_τ) or between muon and electron neutrinos, as it was the ratio that was in deficit: one could not be sure whether there was an excess of electron neutrinos, a deficit of muon neutrinos, or some of both. This lead to suggestions of other "physics" causes, such as nucleon decay favoring electron modes (since the anomaly was not detected above the nucleon mass), or an excess of extraterrestrial electron neutrinos. See Table 1 for a graphical summary of the situation. There were also suggestions of systematic problems, such as problems in muon identification, something wrong with flux calculations or neutrino interaction cross sections, entering backgrounds, or with the water Cherenkov detectors.

Over the intervening years between the emergence of this "atmospheric neutrino anomaly", as it became known, and last year's announcement, a great deal of effort went into study of the possible systematic causes of the anomaly. One troubling problem was that two European experiments, the NUSEX and the Frejus Detectors, did not observe the same anomaly as IMB and Kamioka, and hence some people suspected a peculiarity of water as a target or with the employment of the Cherenkov radiation in vertex location. As it later turned out, not only were the statistics of the European detectors rather small, but as indicated by more recent work from the similar type of instrument in the US, the Soudan II detector, the presence of a surrounding veto counter is vital for the more compact type of slab detectors. As well, the MACRO experiment has elucidated the production of low energy (hundred MeV) pions by nearby cascades in rock, which particles enter cracks in non-hermetic detectors and appear to be neutrino interactions. In any case the Soudan II with now significant exposure (several kiloton-years) finds an R value close to that of SuperK (and IMB and Kamioka)[4].

2 The Super-Kamiokande Revolution

We now proceed to summarize the new evidence for oscillations which comes from the SuperK experiment. Before proceeding it may be worthwhile to point out what permitted the big break-through, which is not so obvious. The increase in size of detector, from near kiloton fiducial volumes for Kamioka and Soudan, and three kilotons for IMB to the twenty two kilotons of SuperK is not the whole story. As will be seen below, the progress comes from the recording of muon events with good statistics in the energy region above 1 GeV. This is due to detector linear dimensions as well as gross target volume: muon events with energy more than 1 GeV and thus 5 m range were not likely to be fully contained in the Kamioka detector (or the IMB detector). SuperK in contrast has decent muon statistics up to almost 5 GeV, and this is turns out to be crucial.

Most of the data to be discussed below is the "fully contained event" (FC) sample, consisting of those events in which both vertex and track ends remain in the fiducial volume. There are also "partially contained" (PC) events, in which a muon may exit the fiducial volume from a contained vertex location. Such events are useful even though the total deposited energy is not known, the energy observed being a lower limit. Of course this is the case even with FC events, though to a lessor degree, because the observed particles are not of the same energy (or direction) as the incident neutrino, which of course is what one would desire to observe.

The particle types are identified by pattern recognition software, now well tested and verified by experiment with known particle beams at the accelerator. Fortunately most of the events (roughly 2/3) are single (Cherenkov radiating) tracks, in which the identification is quite clean (at the 98% level). To be clear and cautious we usually refer to the reconstructed events as "muon-like" and "electron-like", though a safe approximation is that these represent muon and electron neutrino charged-current interactions.

The other two categories of events of which we shall report are the through-going upwards moving muons (UM), produced by neutrino interactions in the rock or outer detector, and which are coming from directions below the horizon (as those from above the horizon can be confused with down going muons from cosmic ray interactions in the atmosphere near overhead). Another category of event is the entering-stopping muon (SM). It is useful that these event categories probe approximately three different energy ranges of neutrinos: $FC \simeq 1\ GeV$; PC and $SM \simeq 10\ GeV$; $UM \simeq 100 GeV$. It should be understood that as far as we know, these neutrinos are all produced in the upper atmosphere by cosmic ray interactions, and are reasonably well described by models in content, energy, and angular dependence (to a few percent)[6].

We shall not take up limited space here with the description of the SuperK detector, which is well documented elsewhere. The interested reader would do well to look at some of the theses from SuperK, which, though large files, are available on the web[7]. The short summary is that the SuperK detector consists of a large stainless steel cylinder (37 m high by 34 m diameter inside the inner detector) with extremely high photo-tube coverage (40%), ten times more pixels than any earlier instrument, and a remarkable sensitivity of roughly eight photoelectrons per MeV of deposited (Cherenkov radiating) energy. The latter permits detection of events down to about 5 MeV, so for the present discussion detection efficiency versus energy is not important because the events we are discussing are all above $\simeq 100\ MeV$. The inner volume is also well protected by a 2 m thick, fully-enclosing veto counter, populated by the recycled IMB photomultipliers and wavelength shifters. The inner "fiducial" volume is further taken as 2 m inside the photomultiplier surface, resulting in the 22 kiloton volume used for most reported data.

The SuperK oscillations claim was first formally presented to the physics community in June 1998 at the $NEUTRINO98$ meeting in Takayama, in a talk by Professor Takaaki Kajita, leader of the on-site contained event analysis group. The data was presented in several papers to the community[8,9,10], building upon past data from Kamioka[3] and IMB[2], and culminating the claim of observation of oscillations or muon neutrinos, published in Physical Review Letters in August 1998[11]. We now proceed to review the evidence, which has changed little except for new indications that the ν_μ oscillating partner is probably the ν_τ, and not a sterile neutrino.

2.1 Up-Down Asymmetry

One way to look at the FC (and PC) data is in terms of a dimensionless up-to-down ratio, difference over sum (which has symmetrical errors in contrast to just up/down)[14]. This quantity is exhibited as a function of charged particle momentum in Figure 1, for both electrons and muons, with the PC data shown as well (for which we know only a minimum momentum). One sees that the electron data fits satisfactorily to no asymmetry, whilst the muon data shows strong momentum dependence, starting from no asymmetry to about -1/3 $(-0.311 \pm 0.043 \pm 0.01)$ above 1.3 GeV.

From this Figure alone, without need for Monte Carlo simulation, assuming the cause to be neutrino oscillations, one can deduce that:

1. The cause of the atmospheric neutrino anomaly is largely due to disappearing muons, not excess electrons.
2. There is little coupling of the muon neutrino to the electron neutrino in this energy/distance range.
3. The oscillations of the muon neutrinos must be nearly maximal for the asymmetry to approach one third.
4. The scale of oscillations must be of the order of 1 $GeV/200\ km$, plus or minus a factor of several.

In fact, as seen by the dashed lines overlying the data points, the simulations do produce an excellent fit to the muon neutrino oscillation hypothesis, while the no-oscillations hypothesis is strongly rejected. The latter is so strong that statistical fluctuations are not in question, one must look for systematic problems to escape the oscillations explanation.

One concern for some people[13] has been the fact that the asymmetry is indeed maximal, which makes it appear that we are very lucky that the earth size and cosmic ray energies are "just so" to produce this dramatic effect. This appears to this author to fall in the category of lucky coincidences, such as the angular diameter of the moon and sun being the same from earth. (There is another oscillations related peculiar coincidence that the matter oscillation scale turns out to be close to one earth diameter, and this depends upon the Fermi constant and the electron column density of the earth.) The phase space for "coincidences" is very large, and we humans are great recognizers of such patterns.

2.2 East-West Asymmetry

The effect of the earth's magnetic field is a little complicated. For example for energies to a few GeV, it provides some shielding from straight downwards going charged cosmic rays in regions near the magnetic equator. For higher energies and incoming trajectories near the horizon, the magnetic field still prevents some arrival paths. As the SuperK detector location is not on the magnetic equator the effect is not up-down symmetric, and this spoils the symmetry otherwise expected from the neutrinos about the horizontal plane (where there is some peaking due to longer flight paths for pions in the atmosphere). However, the effects are mostly limited to neutrino energies below about 1 GeV, corresponding to cosmic ray primaries below about 10 GeV. The picture is made a bit more complicated by the earth's magnetic field not being a nice symmetrical dipole. Fortunately there are good models of the magnetic field and the people who have made flux calculations take this into account. The SuperK group has published a paper[10] showing the azimuthal variation of the SuperK data (± 30 deg about the horizon) for intermediate to higher energies ($400 - 3000$ MeV), where the calculations are reliable. (Certain simplifications such as a one dimensional cascade model have been regularly used, which surely is not a good approximation at the lowest energies). The SuperK data exhibit significant variation from uniformity yet fit the flux predictions very well, giving one confidence in the modeling[10].

2.3 Natural Parameters for Oscillations: L/E

In an ideal world, one would assuredly present this data as a function of distance divided by energy, L/E, since that is the parameter in which one expects to see oscillatory behavior. Since we observe only the secondary charged particle's energy and direction, badly smeared at the energies available, plots in which one would wish for visible oscillations can at best show a smooth slide from the no-oscillations region to the oscillating regime. This is illustrated in Figure 2, where the ratio of numbers of events observed to those expected with no-oscillations is plotted versus "L/E"[15], for muon and electron (type) events.

The plot is not "normalized", and since we see somewhat of an excess of electron type events overall, the solid circle indicated electron points are a bit greater than one on average (+14%). This is a little worrisome, but acceptable since (as already noted) the absolute flux is uncertain to a larger value. In contrast to the electron data, the muon points fall with increasing L/E, reaching a plateau at about one half their initial value, again consistent with oscillations. Muon neutrino oscillations in the Monte Carlo simulation are indicated by dotted line, and fit the data reasonably well.

As noted, the data does not show oscillations, presumably due to convolutions washing out the oscillatory behavior. It was this smooth falloff that got us to wondering if another model might fit the data, one in which one component of the muon neutrino decays with distance. We wrote a paper[16], and a second version[17], suggesting decay to explain the atmospheric neutrino anomaly. I will not discuss details here, but note that in order to get a model that fit the available facts we had to push on all available limits, and we invoke neutrino mass and mixing in any case. Consequently such models do not pass the economy test of Occam's Razor, though most annoyingly they remain not ruled out as yet.

Considering future experiments, this is one area in which improvement may indeed be made. A hypothetical detector, such as a megaton version of the Aqua-RICH instrument studied by Ypsilantis and colleagues could have the resolution to see a multi-peaked L/E plot[18].

2.4 Fits in Energy and Angle

The SuperK Collaboration's preferred method of fitting the ensemble data is to employ a χ^2 test to numbers of events binned by particle type, angle, and energy, a total of 70 bins. The bin choices may seem a bit peculiar, but they have historical precedent (they are as employed for Kamiokande) and though not optimal for the new data set, this choice allows us to avoid paying any statistical (or confidence) penalty for choosing arbitrary bins. The fit employs a set of parameters to account for potential systematic biases. Details cannot be presented here, but it has been shown that the numerical results are quite insensitive to the selection of the parameters or their supposed "errors" (except for the overall normalization)[19].

Figure 3 illustrates the data plotted for two energy intervals ($sub - GeV$ and $multi - GeV$, more or less than 1.3 GeV) for single track events identified as either electron-like or muon-like. The partially contained data is added to the multi-GeV muon data. The data is shown as a function of the cosine of the zenith angle, with +1 being down-going. One sees that the data very well fits the curves gotten from the Monte Carlo simulation, at the values gotten from the grand ensemble fit ($\delta m^2 = 0.003\ eV^2$ and $sin^2(2\theta) = 1.0$).

The results of the fits are often presented in terms of an inclusion plot, showing an acceptable region(s) in the space of mixing angle ($\sin^2 2\theta$) and mass squared difference (δm^2), as presented in Figure 4. The minimum in δm^2 has moved a little upwards with accumulated statistics, though not much, (good news for long baseline experiments anyway) but remains uncertain to about a factor of two.

2.5 Muon Decay Events

It is not often emphasized, but the original indication of the anomaly, a deficit in stopped-muon decays ($\simeq 2.2\ \mu sec$ after the initial neutrino event), remains with us, and constitutes a nice alternate, almost independent sample, with quite different systematics. It is not so clean a sample and the statistics are lower, but the complete consistency of the muon decay sample remains a reassuring complement to the energy and angle analysis employing track identification.

2.6 *Through-going and Entering-Stopping Muons*

Another cross check comes from the UM and SM samples, which is particularly nice as the source energies are a factor of $10 - 100$ higher and the detector systematics rather different (for example, the target is mostly rock not water). In going from the earlier instruments to SuperK, however, the gain is not so great (about a a factor of 2.5 times over IMB, for example), since the rate of collection of through-going muons depends upon area not volume. However, the much greater thickness of the detector and the good tagging of entering and exiting events in the veto layer yields many more stopping (SM) events.

The angular distribution for UM events from below the horizon is shown in Figure 5, where one sees that the angular distribution is nicely consistent with oscillations and not with no-oscillations. However, since much of the effect is close to the horizon, where oscillations for the energies in question are just setting in, one worries about contamination of the near horizon events with in-scattered events from the much greater numbers of down going muons. There is no room for details here, but we find no evidence for significant contamination[12].

The SM sample was predicted to be 35-40% of the UM sample, as indicated in Figure 6, yet in fact we see only about $24\% \pm 3\%$. Fitting the data to the oscillation hypothesis one can make the now usual inclusion plot, which shows that the UM and SM results are completely in accord with those from the FC and PC data. However, as the statistics are smaller and the physics leverage not as great, the muon result does not add much to the FC and PC constraints, though it does stiffen the lower bound in δm^2.

There is a lengthy tale about an SM/UM analysis from the IMB experiment[2], which claims an exclusion region very close to the now preferred solution. This result. seems to have been flawed due to older flux models and Monte Carlo simulations. Work is in progress to reassess the old data with new flux calculations and an updated quark model. Thus there remains a small cloud upon the horizon, but one which appears to be fading.

2.7 *The Muon Neutrino's Oscillation Partner*

Given that the muon neutrino is oscillating, is it oscillating with a tau neutrino or a new sterile neutrino which does not participate in either the charged (CC) or neutral current (NC) weak interaction? Fortunately we have several means to explore this with SuperK data. The NC events should show an up-down asymmetry for sterile neutrinos but not for tau neutrinos (as the NC events for all ordinary neutrinos are the same). Another avenue for discrimination is that sterile neutrinos would have an additional oscillation effect due to "matter oscillations". The consequence would be a unique signature in the angular distribution of intermediate energy muons.

Early SuperK efforts focussed upon the attempt to collect a clean sample of π^o events. As it turns out this has been frustrating as the rings (from the two decay γs) cannot be separated at energies above $\simeq 1\ GeV$, and in net there are not so many reconstructed events as to permit a good discrimination. In fact the absolute rate is consistent with expectations, but the cross section is uncertain to about 20% making the hint at tau coupling not significant.

More recently, tests have been devised employing the PC event sample and the UM sample. The PC sample can be cut on energy to yield a somewhat higher mean source energy, and the upwards going number compared to downwards number of events. For the muons a near horizontal number can be compared to number of nearly straight upcoming events. Preliminary results from SuperK give no encouragement for sterile neutrino model builders. It appears that the tau neutrino hypothesis fits the data, while the sterile neutrino hypothesis is rejected at about the 2 standard deviation level. A publication will be forthcoming from SuperK.

2.8 Hypotheses to Explain Anomaly

We conclude with a summary Table 1 of all hypotheses put forth to explain the atmospheric neutrino anomaly. Space does not permit a full discussion here, but it is the case that with the SuperK data we now have eliminated almost all alternate hypotheses to explain the results. The only exception of which the author is aware involves the peculiar decay model, but it is one that nobody takes very seriously (including the author who is co-author of the model). The only hypothesis which fits the evidence, and it fits very well, is that muon neutrinos maximally mix with tau neutrinos with a δm^2 in the range of $2 - 5 \times 10^{-3}\ eV^2$.

3 Implications

The ramifications of the atmospheric neutrino anomaly are great and span the known realms of fundamental physics from large to small. We have not discussed in this short paper the links to solar neutrinos[20], nor the LSND results[21]. Certainly there is no conflict between the atmospheric muon neutrino results and the possible (nay likely) solar oscillations. If, however, the LSND results are correct, then we have surely some interesting physics to untangle, as it is generally admitted that no simple three neutrino model can incorporate all three neutrino anomalies, and that new degrees of freedom would be required.

Table 1: List of hypotheses invoked to possibly explain the atmospheric neutrino anomaly. The first 3 columns are criteria available prior to SuperK, and the last 4 after the 1998 SuperK publication. The hypotheses divide into 5 systematics issues and 7 potential physics explanations. As indicated in the text, the only remaining likely hypothesis is the oscillation between muon and tau neutrinos. The "x" schematically indicates which evidence rules out the hypothesis in that row.

Evidence		*Old*				*New*	
Hypothesis	R $(E<1$ $GeV)$	μ *decay* $Frac$	Vol $Frac$	R $(E>1$ $GeV)$	A_e $\simeq 0$	A_μ >0	$R(L/E)$ $\simeq 0.5$
Atm. Flux Calc.	xx			x		x	x
Cross Sections	xx			x		x	
Particle Ident.		xx	xx				
Entering Bkgrd.			xx			x	
Detector Asym.			xx				
X-Ter. ν_e						x	x
Proton Decay				x		x	x
ν_μ Decay							$\tilde{x}$
ν_μ Abs.							x
ν_μ - ν_e					x		
ν_μ - ν_s						x	
ν_μ - ν_τ							

3.1 Astrophysics and Cosmology

The implications of the oscillations results have been explored in other talks at this meeting as well. First, it appears that neutrinos with summed masses of the order of 0.1 eV will not make any major contribution to resolving the dark matter quandary. Nonetheless with a ratio of 2 billion to one for photons (and neutrinos) to nucleons from the Big Bang, even such a small neutrino mass may be greater in total than all the visible stars in the sky. Hence one must account for neutrino mass in further cosmological modeling, but neutrinos are not likely to constitute the bulk of the "missing matter". However if the neutrinos should be nearly degenerate in mass and all have masses in range of around 1 eV (and hence we are observing only small splittings), then neutrino mass may dominate the universe. While neutrinos are not favored by astrophysical modelers (fitting the spatial fluctuations in the cosmic microwave background for example), large neutrino masses are ruled out. Nearly degenerate neutrino masses would not present a consistent picture with the quark and charged lepton masses, which make large mass jumps between generations. But who knows? We do not have a viable GUT with mass predictions, so an open mind is appropriate.

The other area of significance, perhaps of the deepest significance has to do with baryogenesis, the origin of the predominance of matter by one part per billion over anti-matter at Big Bang time. There are claims that the old idea of accumulation of net baryon number will not survive the early stages of universe expansion[22,23]. If that is indeed the case, it may be that neutrinos provide the avenue for net baryon asymmetry generation, relatively late in the game[24].

3.2 Theoretical Situation: Why So Important

There have been a number of talks at this Workshop about the particle theory situation, so I can add little. In Figure 8, I show the masses of the fundamental fermions in three generations, on a logarithmic scale in mass. One dramatically sees that if the neutrino masses are near the lower bounds (that is at the presumed mass differences from present atmospheric and solar results), they lie 10-15 orders of magnitude below the other fundamental fermions (charged quarks and leptons). Graphically one notes the spacing between the neutrino masses and the charged fermion masses is just about the same as the distance (on the log scale) to the unification scale. This is a pictorial representation of the see-saw prediction, as we foresaw more than ten years ago[5]. This points up the task for grand unification, and highlights the deep link between neutrino masses and nucleon decay.

3.3 Future

During the last year the physics community seems to have largely accepted the inevitability of neutrino mass and oscillations[25]. Of course the game has hardly begun and many a trick may await our exploration. But if the LSND claims will go quietly away, the mass and mixing may settle into the simple hierarchical pattern explored in the bi-maximal mixing scenario (or similar versions). To my taste this highlights the importance of experiments to follow up on the LSND results as one of the first agenda items in the neutrino business.

Given present indications, it would seem that the K2K and MINOS experiments should confirm the SuperK results and make the oscillation parameters more precise. Of course, one would really like to see tau appearance, not just muon disappearance to be sure we are not being misled. There are many arguments in the community as to what constitutes appearance (see Goodman's talk at this Workshop). Because of the complexity of tau final state identification, this author would prefer to see a real tau track recorded. In any case plans are in progress in the US, Japan and Europe for the obvious follow up experiments to nail things down.

More interesting for the long range physics is filling in the MNS matrix (lepton equivalent of quark CKM matrix) for neutrinos. This is not an easy business. The atmospheric neutrino measurements really are only defining, at best, three of the nine elements! Solar neutrinos get us another, perhaps a constraint on two. Measuring the tau related components directly seems pretty hopeless. Of course if we can assume the matrix to be unitary and real we are, or soon will be, in good shape as there are then only three independent parameters (plus the masses). But we do not know this, and if there exist CP violations we then have a total of three angles and two phases (but only one measurable). If there are more (heavy or sterile) neutrinos, then things could be much more complicated (as the 3 by 3 sub-matrix will not be unitary). By analogy with the quarks (where the 3 by 3 CKM with small mixing angles and one CP violating phase seems to do the job), perhaps we should not worry too much, except for lack of any guidance whatsoever from theory. CP violation is only very weakly constrained experimentally in the neutrino sector at present, so we could be in for big surprises, and given the neutrino connection with cosmology and baryogenesis, one should indeed be suspicious, I believe. As a whole, the particle physics community is just beginning to explore this avenue, but it looks as though muon colliders may provide our best route for exploring this new realm. Measuring absolute mass remains a frustrating problem, which will not be resolved in the near future it seems. In any case we can expect a long and interesting exploration into neutrino mass and mixing now that the door has been opened.

Acknowledgments

The author wishes to thank Susan and Gabor Domokos and their colleagues for organizing an excellent meeting and for their generous support. Such small workshops are wonderful opportunities for the exchange of detailed information and ideas, not possible in larger more impersonal venues.

References

1. "Calculation of Atmospheric Neutrino-Induced Backgrounds in a Nucleon-Decay Search", IMB Collaboration (with T. J. Haines et al.), *Phys. Rev. Lett.* **57**, 1986 (1986).
2. R. Clark, *et al.*, (IMB Collaboration), Phys. Rev. Lett., **79**, 345 (1997); R. Becker-Szendy *et al.*, Phys. Rev. **D46**(1992) 3720; D. Casper *et al.*,, Phys. Rev. Lett. **66**(1991) 2561.
3. Y. Oyama, *et al.*, (Kamiokande Collaboration), hep-ex/9706008 (1997); K.S. Hirata *et al.*, Phys. Lett. **B205**(1988) 416; K.S. Hirata *et al.*, Phys. Lett. **B280**(1992) 146; Y. Fukuda *et al.*, Phys. Lett. **B335**(1994) 237.
4. See talk by Maury Goodman in these Proceedings.
5. "Neutrino Mass and Mixing Implied by Underground Deficit of Low-Energy Muon-Neutrino Events", John G. Learned, Sandip Pakvasa, and Thomas J. Weiler, *Phys. Lett.* B **207**, 79 (1988). Also see V. Barger and K. Whisnant, Phys. Lett. B **209**,365 (1988); K. Hidaka, M. Honda and S. Midorikawa, Phys.Rev.Lett., **61**, 1537 (1988).
6. See paper by Todor Stanev in these proceedings, and also "Geomagnetic Effects on Atmospheric Neutrinos", Paolo Lipari, T.K. Gaisser, Todor Stanev *Phys. Rev.* D **58**, 73003 (1998), preprint astro-ph/9803093, and references therein.
7. http://www-sk.icrr.u-tokyo.ac.jp/doc/sk/pub/
8. "Measurement of a Small Atmospheric ν_μ/ν_e Ratio", The Super-Kamiokande Collaboration: Y.Fukuda *et al.*, *Phys. Lett.* B **433**, 9 (1998), preprint hep-ex/9803006.
9. "Study of the atmospheric neutrino flux in the multi-GeV energy range", The Super-Kamiokande Collaboration, Y. Fukuda, *et al.*,, *Phys. Lett.* B **436**, 33 (1998), hep-ex/9805006.
10. "Observation of the East-West Anisotropy of the Atmospheric Neutrino Flux", The Super-Kamiokande Collaboration: T.Futagami, *et al.*, PRL **82**, 5194 (1999), preprint astro-ph/9901139.

11. "Evidence for Oscillation of Atmospheric Neutrinos", The Super-Kamiokande Collaboration, Y. Fukuda, *et al.*, *Phys. Rev. Lett.* **81**, 1562 (1998), preprint hep-ex/9807003.
12. "Measurement of the flux and zenith-angle distribution of upward through-going muons by Super-Kamiokande", The Super-Kamiokande Collaboration: Y. Fukuda, *et al.*, *Phys. Rev. Lett.* **82**, 2644 (1999), preprint hep-ex/9812014.
13. See paper by J. LoSecco at this Workshop. We note that some of his earlier his conclusions about this plot are drawn from a monochromatic neutrino model which does not reproduce the situation adequately as it is necessary to employ a wide range of neutrino energies and take account of the smearing of observed energies and angles from those of the parent neutrinos. Figure 7 demonstrates that there is no apparent significant variation in the ratio of muon to electron recorded event rates versus calendar time.
14. "Up-Down Asymmetry: A Diagnostic for Neutrino Oscillations" John W. Flanagan, John G. Learned, and Sandip Pakvasa, *Phys. Rev.* D **57**, 2649 (1998), preprint hep-ph/9709438. Also see J.W. Flanagan dissertation, UH 1997, available at the SK web page.
15. See PhD dissertation of M. Messier, BU 1999 for derivation of the correction used for translating observed energy into L/E. This author has used a different form and definition in the past. The figure presented is the SuperK official plot.
16. "Neutrino Decay as an Explanation of Atmospheric Neutrino Observations", V.Barger, J.G. Learned, S. Pakvasa, T.J. Weiler, *Phys. Rev. Lett.* **82**, 2640 (1999), preprint astro-ph/9810121.
17. "Neutrino Decay and Atmospheric Neutrinos", V. Barger, J.G. Learned, P. Lipari, M. Lusignoli, S. Pakvasa, T.J. Weiler, *Phys. Lett.* B **462**, 109 (1999), preprint hep-ph/9907421
18. "The AQUA-RICH atmospheric neutrino experiment", P. Antonioli, it et al., to be published in NIM in the *Proceedings of the RICH98 Workshop in Israel*, preprint CERN-LAA/99-03, 5/5/99.
19. Nonetheless, this author (personal opinion, not collaboration) suspects that the process pulls the minima slightly towards lower values of δm^2. The reasoning is that the introduced parameters un-weight the effect of R on the fitting, which pulls upwards, while the shape pulls downwards. The author's bet remains that δm^2 settles at around $5 \times 10^{-3}\ eV^2$, whereas the official fits give 2.5 *to* 3.5.
20. See the paper by J. Bahcall in these Proceedings.
21. See the paper by D. Caldwell in these Proceedings.

22. A.D. Sakharov, JETP Lett., **5**, 24 (1967).
23. V.A. Kuzmin, JETP Lett. **12**, 228 (1970).
24. "Baryogenesis via neutrino oscillations", E.Kh. Akhmedov, V.A. Rubakov, A.Yu. Smirnov, *Phys. Rev. Lett.* **81**, 1359 (1998). There are many papers dealing with this topic, as a search on "baryogenesis" in the preprint server will reveal.
25. See the paper of L. Wolfenstein in these Proceedings.

Appendix - Super-Kamiokande Authors

Super-Kamiokande Collaboration

- **Institute for Cosmic Ray Research, University of Tokyo** Y. Fukuda, K. Ishihara, Y. Itow, T. Kajita, J. Kameda, S. Kasuga, K. Kobayashi, Y. Kobayashi, Y. Koshio, M. Miura, M. Nakahata, S. Nakayama, Y. Obayashi, A. Okada, K. Okumura, N. Sakurai, M. Shiozawa, Y. Suzuki, H. Takeuchi, Y. Takeuchi, Y. Totsuka (spokesman),S. Yamada
- **Gifu University** S. Tasaka
- **National Laboratory for High Energy Physics (KEK)** T. Ishii, H. Ishino, T. Kobayashi, K. Nakamura, Y. Oyama, A. Sakai, M. Sakuda, O. Sasaki
- **Department of Physics, Kobe University** S. Echigo, M. Kohama, A.T. Suzuki
- **Department of Physics, Kyoto University** T. Inagaki, K. Nishikawa
- **Niigata University** W. Doki, M. Kirisawa, S. Inaba, K. Miyano, H. Okazawa, C. Saji, M. Takahashi, M. Takahata
- **Department of Physics, Osaka University** K. Higuchi, Y. Nagashima, M. Takita, T. Yamaguchi, M. Yoshida
- **Bubble Chamber Physics Laboratory,Tohoku University** M. Etoh, A. Hasegawa, T. Hasegawa, S. Hatakeyama, K. Inoue, T. Iwamoto, M. Koga, T. Maruyama, H. Ogawa, J. Shirai, A. Suzuki, F. Tsushima
- **The University of Tokyo** M. Koshiba
- **Tokai University** Y. Hatakeyama, M. Koike, M. Nemoto, K. Nishijima
- **Department of Physics, Tokyo Institute of Technology** H. Fujiyasu, T. Futagami, Y. Hayato, Y. Kanaya, K. Kaneyuki, Y. Watanabe
- **Boston University** M. Earl, A. Habig, E. Kearns, M.D. Messier, K. Scholberg, J.L. Stone, L.R. Sulak, C.W. Walter
- **Brookhaven National Laboratory** M. Goldhaber

- **University of California, Irvine** T. Barszczak, D. Casper, W. Gajewski, W.R. Kropp, L.R. Price, M. Smy, H.W. Sobel, M.R. Vagins
- **California State University, Dominguez Hills** K.S. Ganezer, W.E. Keig
- **George Mason University** R.W. Ellsworth
- **University of Hawaii** A. Kibayashi, J.G. Learned, S. Matsuno, V.J. Stenger, D. Takemori
- **Los Alamos National Laboratory** T.J. Haines
- **Louisiana State University** E. Blaufuss, B.K. Kim, R. Sanford, R. Svoboda
- **University of Maryland** M.L. Chen, J.A. Goodman, G.W. Sullivan
- **State University of New York, Stony Brook** J. Hill, C.K. Jung, K. Martens, C. Mauger, C. McGrew, E. Sharkey B. Viren, C. Yanagisawa
- **University of Warsaw** D. Kielczewska
- **University of Washington** J.S. George, A.L. Stachyra, R.J. Wilkes, K.K. Young
- **Department of Physics, Seoul National University** S.B. Kim

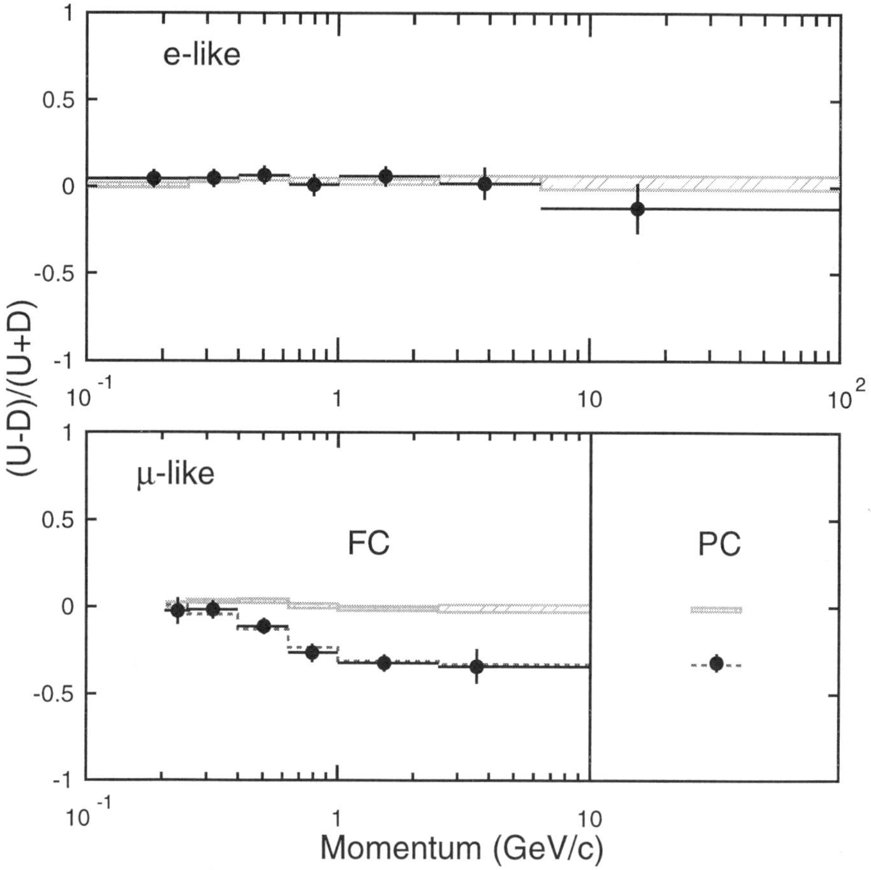

Figure 1: The up-down asymmetry for muon and electron type events in SuperK, from 848 days of live time (analyzed by 6/99), as a function of observed charged particle momentum. The muon data includes a point for the partially contained data (*PC*), which is more than about 1 *GeV*.

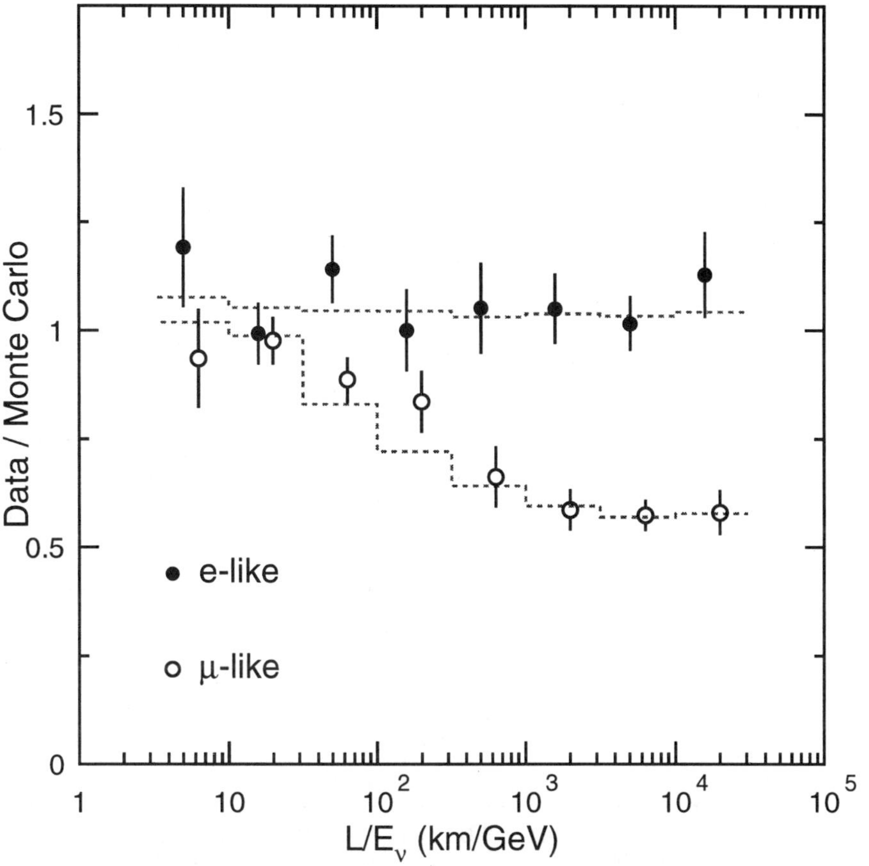

Figure 2: The ratio of numbers of events observed compared to predicted as a function of the natural oscillations parameter, distance divided by energy. The results are not normalized and overall there is a slight excess (not significant) compared to expectations. Electrons show no evidence for oscillations, while muons exhibit a strong drop with L/E. This is consistent with oscillations, as indicated by the dashed lines from simulations.

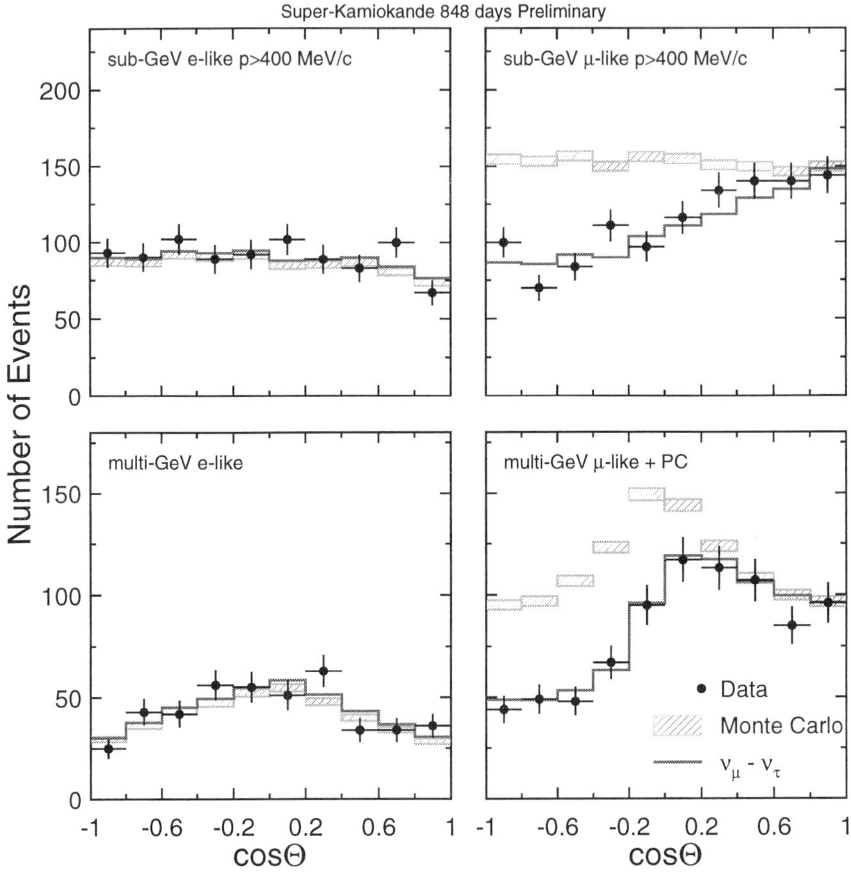

Figure 3: Cosine of zenith angle distributions of the contained event data for two different energy ranges, electron and muon events. Shaded region shows no-oscillations simulation, and heavy line for oscillations between muon and tau neutrinos.

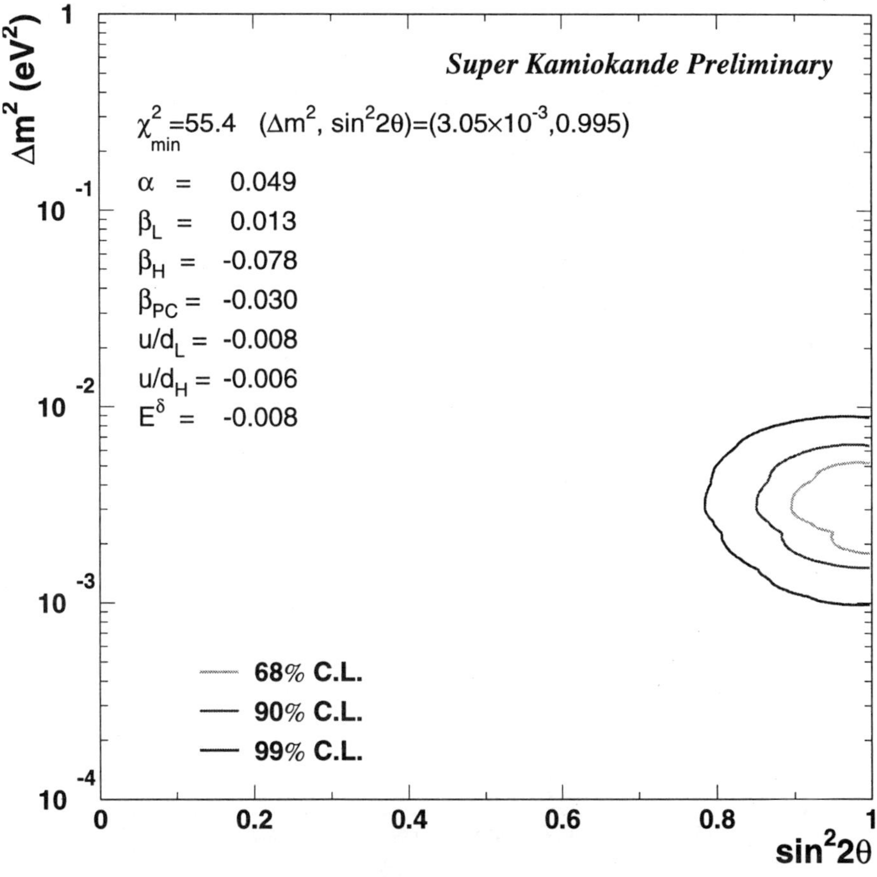

Figure 4: Inclusion plot, showing the regions for various degrees of statistical acceptability in the plane of mixing angle and mass squared difference.

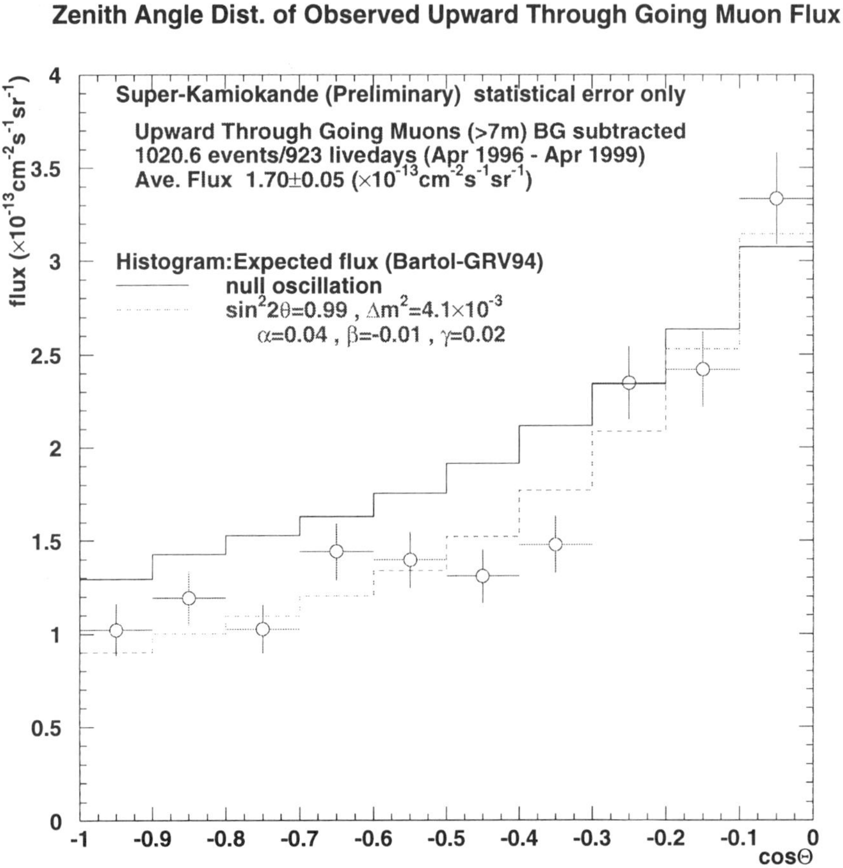

Figure 5: The angular distribution of upcoming muons from 902 days of SuperK data. Expectations for no-oscillations and best fit from contained events are indicated.

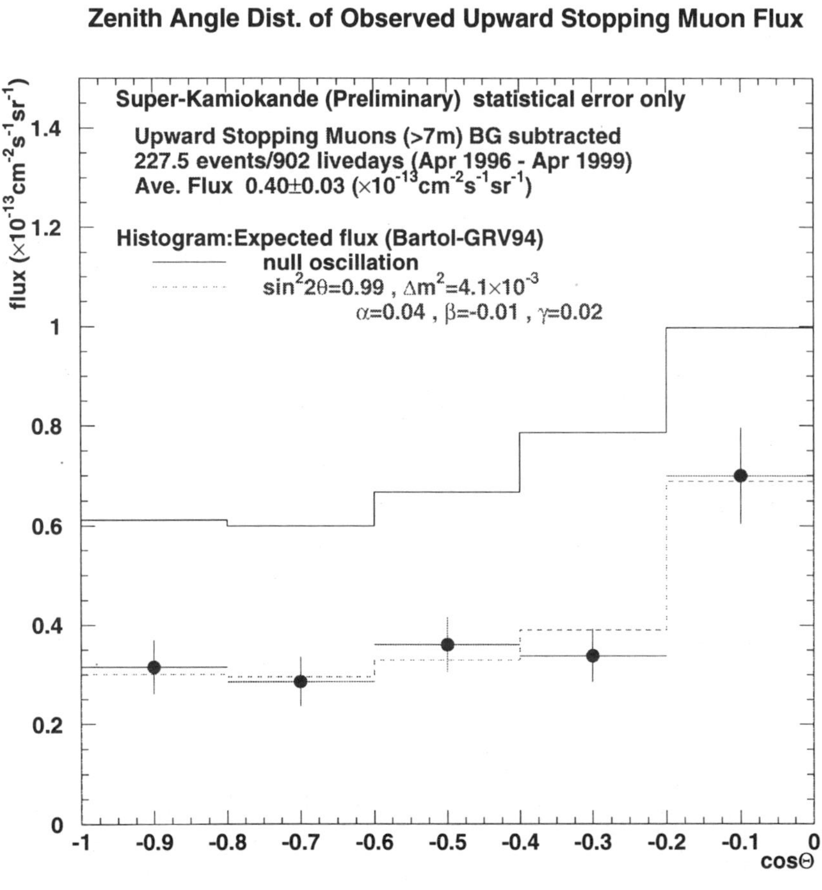

Figure 6: The rate of stopping muons versus zenith angle in 902 days of SuperK data. The large deficit is less significant than it appears because of a 20% uncertainty in absolute flux.

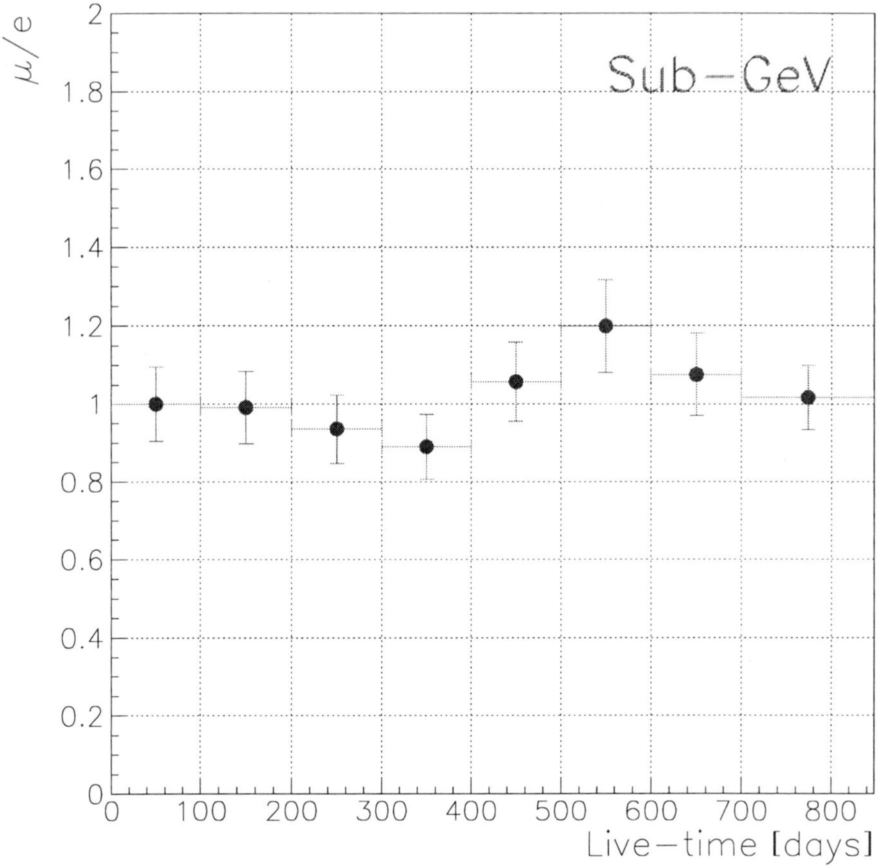

Figure 7: The rate of contained sub-GeV events in SuperK, versus time in 100 day bins. One sees no significant change of ratio with time.

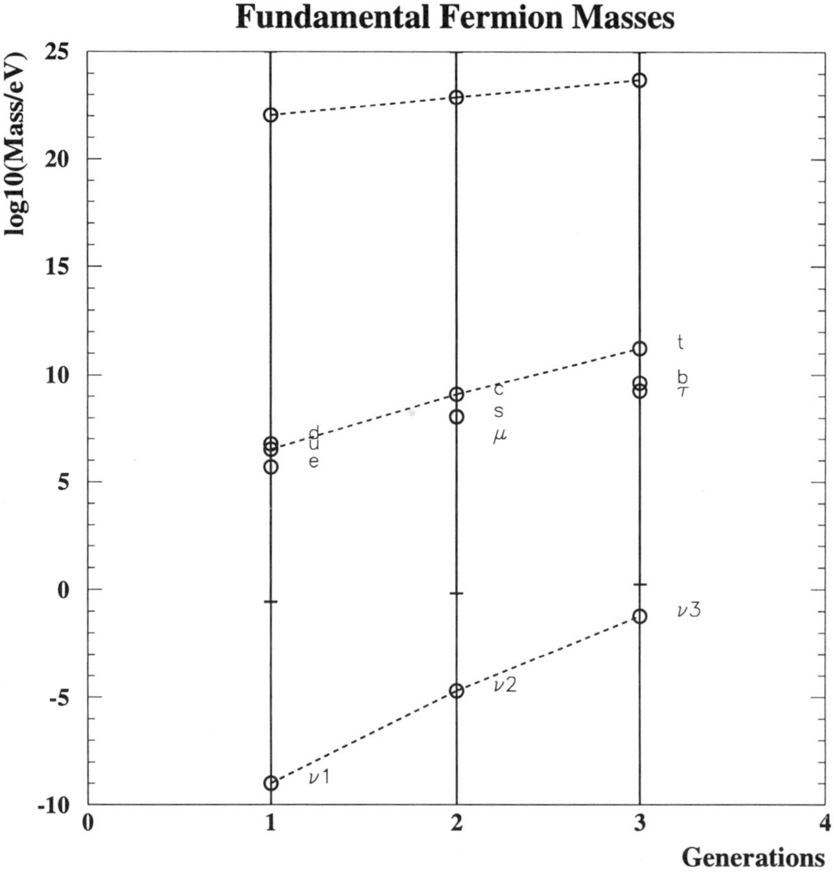

Figure 8: The masses of the fundamental fermions.

ATMOSPHERIC NEUTRINOS: PAST, PRESENT AND ...

JOHN M. LOSECCO
Physics Department, University of Notre Dame
Notre Dame, Indiana 46556-5670 USA
E-mail: losecco@nd.edu

This paper reviews the history of the atmospheric neutrino anomaly. The historical record does not support more recent claims that the anomaly constitutes evidence for a neutrino mass. Most experiments have reported an apparent muon deficiency which is independent of energy and distance. Time dependent variations in the Super Kamioka results are noted.

1 Prehistory

1.1 Opportunity

The early history of the atmospheric neutrino anomaly is closely tied to the growth of grand unified theories. Grand unified theories inspired the experimental search for proton decay which led to large well shielded detectors. The primary background to proton decay was due to atmospheric neutrinos so they were studied in some detail. Neutrino oscillations figured prominently at the first workshop on grand unification. About half of the IMB paper at that meeting [1] was devoted to the question of studying neutrino oscillations with atmospheric neutrinos.

1.2 Early Indications

Early indications that atmospheric neutrinos were not behaving as expected came because many interesting proton decay modes produced unstable particles such as kaons and muons that could be expected to decay in the detector and yield a delayed coincidence with the primary event. In looking for these delayed coincidences it was noticed that there appeared to be too few of them in the global data sample. Based on 148 events Shumard reported [2] that 26±4% of the observed events had a muon decay. Careful studies had indicated that a decay rate of 35±1% was expected. This 2.1 sigma difference was discounted in the thesis.

Confirmation came from the Kamioka experiment in Japan. Kajita's thesis [3] had evidence for a 2.4 sigma deficit of muon decays based on 89 single ring events, But no note of the deficit was made in the thesis.

The atmospheric neutrino statistics rapidly increased and the deficit was noted in a 1986 IMB publication [4]. A 3.3 sigma difference was reported based

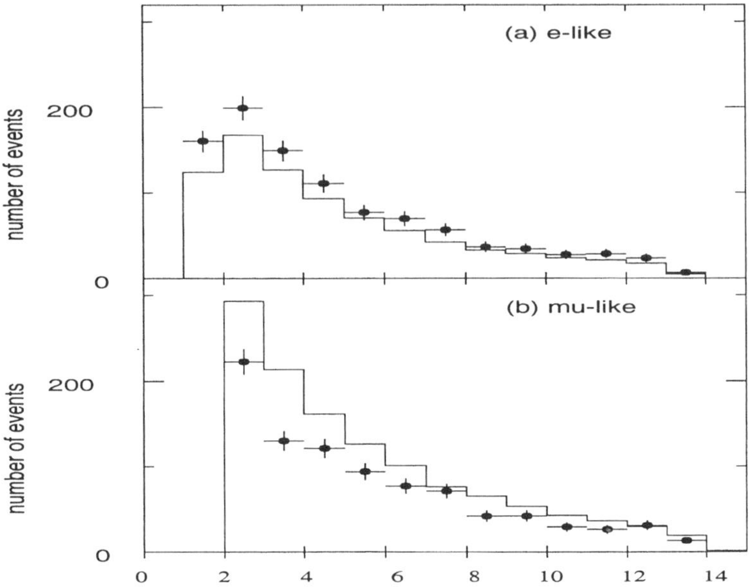

Figure 1: The electron and muon neutrino spectra as reported by the Super Kamioka experiment

on 401 events. 26±2% of the events contained a muon decay when 34±1% was expected.

1.3 Flux Modeling

In all of these cases there were suspicions about the *expected* values. The atmospheric neutrino flux, the neutrino interaction model and the detector response were all required for an estimate. Substantial effort went into modeling the neutrino interactions themselves. (This is because the high multiplicity neutrino interactions were the most serious background to the proton decay signal). Shumard had used Freon bubble chamber data as a model of neutrino interactions. Kajita and Haines had used neutrino interaction models which they compared with deuterium bubble chamber data.

Except for the muon decay rate deficiency the observations were all in good agreement with expectations of atmospheric neutrinos. The rate of interactions, their homogeneity in the detector and the *isotropy* of the signal were as expected [5,6].

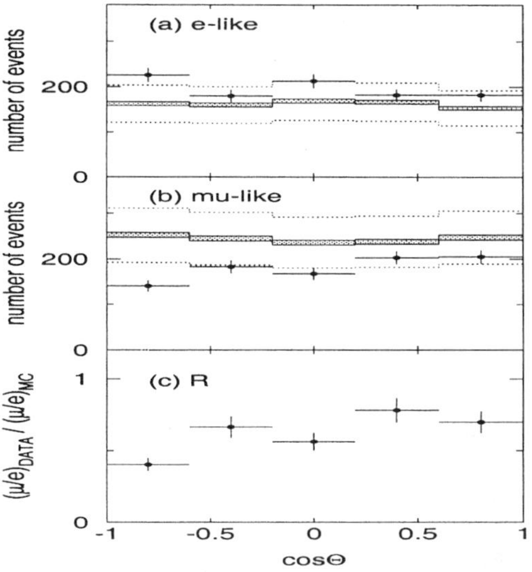

Figure 2: The electron and muon neutrino zenith angle distribution as reported by the Super Kamioka experiment. The bottom plot is the zenith angle distribution for R.

2 The Classical Period

2.1 Rapid Progress

The Kamioka group applied a *revised* form of their particle identification methods to follow up on these indications to establish the general properties of the anomaly [7]. The early work implied that the effect could be due to either a deficit of muon neutrinos or an excess of electron neutrinos. Uncertainties in the flux normalization did not permit one to discriminate between these two possibilities. To reduce the dependence on the absolute flux normalization the Kamioka group introduced a variable called R, $R = (\frac{\mu}{e})_{Obs}/(\frac{\mu}{e})_{Expect}$, which was only sensitive to the relative normalization of the muon and electron neutrino flux.

Typical of the results from this period is an apparent energy independent excess of electron type events and an apparent energy independent deficit of muon type events. This is illustrated in figure 1 which is taken from the Super Kamioka early work [8]. The solid curve is the expected value the points are the data. This confirms prior work [7,9,10] on these properties. These data have $R = 0.61 \pm 0.03(stat.) \pm 0.05(sys.)$ which is well below the expected value of

1. This value of R seems to be energy independent [8] over the range of energies illustrated.

The angular distribution of the effect is very important in testing neutrino evolution effects. Atmospheric neutrinos from above have traveled a few 10's of km. Those from below have traveled on the order of 10,000 km. R as a function of zenith angle from early Super Kamioka work [8] illustrates the most important features of the angular distribution.

While IMB [9] and Kamioka [7,10] have shown purely isotropic distributions the most important feature is still present in figure 2, the Super Kamioka plot. Note that the electrons are high and the muons are low at all angles. The value of R is significantly low over all of the solid angle, even at $0°$. This implies that even atmospheric neutrinos with short path lengths, on the order of 10's of kilometers, are effected by the anomaly.

2.2 Uniformity in Energy and Direction

The isotropic angular distribution could be misleading. The direction reconstructed and plotted is the charged lepton direction emerging from the neutrino interaction. This has a reasonable correlation with the incident neutrino direction, but it is not exact [1]. Two factors mitigate the significance of this scattering effect. The directional correlation improves with energy and the distance scales very slowly with angle except near the horizon. The majority of the atmospheric neutrino events from above have traveled on the order of 10's of km. Those from below have traveled the order of 10,000 km. So scattering within the hemisphere has only a small effect on the distance traveled. With regard to the first factor, the data indicates [11] that the isotropy for both the electron and muon samples is still present at higher energies. See figure 4.

Since neutrino oscillations were always a potential hypothesis for such an effect and since there was no apparent energy or distance dependence in the contained neutrino events another source of signal could be used to extend the sensitivity or corroborate the oscillation hypothesis. Figure 3 shows the exclusion plot for the $\nu_\mu \to \nu_\tau$ hypothesis using upward going neutrinos [12] and the ratio of stopping to through-going upward muons [13]. The Frejus limits [14] are also shown. Most large mixing angle Δm^2 were ruled out by the combination of analyses reviewed in this paper [13]

2.3 Consistency

By early 1998 the anomaly had been consistently observed in about 8 independent measurements. All the observations were consistent with an R value of 0.61.

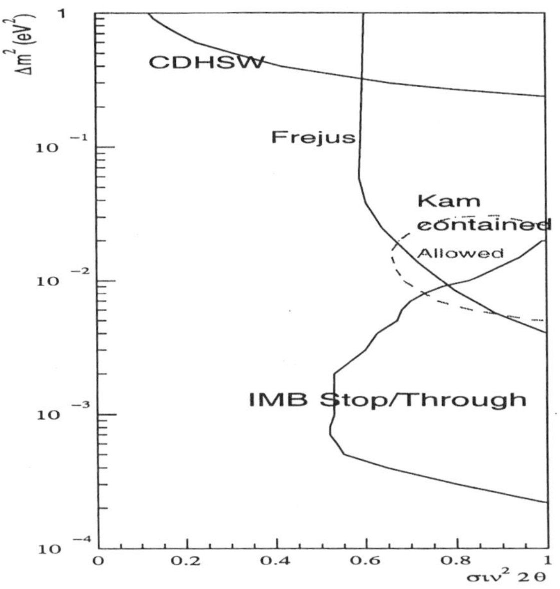

Figure 3: The exclusion plot for $\nu_\mu \to \nu_\tau$ from a number of experiments. The IMB Stop/Through excludes the region favored by recent Super Kamioka analyses and much of the region permitted by the older Kamioka analysis too.

Experiment	Measured R value
Kamiokande Sub-GeV	0.60±0.07±0.05
Kamiokande Multi-GeV	0.57±0.08±0.07
IMB	0.54±0.05±0.12
Frejus	1.00±0.15±0.08
Nusex	0.99±0.29
Soudan	0.64±0.17±0.09
Super Kamiokande Sub-GeV	0.61±0.03±0.05
Super Kamiokande Multi-GeV	0.66±0.06±0.08

2.4 Up/Down Energy Independence

The atmospheric neutrino flux is not truly isotropic. The Earth's magnetic field, which varies from point to point limits the flux of primary cosmic rays which can get close enough to the atmosphere to interact. Fortunately at some sites, such as the IMB site, the effect is small. IMB was located at 81.27° west and 41.72° north where the the downward flux was expected to be about 5%

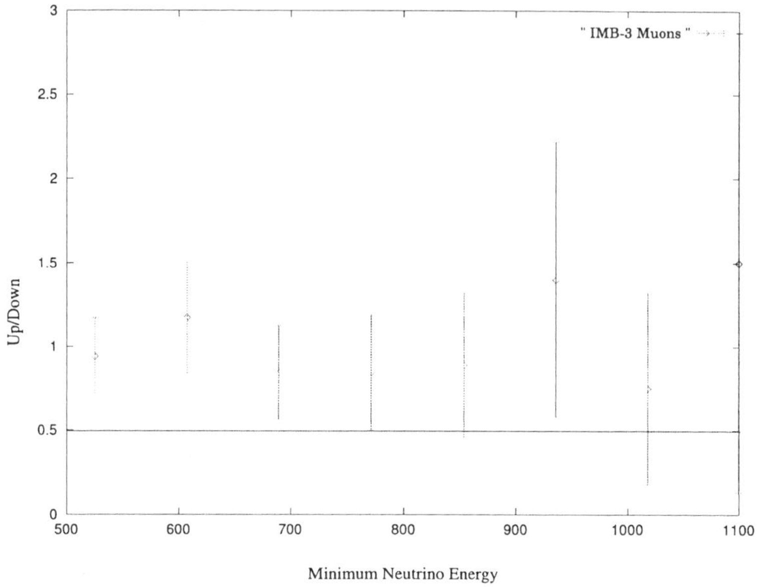

Figure 4: The upward over downward muon rate in IMB3 plotted as a function of the minimum energy. There is no indication of an asymmetry at any energy.

greater than that coming from below. The geomagnetic effects decrease with energy (but are replaced by path length effects at high energy.) In setting limits at IMB it has been customary to neglect the small geomagnetic effects. Any reduction in the upward flux due to geomagnetic effects would be attributed to the phenomena in question. This neglect of geomagnetic effects at IMB makes the limits obtained slightly conservative. In other words in setting limits for neutrino oscillations a larger region could have been excluded if geomagnetic effects were taken into account.

Limits based on the up/down rate for muon neutrino interactions in IMB-3 [9] are shown in figure 5. The up over down ratio is shown as a function of the minimum neutrino energy in figure 4. While the statistical errors grow with energy as fewer events are used, the data are consistent with an energy independent value of 1. There is no evidence for an up-down asymmetry in this data.

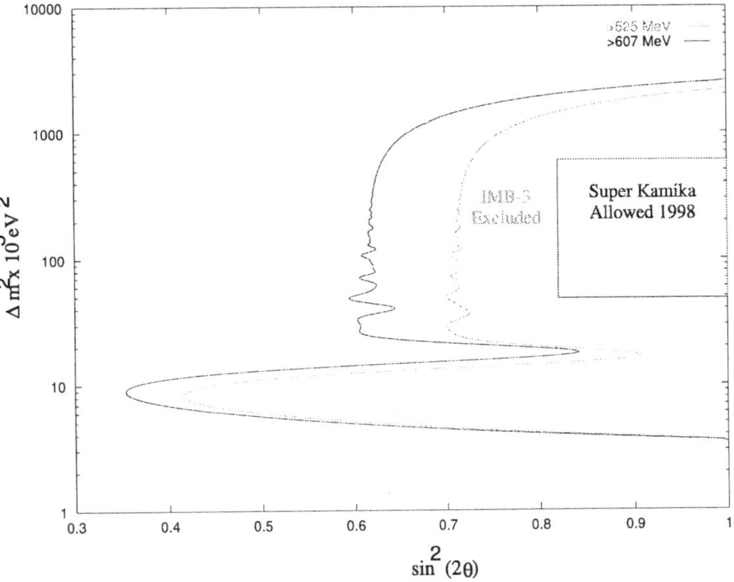

Figure 5: The excluded region extracted from the IMB3 upward over downward muon rate. The region favored by the Super Kamioka experiment is ruled out by this measurement and the one illustrated in figure 3. The larger excluded region uses all of the muon neutrino events. The smaller region uses those events above about 600 MeV.

3 The Modern Era

The hint of a directional modulation, present in figure 2 apparently evolved and in mid 1998 the Super Kamioka collaboration published [15] it as evidence to support the neutrino oscillation hypothesis (figure 6). This result has been widely discussed and interpreted. It is noteworthy that the mass fit to the Super Kamioka data was barely consistent with the earlier Kamioka neutrino analysis. The reason for this difference is that the two data samples are indeed different. Earlier experiments, including Kamioka do not corroborate the Super Kamioka observations of anisotropy. The anomaly is still manifest at $0°$ in this data [15]. It can not be accommodated by the Δm^2 of the fit but is handled by several other fit parameters. Both the μ to e flux ratio and the up to down flux ratio are fit as part of the oscillation analysis.

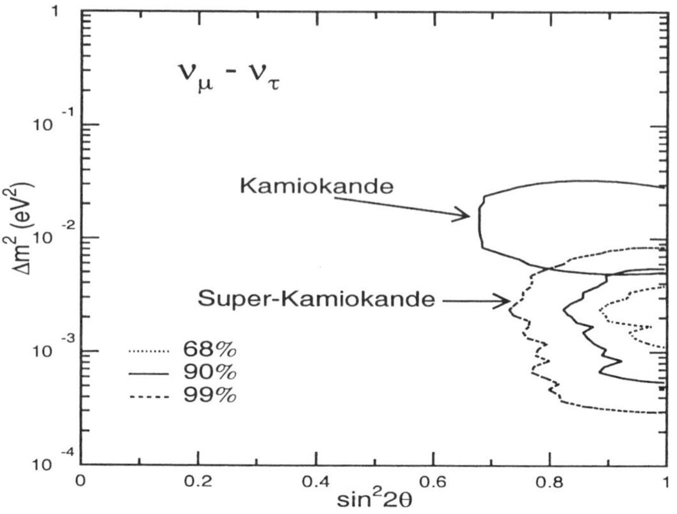

Figure 6: The Super Kamiokande and Kamiokande allowed regions for $\nu_\mu \to \nu_\tau$ oscillations.

4 The Post Modern Era

The modern period did not last long. In January 1999 the Super Kamioka group presented [16] additional data to increase the exposure from 414.4 days to 736 days. The data was presented in a manner to bolster confidence in the neutrino oscillation hypothesis. But under closer examination [17] the additional data appeared to be *inconsistent* with the sample that had already been published.

Without detailed access to the data it is hard to understand the exact nature of the change. Manifestations of the problem are a 12±3% drop in overall event rate for the sub-GeV event sample. Most of the drop seems to be due to a 18±5% drop in the "electron neutrino" interaction rate and a 20±5% drop in the multi-ring rate. The muons manifest a not significant 3±5% rise. There is no significant change in the multi-GeV data sample.

Since the observed "electron neutrino" rate has dropped there is a significant change in R over values reported earlier. The value of R reported in the new sub-GeV sample was, $0.76 \pm 0.04(stat.) \pm 0.06(sys.)$. The earlier value was $R = 0.61 \pm 0.03(stat.) \pm 0.05(sys.)$. Only the statistical error is relevant in comparing these two numbers since they are taken from the same experiment and share the same systematic error. To the extent that the atmospheric anomaly is $R \neq 1$ this variation is rather shocking.

5 What Next?

The apparent temporal variation may be a significant factor in understanding what is actually happening with the anomaly. We have been assured that the temporal effect is not a result of systematic error. If it were attributable to systematic error it would invalidate much of the Super Kamioka physics but not help in furthering our understanding of the effect. The Super Kamioka group has shown evidence [18] of the stability of their detector. Previous experiments such as Kamioka [11] and IMB have found no indications of temporal variation. There is some evidence [17] that the change in R is continuing and has not yet reestablished a constant value.

Acknowledgments

I am grateful to S. and G. Domokos for their hospitality at this meeting. I would like to thank I. Bigi for encouragement.

References

1. L. Sulak in *First Workshop on Grand Unification*, ed. P. Frampton, S. Glashow and A. Yildiz (Math Sci Press, Brookline, Mass. 1980).
2. E. Shumard, Ph.D. thesis, University of Michigan (1984).
3. T. Kajita, Ph.D. thesis, University of Tokyo (1986).
4. T. Haines *et al.* Phys. Rev. Lett. **57** 1986 (1986).
5. R.M. Bionta *et al.* Phys. Rev. **D38**, 768 (1988).
6. M. Nakahata *et al.* J. Phys. Soc. Jap. **55** 3786 (1986).
7. K. Hirata *et al.* Phys. Lett. **B205** 416 (1988).
8. Y. Fukuda *et al.* Phys. Lett. **B433** 9 (1998).
9. D. Casper *et al.* Phys. Rev. Lett. **66** 2561 (1991).
   R. Becker-Szendy *et al.* Phys. Rev. **D46** 3720 (1992).
10. K. Hirata *et al.* Phys. Lett. **B280** 146 (1992).
11. M. Takita, Ph.D. thesis, University of Tokyo (1989).
12. Y. Oyama *et al.* Phys. Rev. **D39** 1481 (1989).
13. R. Becker-Szendy *et al.* Phys. Rev. Lett. **69** 1010 (1992).
14. Ch. Berger *et al.* Phys. Lett. **B245** 305 (1990).
15. Y. Fukuda *et al.* Phys. Rev. Lett. **81** 1562 (1998).
16. M. Messier, Talk presented at the 1999 DPF meeting, January 1999. http://hep.bu.edu/~messier/dpf/index.html

K. Scholberg, hep-ex/9905016
17. J. LoSecco, hep-ph/9903310
18. S. Kasuga Ph.D. thesis, University of Tokyo (1998).

ATMOSPHERIC NEUTRINO OSCILLATIONS AND COSMIC RAY LEPTONS IN THE ATMOSPHERE

Todor STANEV
Bartol Research Institute, University of Delaware, Newark, DE 19716, U.S.A.
E-mail: stanev@bartol.udel.edu

The experimental results from detection of atmospheric neutrinos have suggested a strong anomaly in the ratio of neutrinos of muon and electron flavor. Recent high statistics results of the Super–Kamiokande detector have confirmed the anomaly. The standard interpretation of this result is that muon neutrinos oscillate in neutrinos of different flavor. We discuss the formation of the atmospheric neutrino flux, the new experimental results and some of the prospects for future work.

1. Atmospheric neutrinos

Atmospheric neutrinos are produced in interactions of cosmic rays in the atmosphere and the subsequent development of atmospheric cascades. The atmospheric neutrinos in the GeV range come predominantly from the pion decay chain: $\pi^{ch} \rightarrow \mu + \nu_\mu$; $\mu \rightarrow e + \nu_e + \nu_\mu$. The decay length for 1 GeV muon is about 6 km, while the characteristic length for vertical atmospheric cascades is 20 km. Thus not only all mesons but practically all muons in this energy range decay and the energy spectrum of the resulting neutrinos follows the spectrum of the cosmic ray primaries. In addition, the charged pions decay chain generates a muon neutrino/antineutrino pair $\nu_\mu \bar{\nu}_\mu$ for every electron neutrino ν_e or antineutrino $\bar{\nu}_e$. This rule of thumb is confirmed by different precise numerical calculations. $\frac{\nu_\mu + \bar{\nu}_\mu}{\nu_e + \bar{\nu}_e} = 2$ is only slightly modified by the contribution of other meson decays. The ratio of muon to electron neutrinos, which to the first approximation depends only on the decay kinematics, is known much better than the absolute flux of neutrinos.

With increasing energy the secondary pions are more and more likely to interact because of the relativistic increase of their decay length $\gamma c\tau$ (γ is the Lorentz factor E_π / m_π). In vertical cascades the energy at which pions are equally likely to decay and to interact (*the critical energy* ε_π) is 115 GeV [1]. Above that energy kaon decays become much more important source of neutrinos and dominate the neutrino production by 3:1 at asymptotically high energy. The interaction/decay competition of the secondary mesons steepens the energy spectrum of the neutrinos originating in meson decays by one power of the energy, i.e. if the primary cosmic ray spectrum is $N(E) = A \times E^{-(\gamma+1)}$ the spectrum of the meson decay neutrinos is proportional to $E_\nu^{-(\gamma+2)}$. Another power of the energy is added for neutrinos originating in muon decays, which steepens their spectrum to $E_\nu^{-(\gamma+3)}$. For that reason the spectrum of electron neutrinos, that come mostly from muon decays, is steeper than that of muon neutrinos, which are also generated directly in meson decays. At very high energies, when the atmosphere is too thin for muons to decay, neutral kaon decays are the only source of electron neutrinos.

The ratio of muon to electron neutrinos at high energy depends strongly on the π/K ratio in multiparticle production interactions, which is known only with an accuracy about 30%. This uncertainty also affects the absolute normalization of the atmospheric neutrino flux. The major factor in the absolute normalization of the atmospheric neutrinos is however the magnitude of the galactic cosmic ray flux interacting on the atmosphere.

This major input in every calculation of the atmospheric neutrino flux is complicated by two other factors:

- The solar modulation of the galactic cosmic ray flux introduces a time dependence for rigidities below $\sim$ 100 GV. The cosmic ray flux is inverse proportional to the level of solar activity.

- The deflection of the cosmic rays by the geomagnetic field prevents lower energy particles from reaching the atmosphere and interacting. This introduces a location dependent penetration (and interaction) probability: low energy particles reach the atmosphere only in regions with high geomagnetic latitude, close to the geomagnetic poles. Low geomagnetic latitudes may have 'geomagnetic cutoffs' as high as 50 GV in certain directions. The highest possible cutoff is 59.4 GV, which is the gyroradius of a proton orbiting the Earth at the geomagnetic equator.

The Earth is totally transparent for neutrinos until their mean interaction length $\lambda_\nu = (N_A \sigma_{\nu N})^{-1}$ reaches 10^{10} g/cm^2 – the column density through the center of the Earth. This happens for neutrino energies about 10^5 GeV and does not affect strongly the detection of atmospheric neutrinos because by that energy the atmospheric neutrino flux has decreased by many orders of magnitude.

A neutrino detector thus can detect neutrinos in 4π solid angle. Neutrinos coming through the Earth are produced in the atmosphere of the opposite hemisphere. Local geomagnetic effects are applicable only for 'downward going' neutrinos, while for 'upward going' neutrinos one has to integrate the field values over the whole area of the atmosphere. The flux of upward going neutrinos at different locations thus varies much less than the flux of downward going neutrinos.

Atmospheric cascades initiated by particles hitting the atmosphere horizontally (the direction perpendicular to the local zenith) develop in much greater atmospheric depth (36, 000 g/cm^2 instead of 1000 g/cm^2) and length (500 km). On top of that the atmosphere is very tenuous (10^{-6} – 10^{-7} g/cm^3) for a large fraction of the column density. The decay probability in horizontal cascades is enhanced and this leads to higher neutrino fluxes in horizontal direction. The difference with the vertical fluxes is big for high neutrino energies and relatively small in the GeV range, where most pions and muons decay even in vertical direction. The angular distribution of neutrinos above 10 GeV peaks in horizontal direction ($\cos\theta = 0$) and is symmetric for negative and positive cosines of θ. The convention is that $\cos\theta$ = -1 corresponds to vertical upward going neutrinos and $\cos\theta = 1$ defines vertical downward going neutrinos.

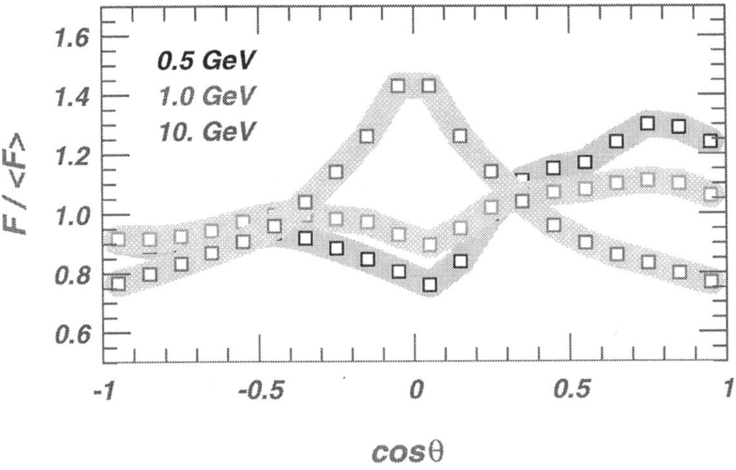

Figure 1: Angular distribution (normalized to the total flux) arriving at the detector at different zenith angle.

Fig. 1 shows the zenith angle distributions of neutrinos of energy 0.5, 1 and 10 GeV at the location of Kamioka, Japan (latitude 34.5°, longitude 137.3° East. The angular distribution of the 10 GeV neutrinos is fully symmetric, while at lower energy the angular distribution is determined by the geomagnetic cutoffs. At Kamioka the vertical geomagnetic cutoff is high and this depresses the flux of downgoing neutrinos.

2. Neutrino detection and experimental definitions

Neutrinos are detected mostly through their charged current interaction

$\nu_i + N \Rightarrow \ell_i + X$ where the lepton number i is transferred to the corresponding lepton. In practical terms this means that an electron or muon track originates inside the detector. Assuming that the atmospheric neutrino flux is of order 1 cm^{-2}s^{-1} one can estimate the necessary detector size. The event rate is

$$
\begin{aligned}
F_\nu \times \sigma_\nu \times N_A & \\
&= 1 \times 10^{-38} \times 6.10^{23} = 6.10^{-15} \textit{ per} \text{ gram } \textit{per} \text{ second} \\
&= 1.8 \times 10^{-7} \textit{ per} \text{ gram } \textit{per} \text{ year} \\
&= 180 \textit{ per} \text{ Kton } \textit{per} \text{ year}
\end{aligned}
$$

Detectors have to have size of thousands of tons (Kt) and to operate for years to collect reasonable atmospheric neutrino statistics. They have to be shielded from downward going atmospheric muons, which places them deep underground. A large fraction of the total volume of the detector (close to the walls) is not used in neutrino detection because it is often polluted by natural radioactivity. It is often used as anticoincidence counter to reject entering tracks. The inner part of the detector is its fiducial volume.

There are two types of detectors:

- traditional calorimeters, where layers of particle detectors are interspaced by layers of target material, e.g. iron.

- water Cherenkov detectors: phototubes on the walls of a large tank filled with water detect Cherenkov radiation from the charged particles created in the neutrino nucleon interaction.

Each detector type has its advantages: Cherenkov detectors are cheaper and can reach total volumes of tens of Kt. Particle calorimeters can measure the charge particle tracks more precisely.

Lepton identification (electron or muon) is based on the different modes of energy loss for these particles. Electrons suffer bremsstrahlung with the radiation length in water of 36 cm, much less than the detector dimensions. Even low energy electrons thus initiate electromagnetic (electron–photon) cascades. Muons of energy less than 100 GeV lose energy mostly through ionization ($\sim$2 MeV/cm). In a traditional calorimeter muons have a straight track with possibly a little bit of kink before stopping. Electrons, on the other hand, show a bunch of short tracks left from individual cascade electrons. In water cherenkov detectors this translates into well defined Cherenkov rings for the muons and 'filled in' circles for the electrons.

Experimentally neutrino events are classified generally in three groups:

- fully contained events (FC) for which the interaction vertex and all tracks are fully contained in the fiducial volume of the detector.

- partially contained events (PC) where only the interaction vertex is contained.

- upward going muons, which are generated by neutrino interactions in the rock surrounding the detector. Upward going muons can be throughgoing (exit the detector) or stopping inside the detector.

FC and PC events are observed in the full 4π solid angle, while upward going muons may only come from below (2π acceptance). Downward going neutrino produced muons are swamped by the atmospheric muons. Neutrino induced upward going muons are a nice way to detect neutrinos because the effective volume of the detector is increased by pathlength of the muon generated in the neutrino interaction, i.e. by its range R_μ in the rock.

Although the exact energy distribution of the neutrinos responsible for each type of events depends on the detector location, detection conditions, as well as on the neutrino zenith angle, FC events are on the average produced by neutrinos of energy slightly less than 1 GeV, PC events correspond to neutrinos of energy several GeV, as do upward going stopping muons. Throughgoing muons have very broad energy distribution of the neutrino parents peaking at ~100 GeV. The bigger the detector, the higher the average energy is for PC events and upward going muons. The energy ranges corresponding to different experimental types of events define the types of neutrino interactions that dominate each group: most of the FC events are produced in quasi elastic neutrino interactions; upward going muons are generated in deep inelastic scattering. Partially contained events are the most complicated set, where both types of neutrino interactions contribute in addition to the one pion resonances [2].

To decrease the effects of the uncertainty in the absolute normalization of the atmospheric neutrino flux the experimentally studied quantity is the *double ratio* **R** between the measured and predicted ratio of the muon and electron neutrino interactions in the detector.

$$R = \frac{(N_\mu/N_e)_{measured}}{(N_\mu/N_e)_{predicted}}$$

During the first round of experiments (1980's) there were hints of contradiction between different types of experiments. Traditional calorimeters Frejus [3] (exposure 2 Kt.yrs) and NUSEX [4] (exposure 0.8 Kt.yrs) measured R values consistent with the predictions while the larger exposure IMB [5] and Kamiokande [6] measured respectively $R = 0.54 \pm 0.05 \pm 0.11$ and $R = 0.60 \pm 0.06 \pm 0.05$ with exposures of 7.7 Kt.yrs each. The first error is statistical and the second one is systematic. The smaller systematic error of Kamiokande indicates the better sensitivity of the detector. The best interpretation of the lack of muon neutrinos is that these oscillate into a different, not measurable, neutrino flavor, possible τ neutrinos.

3. Neutrino oscillations

Neutrino oscillations were first suggested as a possibility by Bruno Pontecorvo. In a consequence of the assumption that the known neutrino flavor eigenstates are a linear combination of different 'basic' neutrino flavors. In the simple two neutrino case this assumption can be expressed as:

$$\begin{aligned} \nu_\alpha &= \cos\theta \times \nu_1 + \sin\theta \times \nu_2 \\ \nu_\beta &= -\sin\theta \times \nu_1 + \cos\theta \times \nu_2 \,, \end{aligned} \tag{1}$$

where θ is the mixing angle between ν_1 and ν_2. A calculation of the wave function for neutrino flavor α and β shows the development of a phase difference $\Delta\phi = \Delta m^2 T/(2E_\nu)$, where $\Delta m^2 = |m_\alpha^2 - m_\beta^2|$ is the absolute value of the difference in the squared masses of the two flavors. This phase difference defines the conversion (oscillation) probability of ν_α into ν_β

$$P_{\nu_\alpha \to \nu_\beta} = \sin^2(2\theta)\sin^2\left(\pi \frac{L}{L_{osc}}\right) , \tag{2}$$

where the oscillation length L_{osc} is proportional to the neutrino energy E_ν and inversely proportional to Δm^2.

Expressed in units suitable for the energy of the atmospheric neutrinos and for the dimension of the Earth the oscillation probability is

$$P_{\nu_\alpha \to \nu_\beta} = \sin^2(2\theta)\sin^2\left(1.27 \frac{\Delta m^2(\mathrm{ev}^2) L_{\mathrm{km}}}{E_{\mathrm{GeV}}}\right) . \tag{3}$$

Atmospheric neutrinos allow the exploration of a variety of L/E_ν values. Downward going vertical neutrinos are created at distances between 10 and 20 km from the detector, while upward going ones must propagate at distances as large as Earth's diameter, i.e. of order 10^4 km. Since different types of events are generated by neutrinos of energy between 1 and 100 GeV, atmospheric neutrinos can probe L/E_ν from 1 to 10^4 and correspondingly Δm^2 values from 1 to 10^{-4} eV^2. Since the atmospheric neutrino statistics is relatively small, however, only large values of $\sin^2(2\theta)$ could be studied, close to the maximum mixing angle of 45°.

4. The Super-Kamiokande detector and the new results

The Super-Kamiokande detector is a water filled cylinder with a total volume of 50 Kt [7]. The outer layer of thickness ~2.7 meters is optically separated from the inner volume of 32 Kt and is used to identify tracks of externally generated events. The inner volume is observed by 11,146 photomultiplier tubes. The fiducial volume of Super-Kamiokande for atmospheric neutrino detection is two meters inside the inner volume and amounts to 22.5 Kt. All 'inner' photomultiplier tubes are of a special design, with a diameter of 50 cm. 40% of the inner wall surface is photocathode. The detector sensitivity is so large that a 6 MeV electron triggers 30 phototubes within a time window of 200 ns.

The event rate for FC events is 8.2 events per day. The FC events sample in divided in two energy bins: sub–GeV and multi–GeV sample. The sub–GeV sample consists of electrons and muon of visible energy (in the detector, i.e. the energy above the cherenkov threshold) less than 1.33 GeV. This sample is similar to the events detected by the IMB and Kamiokande detectors. The multi–GeV sample of Super-Kamiokande is qualitatively different, because the very large size of the fiducial volume of the detector detector allows for the containment of muons of energy up to 10 GeV. The energy response of the multi–GeV set is similar to this of PC events, which are sometimes included in the same sample.

After 848 days of operation Super-Kamiokande has collected 52.3 Kt.yr of exposure and thus fully dominates the world statistics on atmospheric neutrinos. Table 1 gives the detected and expected rates of electron and muon neutrino events in Super-Kamiokande [8]. Only single track ('one–ring') events, that are relatively easy to analyze, are included here. In the following discussion of the Super-Kamiokande results I follow the data set presented in this reference.

The double ratio R in the sub–GeV sample is $0.68 \pm 0.02 \pm 0.05$. Since 98% of the PC events are generated by muon neutrinos, they could be added to the multi–GeV sample to obtain a double ratio $R = 0.68 \pm 0.04 \pm 0.08$.

The systematic effects in the detector and various possible backgrounds have been carefully investigated. The first suspect is the particle identification of the daughter lepton – electrons and muons leave patterns that are not that different to the uneducated eye. The particle recognition algorithms have been tested in an electron and muon beam experiments and are judged to be better than 2%. So is the energy assignment. The R values are stable in different parts of the detector and at different distances from the walls. The backgrounds due to natural radioactivity and atmospheric muons and neutrons are almost fully eliminated by the 2 m of water outside the fiducial volume and the external

Table 1: Number of electron and muon type one–ring events detected by Super–Kamiokande in exposure of 50 Kt.yr.

event type	detected		expected	
	ν_μ	ν_e	ν_μ	ν_e
Sub–GeV	1852	1826	2618	1754
Multi–GeV	351	439	487	414
Partially contained	470		656	

anticoincidence layer.

If one defines upgoing events as events with $\cos\vartheta$ between -1 and -0.8 and downgoing ones respectively between 1 and 0.8, the ratio between detected and expected muon neutrino events is 0.48±0.1 for upgoing and 0.92±0.1 for downgoing multi–GeV events. These ratios are 0.58±0.04 and 0.87±0.03 respectively for sub–GeV events. This behavior means that at neutrino energy below 1 GeV the oscillations are strong enough to convert a large fraction of the muon neutrinos into neutrinos of a different flavor. At higher energy, however, only neutrinos with long pathlengths are subject of oscillations and the downgoing neutrino rate is almost consistent with no oscillations.

The effect appears to be much stronger for muon neutrinos than it is (if it exists at all) for electron neutrinos. The data sets are divided in five zenith angle ($\cos\vartheta$) bins and the angular distributions are compared to the shape of the Montecarlo predictions. The assumption of no oscillations gives χ^2 of 177 for 69 degrees of freedom. The probability that this hypothesis is true is 5×10^{-9}. In the assumption that muon neutrinos oscillate into tau neutrinos the best fit yields χ^2 of 55.4/(67 d.o.f.) with oscillation parameters of $\sin^2(2\theta)$ $= 1$ and $\Delta m^2 = 3.05\times10^{-3}$ eV^2.

The best visual representation of the general behavior of the atmospheric neutrino data of Super-Kamiokande is the plot of the rates of electron and muon events as a function of L/E_ν, which is shown in Fig. 2 for an earlier, smaller data sample [9]. Electron neutrino events have essentially a flat distribution, while muons agree with the predictions at small L/E_ν values and approach 1/2 of the predictions for L/E_ν larger than 10^3. L is estimated from the muon direction with accounts for the angle between the lepton and the parent

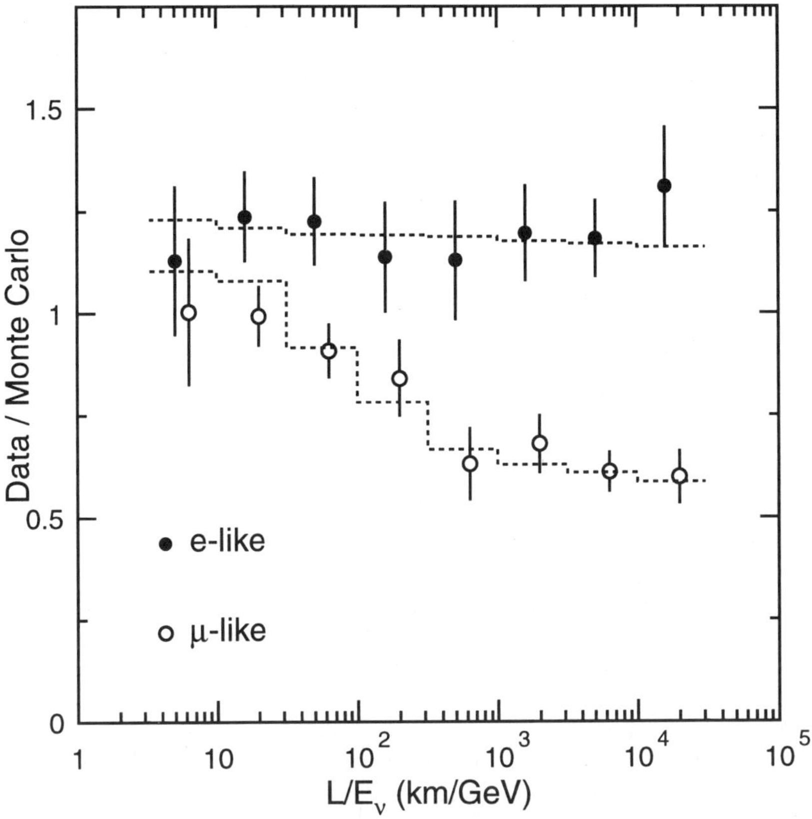

Figure 2: Ratio of data to predictions for fully contained events as a function of L/E_ν. The dashed histograms show the result of the best fit ($\Delta m^2 = 2.2\times10^{-3}$ eV^2, $\sin^2 2\theta = 1$) where the overall normalization of the neutrino flux is also a fit parameter. Adapted from Fukuda et al. (1998).

neutrino and for the height of origin distribution of atmospheric neutrinos in the atmosphere [10].

The consistency of the whole data set with $\nu_\mu \longrightarrow \nu_\tau$ oscillations could be checked independently by a comparison of the fraction of neutral current events to predictions. The rate of neutral current events will not be changed by the oscillations, because neutral current interactions of muon and tau neutrinos are identical. If electron neutrinos are also not involved in the oscillations, the measured ratio $R_{NC/e}$ of neutral current to electron events should correspond to the predictions. In fact, it is $1.00 \pm 0.06 \pm 0.23$ of the predicted value, and is consistent with oscillations.

An independent data set of 1028 upward throughgoing muons and 245 stopping muons is also consistent with the oscillation hypothesis and Δm^2 value derived from the contained event samples. One can form a double ratio for the fraction of stopping muons in the upward going muons. Such a ratio eliminates the uncertainty in the absolute normalization of the cosmic ray flux, although not in the energy spectrum shape. This ratio

$$R_\mu = (N_{stop}/N_{through})_{data}/(N_{stop}/N_{through})_{MC} = 0.63 \pm 0.05 \pm 0.05$$

is also 3.6 σ below the predictions in the absence of oscillations.

Two other recent results are consistent with muon neutrino oscillations. The Soudan 2 detector reports a double ratio of $0.61 \pm 0.11 \pm 0.06$ from the analysis of 3.9 Kt.yr of exposure [11]. The result is important because Soudan 2 is a traditional calorimeter where the distinction between electron and muon tracks is easier and the experimental systematic uncertainties are different from those of the water cherenkov detectors. The MACRO detector [12] presents atmospheric neutrino detection in three different event samples: throughgoing muons, upward stopping + downward partially contained, and upward partially contained events. The rates of all three samples are consistent with muon neutrino oscillations with maximum mixing and Δm^2 in the range of 10^{-2} - 10^{-3} eV^2.

The world set of atmospheric neutrino events confirms the atmospheric neutrino anomaly. The number of muons of atmospheric neutrino origin is much smaller than predicted. Only two out of 14 available independent data sets are statistically inconsistent with this statement: the Frejus 2.0 Kt.yr contained events data set and the Baksan [13] upward going muon sample. The best explanation for the missing muons is that muon neutrinos oscillate in either tau or sterile neutrinos. Neutrino oscillations are possible only if neutrinos have mass. A very important measurement is the angular distribution of the Super-Kamiokande multi-GeV muon sample, which is sensitive to Δm^2 values between 10^{-2} and 10^{-3} eV^2. The best fit value for Δm^2 is 3.05×10^{-3} eV^2.

The muon to electron neutrino oscillations are not only a worse fit to the atmospheric neutrino data, but the relevant Δm^2 range is almost fully excluded by the results of the CHOOZ long baseline reactor neutrino experiment [14]. CHOOZ excludes Δm^2 values greater than 9×10^{-4} eV^2 for maximum mixing.

The final interpretation of the atmospheric neutrino anomaly will require more work. Not all separate data sets are fully consistent with each other. For example the double ratio of the Super-Kamiokande sub–GeV set suggests a Δm^2 value higher than that implied by the angular distribution of the same events. On the other hand, this angular distribution is not confirmed by the Soudan data set. Many of these inconsistencies may be solved with the increase of the experimental statistics. The most important questions, however, concern the absolute normalization and the angular distribution of the atmospheric neutrinos.

5. The Normalization of the Atmospheric Neutrino Flux

The importance of the absolute normalization of the atmospheric neutrino flux is easy to see in Fig. 2. The best fit to the data in the assumption of $\nu_\mu \longrightarrow \nu_{\tau,s}$ oscillations requires neutrino fluxes higher than predicted. At the time of Ref. [9] the ratio of the detected to predicted number of electron events was $\sim$ 1.2. The current difference is about 10%. One the other hand the predictions used in the fit [15,16] are the highest for the location of Kamiokande. All other calculations give smaller neutrino fluxes. The use of a lower neutrino flux would require a more complex disappearance mechanism for the muon neutrinos, where a fraction of the muon neutrinos convert into electron neutrinos.

The dispersion of the predictions of atmospheric neutrinos come mainly from two sources: the uncertainty of the primary cosmic ray flux and the uncertainty in the pion production in nucleon interactions on light nuclei. Until recently there were only two statistically precise measurements of the cosmic ray flux up to about 100 GeV: those of Webber [17] and of the LEAP instrument [18]. These two measurements differ by as much as 30–40% at 10 GeV. This uncertainty is not as large now, when several recent experiments IMAX [19], CAPRICE [20], MASS [21] and BESS [22] give flux values very close to LEAP with a dispersion of not more than 10%.

Different event generators that are used in the calculation of the atmospheric neutrino flux exhibit a variety of secondary charged pion spectra that are produced in nucleon interactions on light nuclei. Gaisser [1] has shown that the Z factors $Z_{p\pi^{ch}} = \int_0^1 x^\gamma f(x) dx$, where x is the ratio of the pion to the proton energy and $\gamma = 1.70$ is the integral energy spectrum of cosmic rays describe quite well the efficiency for neutrino production of a hadronic interaction model. Expressed in terms of these Z factors the differences between different models could be as large as 25% [23]. The important x region 0.1 – 0.2 is largely not measured experimentally in the important energy region 10 t0 100 GeV.

Bartol's atmospheric neutrino flux [24,25] is calculated using a primary cosmic ray spectrum intermediate between the Webber and LEAP parametrizations, but much closer to the latter one (about 10% higher in the 10 - 100 GeV range). It also uses the event generator TARGET that predicts relatively high

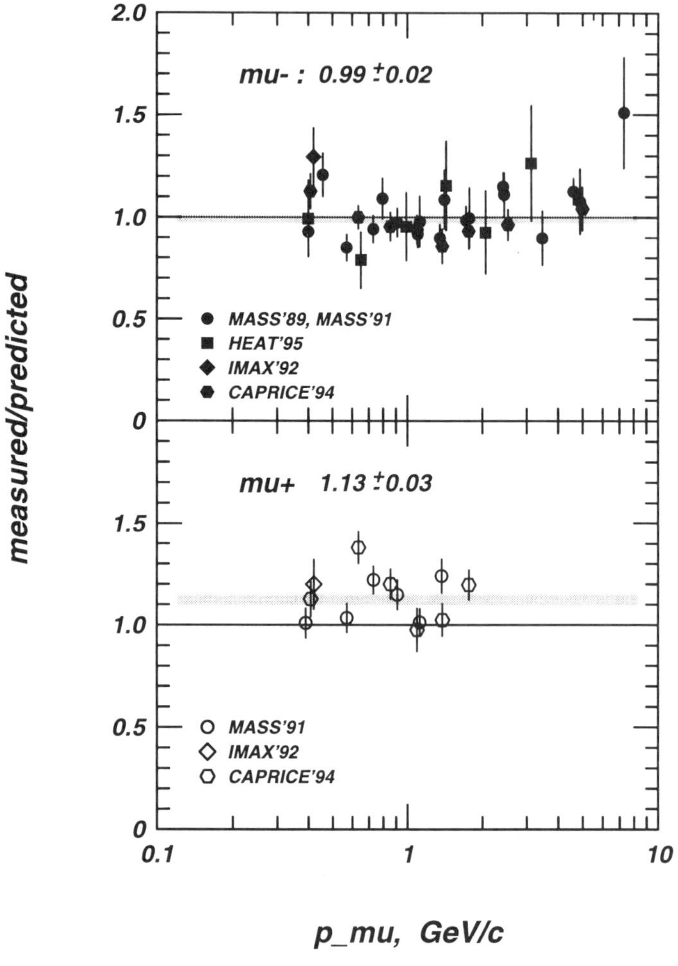

Figure 3: Ratio of data to predictions for positive (bottom) and negative (top) muons measured 'at float' by several different magnetic spectrometers.

$Z_{p\pi^{ch}}$ values. For these two reasons the Bartol neutrino flux prediction should be considered high. Predictions that use either lower cosmic ray flux or have lower charged pion production efficiency would have lower overall normalization. The predictions of Honda[15,16] used in Fig. 2 are about 2–3% higher than Bartol's after accounting for the geomagnetic field effects for Kamioka.

The only way to normalize the calculations of the atmospheric neutrino flux is to compare to measurements of other particle fluxes in the atmosphere.

The most relevant measurement is that of GeV muons. Some authors have recently compared muon fluxes predicted with one dimensional simulations, such as Refs. [24,25,15,16] and concluded that these predictions overestimate the fluxes of atmospheric muons and possibly neutrinos.

Two 3D calculations of the atmospheric neutrino flux [26,27] have been recently performed. The general conclusions are that the simulations in three dimensions do change the predicted angular distribution but not significantly enough to require a change in the interpretation of results. The smearing caused by the angle between the parent neutrino and the generated lepton masks the improvements in the predicted neutrino angular distribution. On the other hand the calculation of atmospheric muons is much more influenced by a three dimensional calculation. The reason is simple: muons generated at certain zenith angle and still within the acceptance of the detectors have longer pathlengths, and thus decay probability. The account for the muon bending in the geomagnetic field emphasizes these 3D effects [28].

Comparisons with 1D calculations can only be made for muons fluxes measured at 'float' (very high altitude corresponding to atmospheric depth less than 5 g/cm^2). Fig. 3 compares the fluxes measured by five different balloon experiments at float with predictions (Bartol) that account for the modulation of the cosmic ray flux for the particular time of the flight. The ratio of measured to predicted muon fluxes is shown separately for positive and negative muons. The average ratios in the momentum range 0.3 to 10 GeV/c are 1.13±0.03 and 0.99±0.02 respectively. Although this comparison suggests that the calculation underestimates the charge ratio, which has to be somewhat increased in TARGET, it supports the overall normalization of the pion production in the model. Note that the muon flux at such high altitude is a simple folding of the primary cosmic ray flux and the pion production features. All the muons are generated in the first cosmic ray interaction and the details of atmospheric cascading are not very important.

6. Checks and Balances

Kamioka is located at low geomagnetic latitude with correspondingly high geomagnetic cutoffs. There should be more upward than downward GeV neutrinos at such a geomagnetic location. The exact value of this asymmetry is still disputed, but the geomagnetic field models are very precise and an agreement should be reached soon. The geomagnetic effects in the zenith angle distribution of the neutrino events in Super–Kamiokande create up/down asymmetry which is the opposite of the expectations from neutrino oscillations. The data of the presumably 'non oscillating' neutrino species should confirm this angular distribution and aid the more precise estimates of the efficiency and effective area as a function of the zenith angle – a very essential part of the neutrino oscillation analysis.

The geomagnetic effects were widely used in 1930's for studies of the nature of the cosmic rays. At was at that time when the 'East–West' effect was experimentally proven: positively charged particles penetrate the geomagnetic field much easier entering from the West. The cosmic ray intensity is much higher looking West than looking East and the effect for large zenith angles at Kamioka is very strong. Atmospheric neutrino fluxes have the same azimuth angle dependence which is to a large degree preserved even after accounting for the large neutrino–lepton scattering angle at low energy. [29].

The experimental study of the East–West effect in the Super-Kamiokande neutrino data[30] demonstrated the good efficiency of the experiment as a function of zenith and azimuthal angles as well the theoretical understanding of the geomagnetic effects on the neutrino flux.

In spite of the intrinsic uncertainties related to the cosmic ray flux and its time variations, atmospheric neutrinos provide a very wide L/E range for the investigation of neutrino oscillations with an accuracy that in the case of full mixing challenges accelerator long baseline setups[31]. These of course may have the advantage to be 'appearance' experiments, that can directly detect ν_τ interactions.

7. Tau appearance experiments with atmospheric neutrinos

This could be also done with the atmospheric neutrino flux. There are at least two suggestions that could detect τ neutrinos in the case of $\nu_\mu \to \nu_\tau$ oscillations. One of them is the construction of a large mass and high density tracking calorimeter with a good energy and direction reconstruction [32]. The other one[33] is based on the properties of the ν_τ interactions and of the τ meson that could be observed with a relatively crude detector of large size.

The point is that in the 10 - 100 GeV energy range a charge current ν_τ interaction and the subsequent τ decay will not be resolved as separate events because of the very short τ decay length. A very large fraction of the τ energy will be deposited in hadronic and electromagnetic particles which will generate shower in the detector, presumably a water (or ice) cherenkov detector. Thus, if one studies the energy spectra of shower events caused by cosmic ray neutrinos, there will be a strong τ signature that will reflect the values of the oscillation parameters. Fig. 4 shows the expected spectrum of neutrino induced showers in the absence of oscillations and for $\nu_\mu \to \nu_\tau$ oscillations with Δm^2 of, 0.01,

0.005 and 0.002 eV^2 and full mixing. These events come from the lowest π solid angle, i.e. between zenith angles of 180 and 120°.

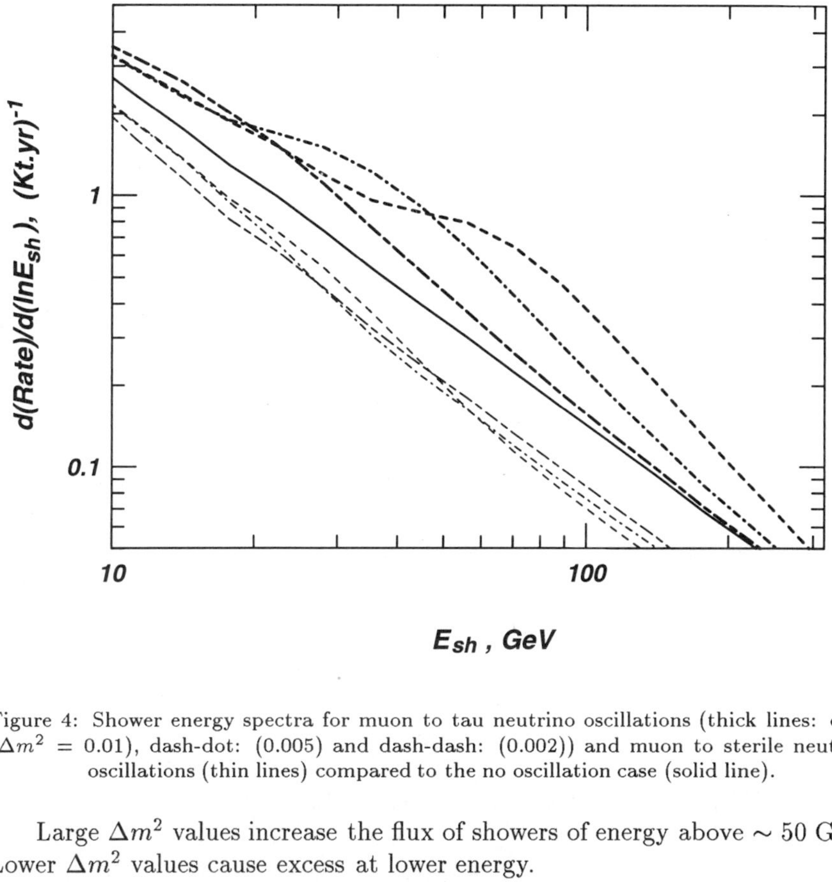

Figure 4: Shower energy spectra for muon to tau neutrino oscillations (thick lines: dash (Δm^2 = 0.01), dash-dot: (0.005) and dash-dash: (0.002)) and muon to sterile neutrino oscillations (thin lines) compared to the no oscillation case (solid line).

Large Δm^2 values increase the flux of showers of energy above $\sim$ 50 GeV. Lower Δm^2 values cause excess at lower energy.

The shower spectrum measurement is also sensitive to $\nu_\mu \to \nu_s$ oscillations. In such a case the overall rate of showers is decreased and the features in the shower energy spectrum are most likely unobservable.

One can easily estimate the expected shower rate for a detector that can measure the shower energy with an error of 30% and its direction within 10°. One expects 425 showers of energy above 10 GeV per year in a 100 Kt detector in the case of no oscillations. In the case of $\nu_\mu \to \nu_\tau$ oscillations with Δm^2 of 0.01 (0.002) eV^2 the rate will increase to 670 (604) events, i.e. with more than 8σ. Systematic uncertainties will somewhat decrease this statistical significance. The increase of the rate, however, is accompanied by substantial changes of the shape of the shower energy spectrum. The excess showers are centered around energies of maximum $\nu_\mu \to \nu_\tau$ transitions corresponding to the Δm^2 value. Distributions are wide because of the large angular (path-length) range of π steradian. A narrower range makes these excesses more prominent but correspondingly decreases the statistics. For δm^2 smaller than 0.002 eV^2 the sensitivity decreases very fast. In case of $\nu_\tau \to \nu_s$ oscillations there is no sensitivity to Δm^2 but the shower rate decreases by 6σ for values between 0.001 and 0.01.

Acknowledgments. My talk is based on work performed in collaboration with T.K. Gaisser, P. Lipari and others. I appreciate the cooperation of colleagues involved in the measurements of muon fluxes in the atmosphere and specifically of Marco Circella, Stephane Coutu and John Krizmanic. Thanks to Gabor & Susan Domokos for organizing the excellent workshop. This research is supported in part by DOE contract DE FG02 01 ER 40626.

1. T.K. Gaisser, Cosmic Rays and Particle Physics, Cambridge University Press (1990).
2. P. Lipari, M. Lusignoli & F. Sartogo, *Phys. Rev. Lett.* **74**, 4384 (1995).
3. Ch. Berger *et al. Phys. Lett. B***227**, 489 (1989); *Phys. Lett. B***245**, 305 (1990); K. Daum *et al. Z. Phys C***66**, 417 (1995).
4. M. Aglietta *et al., Europhys. Lett.* **8**, 611 (1989).
5. D. Casper *et al., Phys. Rev. Lett.* **66**, 2561 (1991); R. Becker–Szendy *et al., Phys. Rev. D***46**, 3720 (1992).
6. K.S. Hirata *et al., Phys. Lett. B***205**, 416 (1988); *Phys. Lett. B***280**, 146 (1992).
7. Y. Fukuda *et al., Phys. Lett. B***433**, 9 (1998); Y. Fukuda *et al., Phys. Lett. B***436**, 33 (1998); Y. Fukuda *et al., Phys. Rev. Lett.* **82**, 2644 (1999)
8. M. Nakahata (for the Super–Kamiokande collaboration), Talk at the TAUP'99 (Paris) (1999).
9. Y. Fukuda *et al., Phys. Rev. Lett.* **81**, 1562 (1998).
10. T.K. Gaisser & T. Stanev, *Phys. Rev. D***57**, 1977 (1998).
11. E. Peterson *et al.,* (Soudan 2 collaboration) *Nucl. Phys. B* (Proc. Suppl.) **77**, 111 (1999).
12. M. Ambrosio*et al., Phys. Lett. B***434**, 451 (1998).
13. M.M. Boliev *et al.*, in *Proc. 3rd Int. Workshop on Neutrino Telescopes* (ed. Milla Baldo Ceolin), 235 (1991).

14. M. Apollonio *et al., Phys. Lett. B*420, 397 (1998).
15. H. Honda *et al., Phys. Lett. B*248, 193 (1990).
16. M. Honda *et al., Phys. Rev. D*52, 4985 (1995).
17. W.R. Webber, R.L. Golden & S.A. Stephens, Proc. 20th ICRC, Moscow, **1**, p. 325.
18. Y.S. Seo *et al., Ap. J.* **378**, 763 (1992).
19. W. Menn *et al.*, Proc. 25th ICRC, (Durban) **3**, p. 409 (1997)
20. G. Barbiellini *et al.*, Proc. 25th ICRC (Durban) **3**, p. 369 (1997).
21. M. Circella *et al., Phys. Rev. D*60: 052002 (1999).
22. S. Orito *et al.*, talk presented by T. Sanuki at the Symposium "New era in neutrino physics", Tokyo Metropolitan University, 1998.
23. T.K. Gaisser *et al., Phys. Rev. D*54, 5578 (1996).
24. T.K. Gaisser & T. Stanev, Proc. 24th ICRC, (Rome) **1**, 694 (1995).
25. V. Agrawal *et al.* 1996: Phys. Rev. D**53**, 1314.
26. G. Battistoni *et al., Astropart. Phys.*, in print, *hep-ph/9907408.*
27. Y. Tserkovnyak *et al., Phys. Rev. D*, submitted, *hep-ph/9907450.*
28. T. Stanev *et al.*, Proc. 26th ICRC (Salt Lake City), **2** , 96 (1999).
29. P. Lipari, T.K. Gaisser & T. Stanev, *Phys. Rev. D*58: 073003 (1998).
30. T. Futagami *et al., Phys. Rev. Lett.* **82**, 5194 (1999).
31. see a list of the proposed long baseline oscillation experiments and a collection of documents at *www.hep.anl.gov/ndk/longbnews/*
32. M. Aglietta *et al.*, Presented at the Lepton–Photon Conference, Stanford University (1999).
33. T. Stanev, *Phys. Rev. Lett* (in print) *astro-ph/9907018.*

Recent Results and Future Prospects on Neutrino Oscillations from Reactors

Yi Fang Wang
Department of Physics, Stanford University,
Stanford, CA 94305, USA

Abstract

We review the current status on searching for neutrino oscillations using reactor neutrinos. The initial results from the Palo Verde experiment, together with the results from Chooz, suggest that the "atmospheric anomaly" is unlikely due to $\nu_e \to \nu_\mu$ oscillations. The next generation KamLAND experiment will improve the current sensitivity on Δm^2 by two orders of magnitude, making it the first experiment which can access the solar neutrino problem using man-made sources.

1 Introduction

For more than twenty years nuclear reactors have been used as powerful sources of electron anti-neutrinos in experiments searching for neutrino oscillations [1]. Generally such experiments detect $\bar{\nu}_{\mathrm{e}}$ by the large cross section process $\bar{\nu}_{\mathrm{e}} + \mathrm{p} \to \mathrm{e}^+ + \mathrm{n}$, where the cross-section-weighted energy spectrum of electron anti-neutrinos, peaked at about 4 MeV, can be directly calculated from the measured e^+ spectrum. Any distortions of the $\bar{\nu}$ energy spectrum or flux deficit would indicate oscillations. The low energy of reactor $\bar{\nu}$'s allow these experiments to reach very small masses, albeit with modest mixing-angle sensitivity. While for long time reactor oscillation experiments have been performed following the general idea that neutrino mixing might occur, in the last few years, the atmospheric neutrino anomaly [2] has provided motivation for the Chooz [3] and Palo Verde [4] experiments, designed to be sensitive to Δm^2 as small as $\sim 10^{-3}$ eV^2. The KamLAND experiment now under construction would further improve this sensitivity by another two orders of magnitude, making it the first experiment which can access the "solar neutrino problem" using man-made sources.

2 Recent Results from the Palo Verde Experiment

The Palo Verde neutrino oscillation experiment [4] is located at the Palo Verde Nuclear Generating Station near Phoenix, Arizona. The total thermal power of three reactors is 10.9 GW, giving 2.3×10^{21} electron anti-neutrinos per second.

The detector consists of a 6×11 array of 9 m long cells filled with Gadolinium-loaded liquid scintillator, as shown in Figure 1. Positrons deposit their energies in the scintillator and annihilate, yielding two 511 KeV γ's. This process generates a triple coincidence signal in the detector. Neutrons thermalize and are captured in Gd with a time constant of about $30\mu s$, giving a γ-ray shower of 8 MeV total energy. The 12-ton central detector is surrounded by a 1 m thick water shield and by large cosmic ray veto counters. The detector is located in a shallow underground laboratory with an overburden of 40 MWE for shielding cosmic rays.

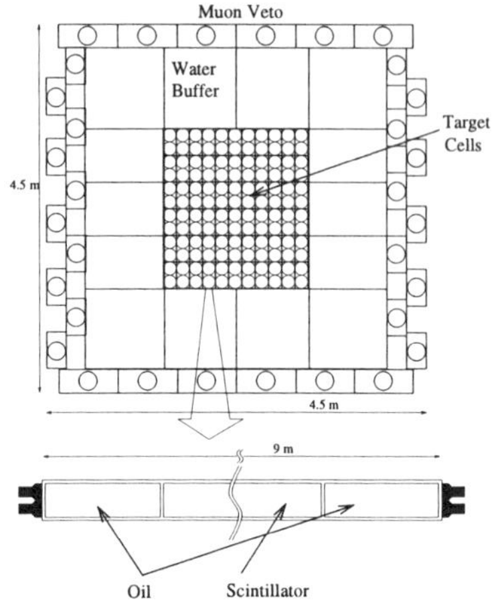

Fig. 1 Schematic view of the Palo Verde detector.

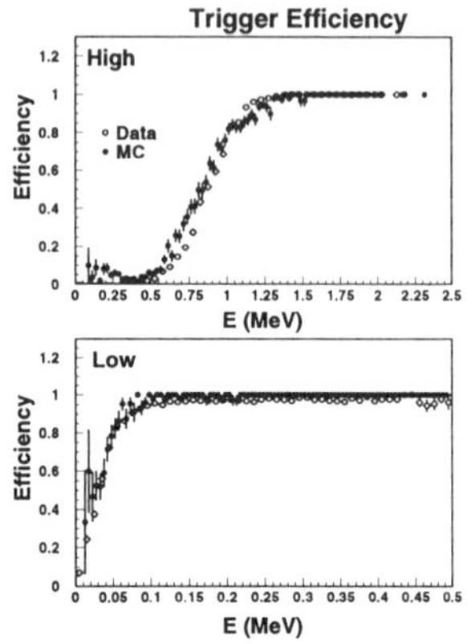

Fig. 2 Measured trigger efficiency at one particular position compared with Monte Carlo simulation.

The neutrino flux at the detector is calculated by tracking the ^{235}U, ^{238}U, ^{239}Pu, and ^{241}Pu fission rates in the three plant reactors, taking into account both power level and fuel age. The neutrino multiplicity and spectrum per fission has been measured [5,6] for all but ^{238}U; the latter contributes 11% of the neutrino signal and theoretical calculations are used for its contribution [7]. The neutrino flux emerging from the reactors, combined with reactor distances, the effective number of target protons and the inverse beta decay cross section then gives the expected neutrino interaction rate. With all three reactors operating at full power there are expected to be approximately 200 neutrino interactions per day in the detector target. The uncertainty in this rate is less than 4%, dominated by the measurement of neutrino flux per fission.

Rigorous calibrations have been taken to ensure the understanding of neutrino detection efficiency:

- Blue LED's at known position to calibrate timing/position;
- Optical fiber scan at different light intensity to monitor PMT's gain and non-linearity;
- Radioactive sources to determine energy scale and liquid transparency;
- Trigger thresholds measured using ^{22}Na source;
- Absolute positron detection efficiency measured using calibrated ^{22}Na source;
- Absolute neutron detection efficiency measured using Am-Be neutron source triggered by accompany γ's using a NaI detector.

Figure 2 shows the measured trigger efficiency compared with detailed Monte Carlo simulation. Both the prompt positron and the delayed neutron are triggered by the same condition: at least one cell above HIGH threshold(for e^+ ionization and neutron shower core) and at least two cells above LOW threshold(511 KeV γ's or neutron shower skirt). Each trigger is a half event(either e^+ or n, called bank) and the time correlation of two banks is formed offline.

There are two main backgrounds in our experiment: uncorrelated from natural radioactivity and correlated from cosmic-muon-induced neutrons. The first one can be seen in Fig. 3 where the time difference between prompt and delayed banks are shown. The exponential falling part shows the neutron capture time which includes neutrino signals and neutron-related background. The constant part is the uncorrelated background coming from the random coincidence of natural radioactivity. The correlated background can be seen in Fig. 4 where the time-since-last-veto hits is shown. There are two exponentials: the first one with $\tau = 21\mu s$ is due to neutrons produced from muons passing through the veto. The fact that the capture time is shortened to 21 μs shows that multiple neutrons are produced. The second exponential includes all our neutrino events and neutron background which are produced in lab walls. The exponential manifests the fact that the time interval between two cosmic rays is about 500 μs.

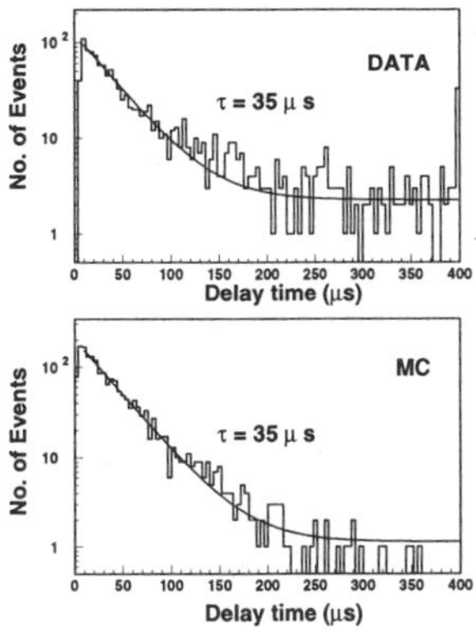

Fig. 3 Time difference between prompt and delayed bank

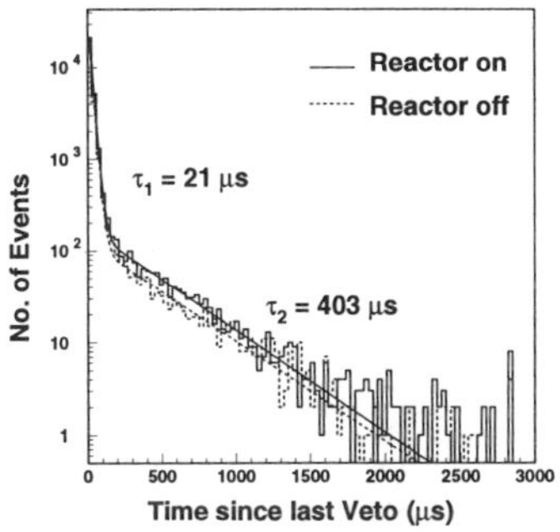

Fig. 4 Time difference between the the central detector event and the last veto event.

We select neutrino candidates using the following criteria:

- the time correlation between two banks should be less than 5 μs to reject muon decays and more than 200 μs to reject random coincidence;
- time-since-last-veto-hit should be more than 150 μs to reject cosmic-ray-induced background;
- two correlated banks should occur within a cell matrix of 5×7 and 100 cm in Z;
- at least one bank should have energy more than 3 MeV;
- triple condition for each bank: the most energetic cell should have energy more than 0.5 MeV and less than 8.0 MeV; the third energetic cell should have energy more than 30 KeV;
- for the positron triple, the second and third most energetic cell should have energy less than 600 KeV and the total energy outside the most energetic cell should be less than 1.2 MeV. This is to ensure compatibility with two 511 KeV γ's.

In 1998, the Palo Verde experiment started to be operational and we have taken data for 37.86 days with full power and 32.74 days with one of the three reactors off for refueling. After the selection, we obtained 39.06 ± 1.00(stat.) events/day for reactor full power and 32.62 ± 1.02(stat.) events/day when one of the three reactors is off. The difference is 6.44 ± 1.43(stat.) events/day. The signal to background ratio can then be calculated to be about 1.22. This is confirmed by the measurement of neutron forward-backward asymmetry with respect to positron, as shown in Fig. 5. The energy spectrum of positron calculated by subtracting the spectrum at 2/3 power from that of full power is shown in Fig. 6. The data is in good agreement with Monte Carlo simulation for no oscillation.

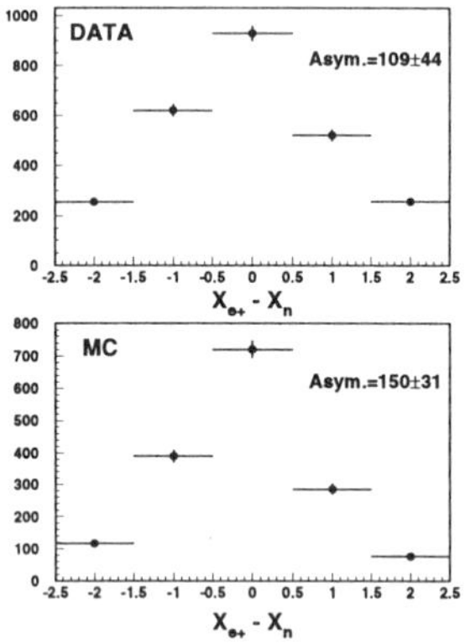

Fig. 5 Asymmetry between forward and backward hemisphere

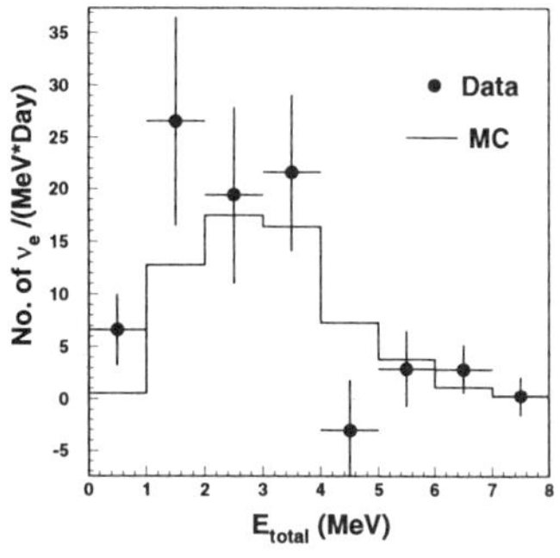

Fig. 6 Total energy measured in our detector compared with Monte Carlo simulation without oscillation.

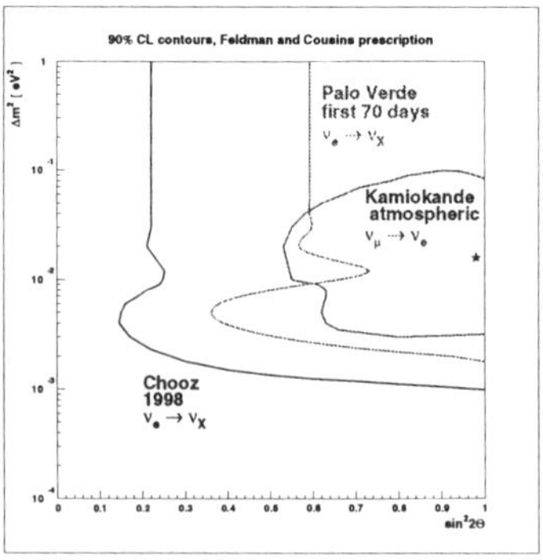

Fig. 7 Exclusion plot of the Palo Verde experiment

Taking into account efficiency corrections, we obtained our measurement for one of the three reactors 77 ± 17(stat.) ± 11(syst.) $\bar{\nu}_e$/day, in good agreement with expectation of 59.6 $\bar{\nu}_e/day$ for no oscillation. The exclusion plot using Feldman and Cousins prescription [8] is shown in Fig. 7. Together with results from Superkamiokande [9] and Chooz [3], the atmospheric neutrino anomaly is very unlikely to be caused by $\nu_\mu \to \nu_e$ oscillations.

3 The Chooz experiment

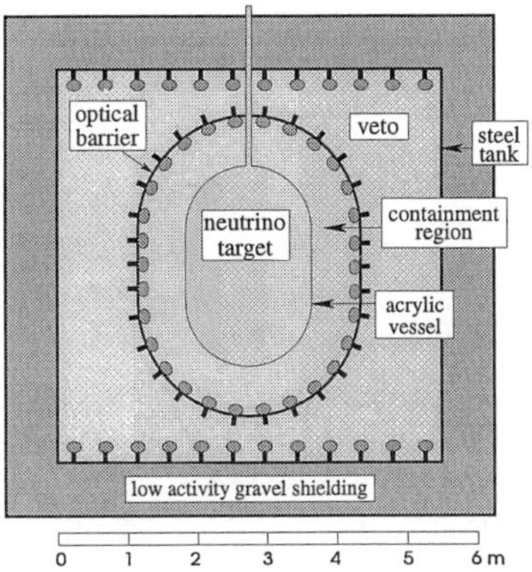

Fig. 8 The schematic view of the Chooz detector

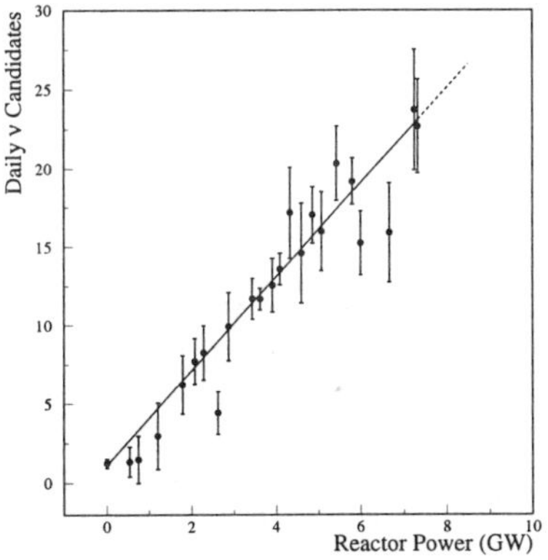

Fig. 9 Number of ν_e candidates per day, as a function of the reactor power.

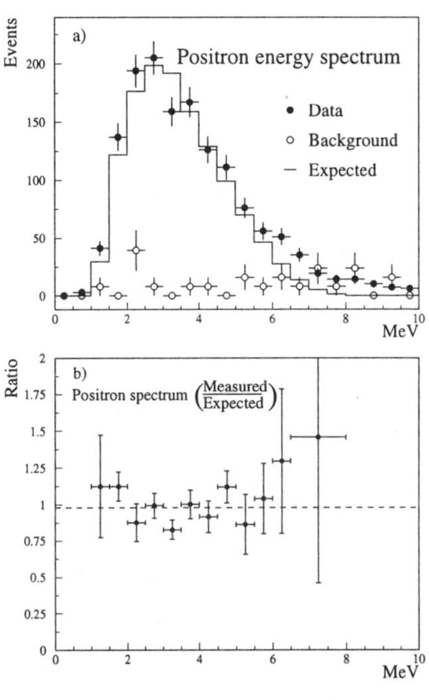

Fig. 10 a) Measured positron energy spectrum compared with the expectation for no oscillations. b) Ratio of the measured(background subtracted) to the expected positron spectrum.

The Chooz experiment is located near the village of Chooz in Ardennes region of France. It has two reactors with a full power of 8.5 GW(thermal). The detector is also based on Gd-loaded liquid scintillator as shown in Fig. 8. It is located underground with a 300 MWE overburden, greatly reducing cosmic-muon-induced backgrounds. Since no segmentation needed to reject neutron background, as opposite to the Palo Verde experiment, the Chooz detector has higher efficiency for neutrinos and better energy resolution for neutron-capture showers.

The data taking was started in March 1997 and it is now finished. Fig. 9 shows the observed number of ν_e candidates per day as a function of the reactor power. From the fit they obtained the neutrino signal 25.5 ± 1.0 /day and the background 1.1 ± 0.25/day. The uncorrelated background is actually estimated to be 0.23 ± 0.05/day and correlated background 0.8 ± 0.2/day, in good agreement with results from the fit. The measured positron energy spectrum is shown in Fig. 10 compared with Monte Carlo expectation for no oscillations. The ratio of the measured to the expected neutrino signal is

$$\mathrm{R}_{\mathrm{measured/expected}} = 0.98 \pm 0.04(\mathrm{stat.}) \pm 0.04(\mathrm{syst.})$$

No deviation from no-oscillation scenario is observed. A 90% CL exclusion region for $\nu_e \to \nu_x$ oscillation is shown in Fig. 7.

4 Future Prospects of the KamLAND Experiment

It is naturally to ask "what is next" after Chooz and Palo Verde since the $\nu_\mu \to \nu_e$ oscillation is still an open question and the solar neutrino puzzle is still there. A new experiment call KamLAND [10] is now under construction to improve the sensitivity to neutrino mass differences as low as $\Delta m^2 \simeq 7 \times 10^{-6}$ eV^2, well cover the MSW Large Mixing Angle solution of solar neutrinos.

The KamLAND detector, situated in the old Kamiokande cavern, is shown in Fig. 11. It consists of 1000 tons liquid scintillator surrounded by a 2.5 m buffer liquid for reducing backgrounds, and a water Čerenkov radiator to veto cosmic muons. The central detector will be viewed by 17" photomultipliers from Hamamatsu and the veto will be viewed by old 20" PMT's from the Kamiokande experiment. A 13 m diameter Balloon will separate the scintillator from the buffer and act as a Rn barrier.

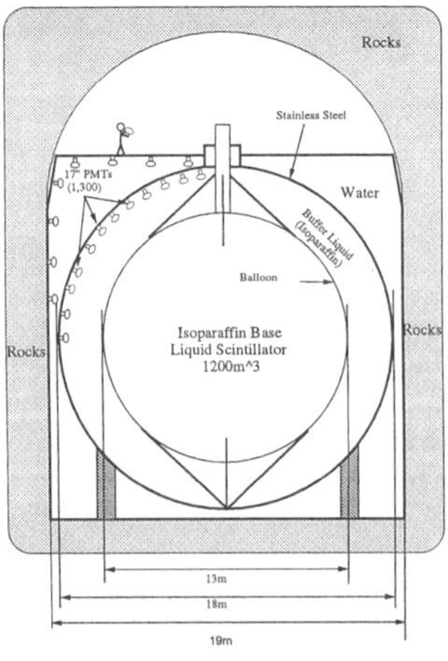

Fig. 11 Schematic view of the KamLAND detector

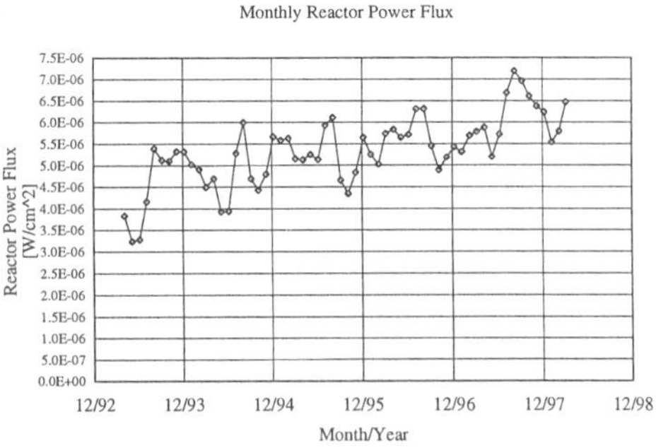

Fig. 12 Monthly reactor power flux as a function of time.

The mechanism to detect reactor $\bar{\nu}_e$ is the same as before except that the scintillator does not contain Gd, therefore neutrons can only be captured by proton, giving a 2.2 MeV γ-ray with capture time of 170 μs. There are 17 commercial nuclear power plants in Japan, supplying a total of 130 GW thermal power. The total number of $\bar{\nu}_e$ events expected from reactors is 750 kton^{-1}year^{-1} for a C_nH_{2n+2} target. The main advantage of using man-made source is that you can vary the source of signals and control the experimental phenomena. The seasonal variation of electricity consumption provides us an excellent opportunity to vary the neutrino flux and measure the background. Fig. 12 shows that a 40% variation of the total thermal power flux is expected at Kamioka during a typical year.

The background due to natural radioactivity and cosmic muons has been carefully studied and simulated using the GEANT Monte Carlo program. Table 1 lists all backgrounds for single (E> 1MeV) and double hit(Correlation time < 500μs) events. Keep in mind that we expect 2 ν/day and the total background is only 0.1 event/day.

Backgrounds for singles	1 kton back.	600 ton back.
neutrons from muon spallation	2.0 day^{-1}	1.2 day^{-1}
activation from n capture	0 day^{-1}	0 day^{-1}
activation from n spallation	<1800 day^{-1}	<1200 day^{-1}
activation from μ capture	0 day^{-1}	0 day^{-1}
activation from μ spallation	< 1800 day^{-1}	< 1200 day^{-1}
Natural radioactivity	1.02 Hz	<0.012 Hz
Backgrounds for doubles		
Cosmic muons induces neutrons	0.05 day^{-1}	0.03 day^{-1}
Natural radioactivity (random)	0.05 day^{-1}	0.00 day^{-1}
Natural radioactivity (correlated)	0.00 day^{-1}	0.00 day^{-1}

Table 1: Background estimate for single and double hit events.

The KamLAND experiment can reach a sensitivity of $\Delta m^2 \simeq 7 \times 10^{-6}$ eV^2, improving the present sensitivity by over two orders of magnitude as shown in Fig. 13. This curve also covers completely the LMAS for the "solar neutrino anomaly". Since most of the evidence for the neutrino oscillations were derived from natural neutrino sources outside the control of experimenters, "laboratory style" experiments are desired to give a firm conclusion. The very low energy of anti-neutrinos from reactors and the very long baseline ($\sim$ 170 km) of the KamLAND experiment provide the first opportunity to access the "solar neutrino anomaly" using man-made neutrino sources.

5 Other Physics Opportunities with the KamLAND Experiment

Besides reactor neutrinos from inverse-β decays, the KamLAND detector can also measure neutrinos directly from the sun via $\nu - e$ scattering. By carefully controlling the purity level of liquid scintillator [10], we can measure the ^7Be solar neutrinos as shown in table 2. The "day-night effect" and "seasonal effect" can be also studied as shown in Fig. 14 together with various solutions to the solar neutrino problem and the sensitivity of KamLAND experiment.

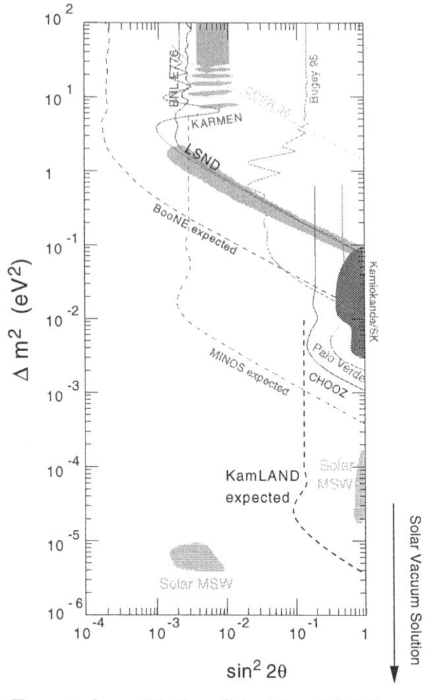

Fig. 13 Expected sensitivity of the KamLAND experiment.

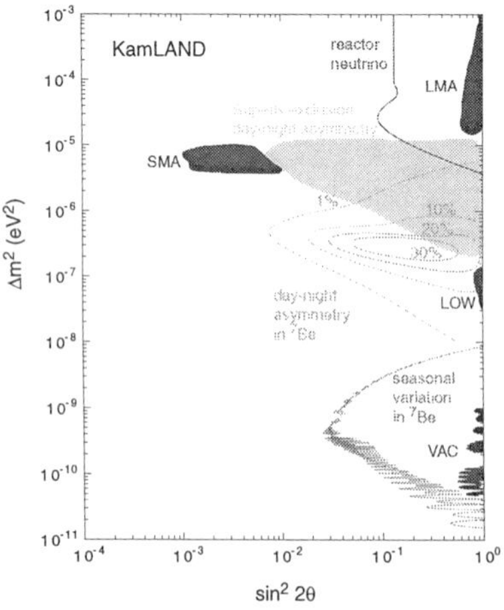

Fig. 14. Sensitivity of the KamLAND experiment to the solar neutrino problem together with various solutions.

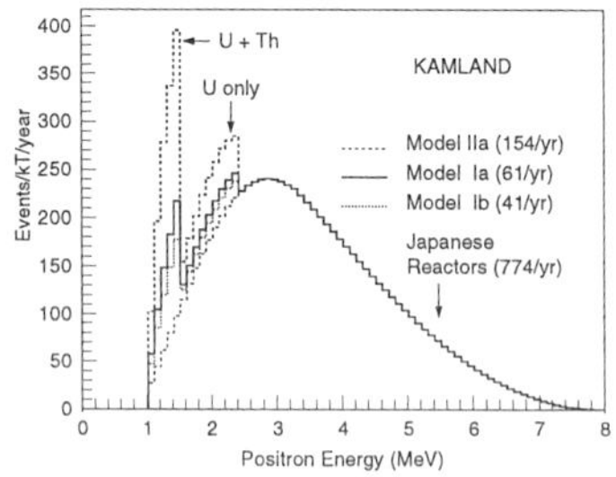

Fig. 15. Energy spectrum from terrestrial antineutrinos compared with reactor signal as expected in KamLAND. Models Ia and Ib correspond to two different geophysical models with different heavy elements concentration in the oceanic and continental crusts [11]. The model IIa shows when the entire 40 TW of heat escaping from the Earth's interior was generated by the Th and U decay chains.

Source	Rate (280-800 keV) [Counts/(kt day)]	
	$\nu_e \to \nu_{\mu,\tau}$	$\nu_e \to \nu_s$
SSM	466	
LMAS	262	207
SMAS	98	2
Background		
without SS	438	
with SS	217	

Table 2: Solar neutrino events in KamLAND

Terrestrial neutrinos from U/Th decays can provide us valuable insights on early processes in the formation of our planet that gave origin to the core, mantle and crust. It will provide a better understanding of the present planetary dynamics and heat flows, helping in calibrating models of partial melting in the mantle. The energy spectrum of neutrinos from U/Th decays has a characteristic double-hump structure as shown in Fig. 15 on top of reactor neutrinos. With five years data taking, we should be able to differentiate different models at 3σ level.

The KamLAND detector can also study supernova neutrinos, solar and astrophysical anti-neutrinos and possibly, double beta decays. In summary, the KamLAND experiment has a very rich physics opportunities and it will certainly provide us insight knowledge of neutrinos and physics beyond the Standard Model.

References

1. G. Zacek *ae al.* Phys. Rev. D34 (1986) 2621; Y. Declais *et al.* Phys. Lett. B338 (1998) 383;
2. For a review, see Particle data group;
3. M. Apollonio *et al.*, *Phys. Lett.* B **420**, 397 (1998).
4. For information, see http://citnp.caltech.edu/PV/Palo-Verde.html.
5. A.A. Hahn *et al.*, Phys. Lett. **B218**, 365 (1989).
6. K. Schreckenbach, *et al.*, Phys. Lett. **160B**, 325 (1985).
7. P. Vogel, Phys. Rev. C24(1981)1543.
8. G.J. Feldman and R.D. Cousins, *Phys. Rev.* D **57**, 3873 (1998).
9. Y. Fukuda *et al.*, *Phys. Lett.* B **436**, 33 (1998).
10. P. Alivisatos *et al.*, "KamLAND Proposal", Preprint Stanford-HEP-98-03, Tohoku-RCNS-98-15. Revised version, March 1999.
11. R.S. Raghavan *et al.*, *Phys. Rev. Lett.* **80**, 635 (1998)

THE NEUTRINO PHYSICS POTENTIAL OF THE ICARUS DETECTOR PROGRAM

DAVID B. CLINE (for the ICARUS Collaboration)
University of California Los Angeles, Department of Physics and Astronomy
Box 951547, Los Angeles, California 90095-1547 USA
E-mail: dcline@physics.ucla.edu

We describe the progress in the construction of the first 600-ton module of the ICARUS liquid-argon detector and the possible next step to a much larger detector for the NGS neutrino beam to the Gran Sasso Laboratory. We then discuss three specific neutrino physics objects: (1) The study of the high energy part of the ^{8}B spectrum for solar neutrinos to determine the "Just So" solar neutrino solution; (2) Possible detection of the ν_e collect flux from all past Supernova IIs; (3) Possible detection of the $\nu_\mu \to \nu_e$ transition for the $\delta^2_m \times 10^{-3}$ eV^2 region.

1 The First 600-Ton Module for the Gran Sasso

The ICARUS Collaboration is constructing the first 500-ton module at Pavia. The UCLA group is providing the high voltage system for this detector. Figure 1 shows a schematic of this detector, as well as the future location of the detector in Hall C at the Gran Sasso. It is expected that the detector will be installed in the middle of 2000.

2 Possible Future Program for ICARUS - ICANOE

There are several options for the future of the ICARUS program, as outlined in Table 1. The ICARUS and NOE groups have joined forces for a proposal to construct a 10-kiloton (kt) detector for Hall B of the Gran Sasso (see schematic of ICANOE, Fig. 2). In addition to the new accelerator physics program of ICARUS (solar neutrinos, proton decay, etc.), this detector will be optimized to detect neutrinos from the long-baseline beam, NGS (neutrinos to Gran Sasso from CERN).

3 The Study of the High Energy ^{8}B Flux and the "Just So" Solution

For many years, the study of the solar neutrino puzzle has emphasized the low-energy ^{8}B spectrum (SNO) or the ^{7}Be (Borexino) line. It is ironic that with the most recent SK data, this concept could be out of date and it is really the high-energy region above 10 MeV for the ^{8}B spectrum where the important new physics can be detected. This also

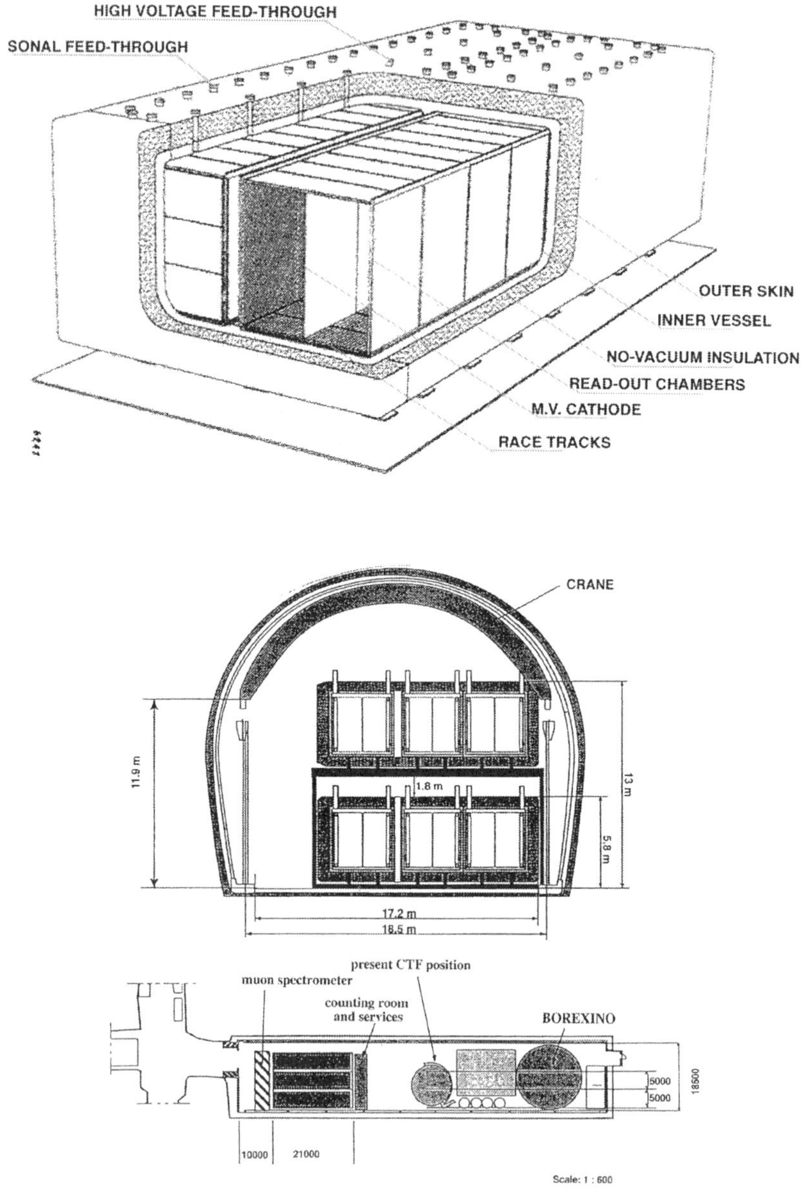

Figure 1: Drawings of the (top) ICARUS 600-ton module and (bottom) 1st 600-ton module in Hall C.

Table 1: Future of the ICARUS program.

There are several possibilities for the future of ICARUS after the 600-ton detector is completed and operational:

1. Produce more 600-ton modules for Hall C or B. (The UCLA group has responsibility only for the first 600-ton detector.
2. Go to an extreme detector, e.g., Super ICARUS of 30 kt outside the Gran Sasso Lab.
3. Construct a new detector in collaboration with the NOE consortium of about 11 kt. This option would go into Hall B and is actually being considered in the proposal stage. The strength of this detector is the ability to detect both electron and μ final states and to measure the charge of the muon. It would be a powerful detector for the NGS, CERN, Gran Sasso neutrino beam, and it could operate in a neutrino factory beam. This will also be a powerful proton decay detector.
4. The future of the NGS program could be a neutrino factory in the U.S. (at BNL or FNAL). Option 3 would be well suited to such a neutrino source.

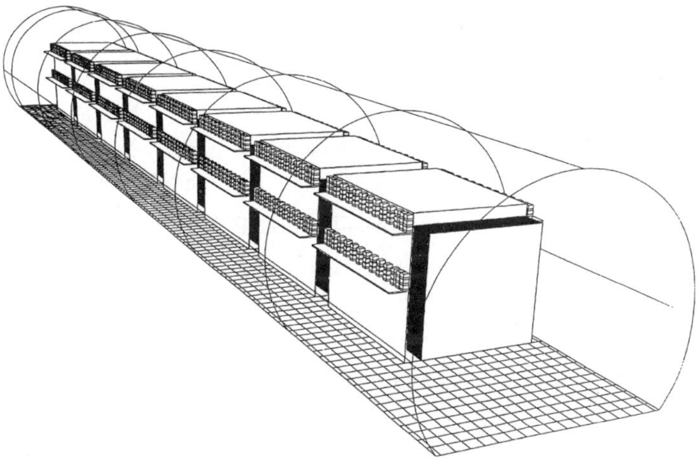

Figure 2: Schematic of the liquid drift detector, ICARUS, being constructed at the Gran Sasso Underground laboratory near Rome, Italy. It comprises a 5,000-ton liquid-argon image changer, with excellent spatial and energy resolution, and an iron plate detector, NOE.

brings up a solution that has been largely ignored – the "just so" or vacuum oscillation solution with $\delta^2_\nu \sim 10^{-10}$ eV^2. Note that this would imply that one neutrino has $m_\nu \sim 10^{-5}$, a value that could be very compatible with the GUT theories and the see-saw mechanism for the m_{ν_μ}

$$m_{\nu_\mu} \sim \frac{(m_{\rm charm})^2}{m_{\rm GUT}} \sim 3 \times 10^{-6}\ {\rm eV} \quad .$$

This would lead to an excess of high-energy ^{8}B neutrinos. Proof of this possibility would be very important. There is a problem – the HEP neutrino flux is uncertain by a factor of 20 to 50 [J. Bahcall, private communication]; therefore, one must separate HEP neutrinos from the effects of the vacuum oscillation possibility.

As of the UCLA DPF meeting (Jan. 1999), the SK data prefers the so-called vacuum oscillation solution. This is largely because of[1]

(1) A hunt of annual variation of the signal beyond the $1/r^2$ dependence of the earth's orbit around the Sun, and

(2) The existence of an excess of high-energy solar-neutrino events (see Table 2 and Fig. 3).[2]

This excess could be due to three things:

(1) The "just so" solution where, at high energy, the oscillations no longer fit into the distance of the earth to the Sun and the flux wants to go back to the normal solar neutrino value;

(2) A larger flux value of the so-called HEP neutrinos;

(3) Effects of energy resolution in the SK detector.

In the latter case, careful study of the energy resolution of the SK detector "appears" to exclude this possibility. It is clear that a different approach to study the high-energy spectrum will be very valuable.

Table 2: Rate of Events (per year) in the ICARUS 600-ton Detector.[2]

	TPC1	TPC1	TPC1	TPC1	Total	
Elastic channel 5 MeV threshold (background)	46 (8)	46 (1)	46 (1)	46 (5)	184 (15)	Contamination of absorption events < 2 events/yr
Elastic channel 6 MeV threshold (background)	40 (3)	40 (<1)	40 (<1)	40 (3)	160 (6)	
Absorption 5 MeV threshold (background)	229 (13)	229 (3)	229 (3)	229 (8)	916 (27)	Contamination of elastic events < 15 events/yr

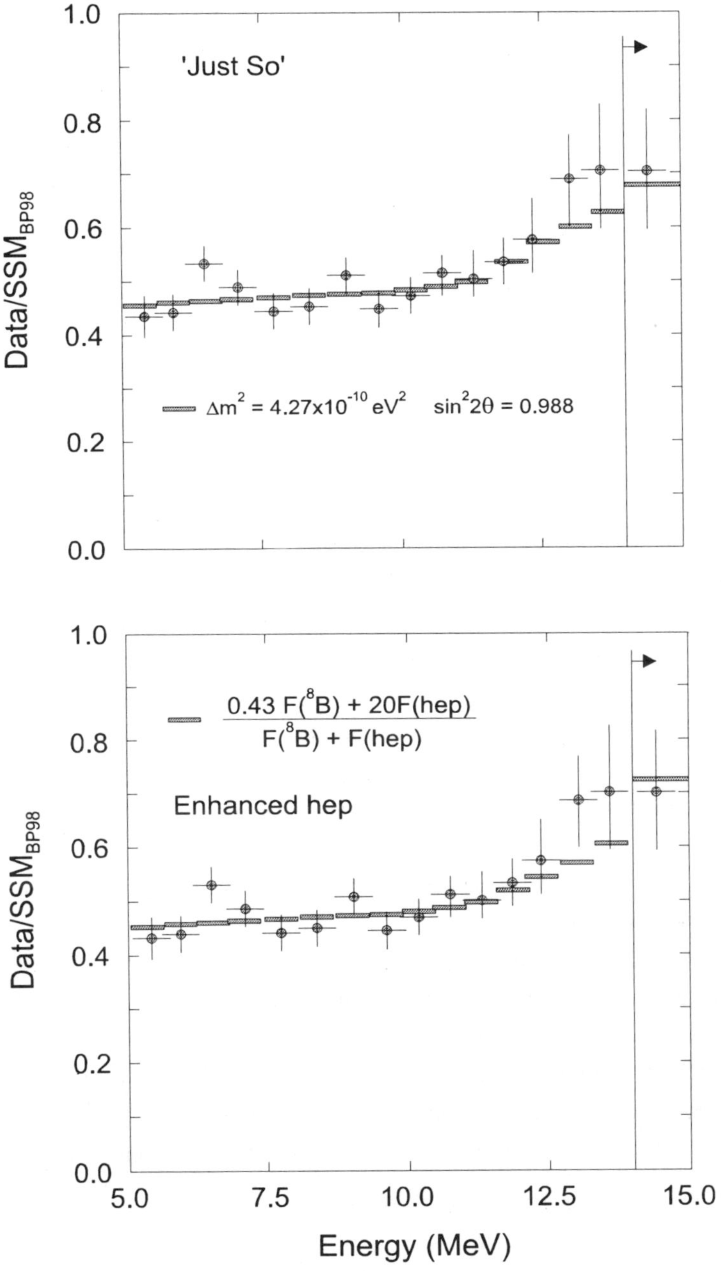

Figure 3: The measured energy spectrum.

Aside from the powerful spatial and time information in ICARUS, the extremely good energy resolution is also very important (see Fig. 4). This good resolution is because of the fact that we are recording ionization and not Cherenkov light like SK or SNO. There is vastly more ionization per charged particle track than Cherenkov radiation. Furthermore, ICARUS is sensitive to **non relativistic tracks**, unlike SK or SNO.

The method of the ICARUS group is to study $\nu_e + {}^{40}Ar \rightarrow {}^{40}K^* + e^-$, where the de-excitation of the K* provides the "signature" to reduce background. In Table 2, we give the expected rate for the 600-ton in one year. We expect about 2000 events with a total neutrino energy of about 9 MeV in two years. In Fig. 5, we show the part of the energy spectrum probed by ICARUS and the flux of HEP neutrinos (including the enhanced flux needed to explain the SK data).[2] Note that ICARUS could observe between 20 and 60 HEP neutrino events in two years. The excellent energy resolution allows ICARUS to distinguish between the three possibilities noted above!

If events from the hep process can be identified in ICARUS ($E_{e^- + K^*} \geq 15$ MeV), then it may still be possible to separate the enhanced hep flux and the vacuum oscillation solution. This is due to the expected distortion of the ^8B spectrum from 10 - 14 MeV and the hep spectrum from 15 - 18 MeV. We expect to carry out Monte Carlo simulation of this effect at UCLA this year.

4 Search for the Integrated Flux of Supernova Neutrinos

Another kind of relic neutrino is that which arises from the integrated flux from all past SNIIs. Figure 6 shows a schematic of these (and other) fluxes. These fluxes could be modified by transmission through the SNII environment, as discussed by G. Fuller and others and as shown in Fig. 7.

The detection of $\bar{\nu}_e$ from the relic supernovas may someday be accomplished by the SK detector; it would be as interesting to detect ν_e with the ICANOE 5-kt detector, as illustrated in Table 3. High-energy ν_e would come from $\nu_{\mu,\tau} \rightarrow \nu_e$ in the supernova. A window of detection occurs between the upper solar neutrino energy and the atmospheric neutrinos, as first proposed by D. Cline and reported in the first ICARUS proposal (1983-1985). The ideal detector to observe this is a large ICANOE liquid-argon detector.

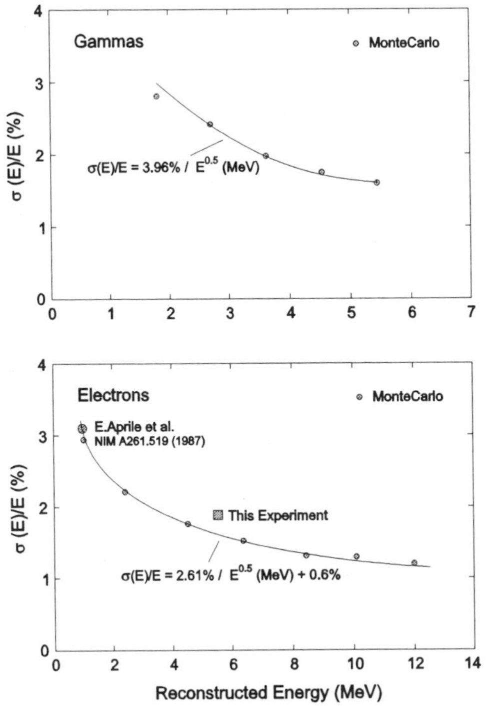

Figure 4: Energy resolution of the ICARUS detector.

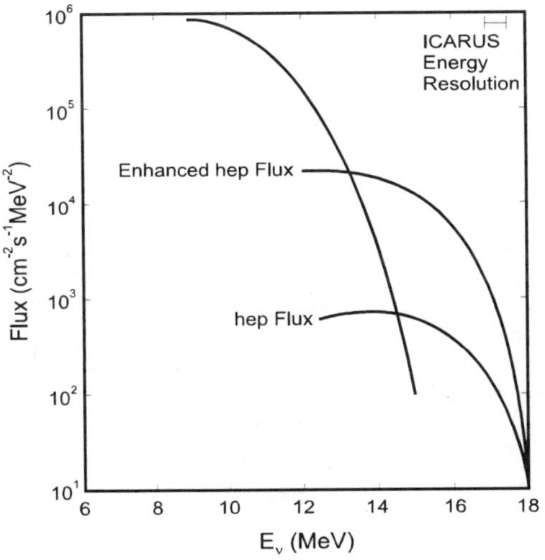

Figure 5: The high-energy ^8B spectrum probed by the ICARUS detector with two possible HEP flux values.

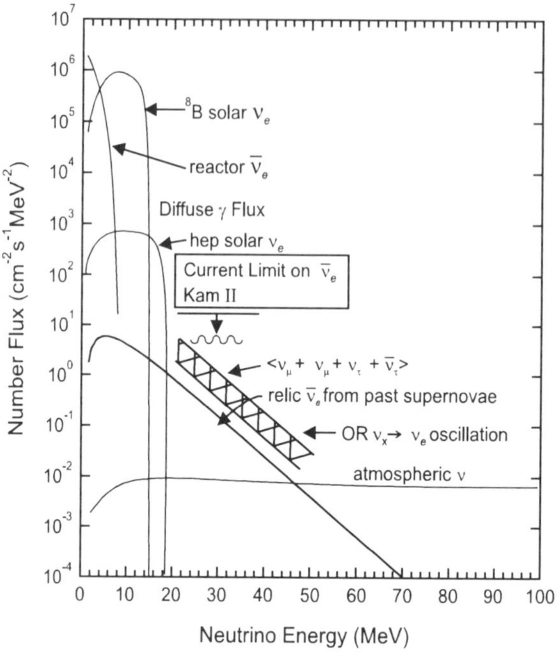

Figure 6: Relic neutrinos from past supernova. Note: $\nu_x \to \nu_e$ in the supernova can boost the energy of the ν_e if we find $<E\nu_e> > <E\bar{\nu}_e>$. This will be a signal for neutrino oscillation in supernovae! and measure $\sin^2 \theta x_e$.

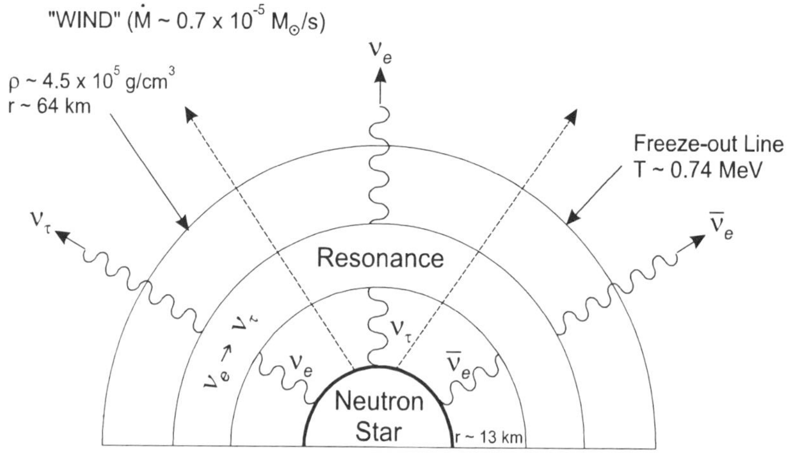

Figure 7: Transmission of various neutrinos through the outer part of the SNII. It is possible that $\nu_\tau \to \nu_e$ conversion can occur in this environment.

Table 3: Detection of $\nu/\bar{\nu}_e$ Relic Neutrino Flux from Time Integrated SNII.

1.	Relic $\nu/\bar{\nu}_e$ from all SNII back to $Z \sim 5$: $<E_\nu> \sim 1/(1+Z)<E_\nu>$
2.	Detection would give integrated SNII rate from Universe - Window of detetion [D. Cline, ICARUS proposal, 1984]
3.	Neutrino oscillations in SNII would give $\nu_x \to \nu_e$ with higher energy than $\bar{\nu}_e$
4.	Detect $\bar{\nu}_e$ with SK or ICARUS. Attempt to detect ν_x/ν_e detection.

5 Possible Detection of $\nu_\mu \to \nu_e$ in the NGS Beam

ICARUS and ICANOE can play a key role in the study of neutrino oscillations. First there will be an independent study of atmospheric neutrino events with the 600-ton ICARUS.[2] This detector is very different from the Kuper Kamiokande water detector and can detect non-relativistic particles. In Table 4 we list the number of different types of atmospheric neutrino events that will be recorded each year in the 500-ton detector.[2]

It is likely that the NAG neutrino beam will be approved in 1999. The role of the ICANOE detector will be to

(A) Provide a unique identification of $\nu_\mu \to \nu_\tau$,[2]

(B) Search for $\nu_\mu \to \nu_e$,[2]

In Table 5 we list the key measurement(s) that can be made for this reactor – namely the θ_{13} angle in the 3ν "CKM" matrix. This angle is a key parameter for any future search for CP violation in the neutrino complex. A method to detect $\nu_\mu \to \nu_e$ has been described for ICARUS (NOE) by the author.[2] It relies on the observation of an enhancement in the low-energy ν_e-like flux, as shown in Fig. 8. We believe that ICANOE can measure the probability $P\nu_e$ to about the 10^{-2} level.

6 Summary

The neutrino physics possible with ICARUS (600 ton) and ICANOE is very rich. From past supernovae, to the solar neutrino puzzle, to the issue of atmospheric neutrino oscillations, and ultimately to the study of the "CKM"-like matrix for 3ν mixing. This is certainly one of the most interesting areas of science in our time.

Acknowledgment

I wish to thank the members of the ICARUS group for help and discussion.

Table 4: ICARUS Neutrino Events

Event Classification		No. Events*
Track (CC; QE; $\nu_e, \bar{\nu}_e$)	$\nu_\mu + n \to p + \mu^-$	80.0
	$\bar{\nu}_\mu + p \to n + \mu^+$	14.0
Shower (CC; QE; $\nu_e, \bar{\nu}_e$)	$\nu_e + n \to p + e^-$	51.0
	$\bar{\nu}_e + p \to n + e^-$	6.3
Multiprong (CC; NC; RS, $\nu_\mu, \bar{\nu}_\mu, \nu_e, \bar{\nu}_e$)	$\nu_{inel} + \ell + X$	33.0
28 1-π reactions	$\nu_{inel} + \nu + X$	12.6
Proton (NC; E; $\nu_\mu, \bar{\nu}_\mu, \nu_e, \bar{\nu}_e$)	$\nu + p \to \nu + p$	81.6
	$\nu + n \to \nu + n$	

*Total Events: CC = 184.3; NC = 94.2

Table 5: "CKM" Matrix for ν Oscillations.

1. $\nu_\mu \to \nu_\tau$ Presumably a large amplitude

   $P_{e\mu}(\nu_\mu \to \nu_\tau) = \cos^4 \theta_{13} \sin^2 2\theta_{23} \sin^2 [1.27\,(L/E)\, \delta^2_m$

   ➔ $\theta_{23} \sim 45^\circ$; $\theta_{13} < 20^\circ$

2. To measure θ_{13}, need a term with a $\sin \theta_{13}$

   $P_{e\mu}(\nu_\mu \rightleftarrows \nu_e) = \sin^2 \theta_{23} \sin^2 2\theta_{13} \ldots$

   $P_{e\tau}(\nu_e \rightleftarrows \nu_\tau) = \cos^2 \theta_{23} \sin^2 2\theta_{13} \ldots$

   If θ_{13} is very small: $P_{e\mu} \sim \theta^2_{13} \ldots$; $P_{e\tau} \sim \theta^2_{13}$

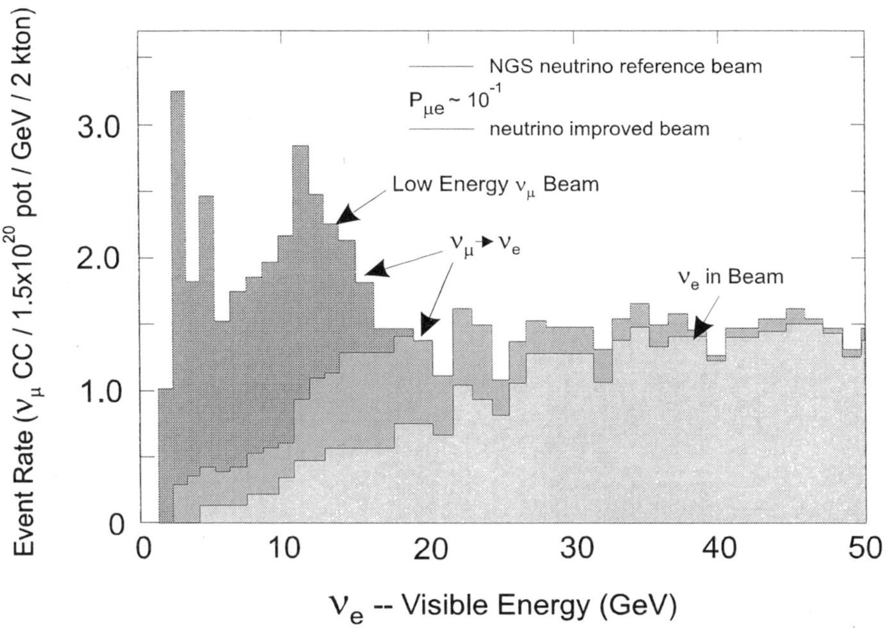

Figure 8: Possible separation of $\nu_\mu \to \nu_e$ from ν_e flux in NAG beam.

References

1. ICCR-Report 442-98-38.
2. ICARUS internal notes.
3. G. Fuller, UCSD, private communication, 1999

HIGH RATE PHYSICS AT NEUTRINO FACTORIES

B.J. KING
Brookhaven National Laboratory, Building 901A, P.O. Box 5000, Upton, NY11973
email: bking@bnl.gov
web page: http://pubweb.bnl.gov/people/bking

Both muon colliders and non-colliding muon storage rings using muon collider technology have the potential to become the first true "neutrino factories", with uniquely intense and precisely characterized neutrino beams that could usher in a new era of high rate and long baseline neutrino physics studies at accelerators. This paper gives an overview of the predicted capabilities of neutrino factories for high rate neutrino physics analyses that will use huge event samples collected with novel, high performance neutrino detectors.

1 Introduction

The idea of using muon storage rings for neutrino physics is an old one [1]. More recent feasibility studies and design work for muon colliders [2,3] led to investigations[4,5] of the exciting neutrino physics possibilities from the uniquely intense neutrino beams they will produce and, as a variation on this theme that is now attracting very considerable interest, it was then proposed[6] to use muon collider technology for non-colliding muon storage rings dedicated to neutrino physics. An extensive literature[7] is now building up on the impressive potential of both types of "neutrino factories" (or "nufacts", for short).

Nufacts will be used for each of the two classes of neutrino experiments at accelerators:

1. high rate (HR) experiments, where the detector is placed close to the neutrino source to obtain the most intense beam possible and hence gather very high event statistics of neutrino interactions.

2. long baseline (LB) experiments, where a very massive neutrino detector is placed far away from the neutrino source, deliberately sacrificing event rate in order to study baseline-dependent properties of the neutrinos and, in particular, whether there are "flavor oscillations" in the types of neutrinos composing the beam.

The large muon currents and tight collimation of the neutrinos from nufacts results in extremely intense beams with several important advantages over the neutrino beams produced today from pion decay at accelerator beamlines:

1. event statistics for HR experiments that might be three or more orders-of-magnitude larger than in today's HR neutrino experiments

2. both higher statistics and longer baselines for LB experiments

3. extremely well understood and pure two-component beams with accurately predictable energy spectra, angular divergences and intensities

4. the first high flux electron-neutrino and electron-antineutrino beams at high energies.

Of the two classes of neutrino experiments at nufacts, LB neutrino oscillation studies are currently attracting an enormous amount of interest, with a large and growing literature [7] that includes another paper [8] in these proceedings. This paper will instead concentrate on the less-developed topic of the potential for a rich and broad-based program of HR neutrino physics at nufacts. It summarizes the material covered in a much longer that is in preparation on this topic [9].

The advantages of neutrino beams from the decays of stored muons over traditional neutrino beams from pion decays are in some ways even more notable for HR experiments than for oscillation studies. In particular, the beam intensity and uniquely small transverse extent close to production invites the use of compact fully-active tracking targets backed by high-rate, high-performance detectors.

In contrast to long baseline neutrino oscillation measurements, the physics interest lies in the interactions of neutrinos rather than in their internal properties. Neutrinos are unique in participating only in the weak interaction and so, for example, they provide a probe of nucleon structure that is intrinsically cleaner than the alternative of charged lepton (electron, positron or muon) scattering. The weak interaction couplings of neutrinos to both quarks and electrons are also interesting in their own right, as will be discussed in the sections 4 and 5 on electroweak measurements and the CKM quark mixing matrix, respectively. Nufacts will also have potential for examining, or searching for, rare and exotic interaction processes. As a bonus outside neutrino physics, HR experiments at nufacts will be impressive factories for the study of charm decays.

The following section presents background material on the expected experimental conditions for HR neutrino physics at nufacts, in preparation for the subsequent physics sections on the topics of nucleon structure and QCD measurements, precision electroweak studies, quark mixing measurements, rare and exotic processes and, last but not least, charm decay physics, before the summary section.

2 Experimental Overview

2.1 The Neutrino Beam

The beam spectrum from muon decays in a monochratic muon beam is a completely pure 2-component mixture; the decays of parent μ^- provide beams of ν_μ plus $\overline{\nu}_e$ and μ^- decays provide beams of $\overline{\nu}_\mu$ plus ν_e:

$$\begin{aligned} \mu^- &\to \nu_\mu + \overline{\nu_{\rm e}} + {\rm e}^-, \\ \mu^+ &\to \overline{\nu_\mu} + \nu_{\rm e} + {\rm e}^+. \end{aligned} \tag{1}$$

These beams will be denoted as $(\nu_\mu\overline{\nu}_e)$ and $(\overline{\nu}_\mu\nu_e)$ in the rest of this paper. The kinematics of a muon decaying to an electron and 2 neutrinos is precisely specified by the electroweak theory and leads to precisely modeled neutrino spectra for HR physics at nufacts. This is a substantial advantage over conventional neutrino beams from pion decays, particularly for the high-statistics precision measurements that will be described in sections 3 through 5.

Reference [9] derives explicit expressions for the beam spectra at high rate experiments in the context of a simplified but relatively realistic model of a thin, monochromatic muon beam in the production straight section of the storage ring. The derivations assume zero net polarization over the data sample; in order to minimize beam modeling systematics in experimental analyses, it is likely that the storage ring will be designed so that the muon polarization precesses to average to zero over each fill of muons.

The derivation in reference [9] begins from the neutrino energy distributions in the rest frame of the decaying muon, E'_ν, where the scaled energy, $x \equiv 2E'_\nu/m_\mu$, has a distribution given by

$$\begin{aligned} \frac{dN_{\nu_\mu}}{dx} &= 6.x^2 - 4.x^3 \\ \frac{dN_{\nu_e}}{dx} &= 12.x^2 - 12.x^3 \end{aligned} \tag{2}$$

for muon-type neutrinos or anti-neutrinos and for electron-type neutrinos or anti-neutrinos, respectively. It is shown that x is simply related to the neutrino energy in the laboratory frame, E_ν, by:

$$E_\nu(x, \theta') = x\frac{E_\mu}{2}\left(1 + \beta\cos\theta'\right), \tag{3}$$

where $\beta = 1$ to a very good approximation and each angle in the muon rest frame, θ' corresponds to an angle θ in the laboratory frame according to:

$$\sin\theta = \frac{\sin\theta'}{\gamma(1 + \beta\cos\theta')}, \tag{4}$$

where $\gamma \equiv \frac{E_\mu}{m_\mu c^2}$. Equation 3 combined with equations 2 and 4 therefore specifies the neturino energy spectra at any position in an experimental target, at least in this simplified model.

Substituting the specific value $\theta' = \frac{\pi}{2}$ into equation 4 shows that the forward hemisphere in the muon rest frame is boosted into a narrow cone of half angle

$$\theta_\nu \simeq \sin\theta_\nu = 1/\gamma = \frac{m_\mu c^2}{E_\mu} \simeq \frac{0.106}{E_\mu[\text{GeV}]}. \tag{5}$$

This characteristic opening half-angle with respect to the muon beam direction will include approximately half of the neutrinos in the thin pencil beams for HR experiments at nufacts and the radius subtended by this angle at the experimental target is a reasonable choice for the radius of the target. This corresponds, for example to a 20 cm radius at 100 m from production in a 50 GeV muon beam or at 1 km from a 500 GeV muon beam.

2.2 Primer on Neutrino-Nucleon Deep Inelastic Scattering

The dominant weak interaction processes for many-GeV neutrinos and anti-neutrinos are charged current (CC) and neutral current (NC) deep inelastic scattering (DIS) off nucleons (N, i.e. protons and neutrons) with the production of several hadrons (X):

$$\begin{aligned} \nu(\overline{\nu}) + N &\rightarrow \nu(\overline{\nu}) + X \qquad (NC) \\ \nu + N &\rightarrow l^- + X \qquad (\nu - CC) \\ \overline{\nu} + N &\rightarrow l^+ + X \qquad (\overline{\nu} - CC), \end{aligned} \tag{6}$$

where the charged lepton, l, is an electron/muon for electron/muon neutrinos. The interaction of neutrinos with electrons is three orders of magnitude less common than neutrino-nucleon DIS. From the property of lepton universality, the DIS interactions ν_e's and ν_μ's, or of their anti-neutrinos, are expected to be identical up to final-state lepton mass corrections that are calculable and almost negligible for multi-GeV neutrinos. This simplifies the physics analyses and allows for a useful check of experimental systematics through the comparision of DIS event samples from the two neutrino flavors. The CC (NC) DIS interactions are well described by the "naive quark-parton model" as quasi-elastic (elastic) scattering off one of the many quarks (q) inside the nucleon through the exchange of a virtual W (Z) boson:

$$\nu(\overline{\nu}) + q \rightarrow \nu(\overline{\nu}) + q \qquad (NC) \tag{7}$$

$$\nu + q^{(-)} \rightarrow l^- + q^{(+)} \qquad (\nu - CC) \tag{8}$$

$$\overline{\nu} + q^{(+)} \rightarrow l^+ + q^{(-)} \qquad (\overline{\nu} - CC), \tag{9}$$

To conserve charge, the initial-state and final-state quarks in equations 8 and 9 differ by one unit of charge and have been labelled accordingly, where $q^{(-)} \in d, s, \overline{u}, \overline{c}$ and $q^{(+)} \in u, c, \overline{d}, \overline{s}$. The final state quark always "hadronizes" at the nuclear distance scale, producing quark-antiquark pairs that arrange into the several hadrons seen in the detector.

The naive quark-parton model will be used in the rest of this paper to give qualitative understanding of physics processes, although it should be understood that the actual analyses will require more sophisticated modeling.

2.3 The Role of High Performance Neutrino Detectors

The huge reduction in the neutrino beam cross section over the conventional meters-wide beams from pion decay allows the use, for the first time in neutrino physics, of compact, specialized targets surrounded by high performance detectors. As an example, figure 1 illustrates the sort of HR general purpose neutrino detector that would be well matched to the intense neutrino beams at nufacts. (Subsection 3.2, on polarized targets, and subsection 4.1, on neutrino-electron scattering will discuss other possible examples of specialized targets and detectors.)

The neutrino target in figure 1 forms a notable contrast with the kiloton-scale calorimetric targets used in today's HR neutrino experiments. Instead, it might comprise a 2 meter long stack of CCD tracking planes with a radius of 20 cm chosen to match the beam radius at approximately 100 meters from production for a 50 GeV muon beam. As a detailed example, it might contain 1500 planes of 300 micron thick silicon CCD's, corresponding to a mass per unit area of approximately 100 g.cm^{-2}, about 5 radiation lengths and one interaction length. (Further parameters for this target will be presented in the following subsection.)

Besides providing the mass for neutrino interactions, a tracking target will allow precise reconstruction of the event topologies from charged tracks, including event-by-event vertex tagging of those events containing charm hadrons or tau leptons and, with the higher energy beams, beauty hadrons. Given the favorable vertexing geometry and the $\sim 3.5\ \mu$m typical[10] CCD hit resolutions, it is reasonable to expect[9] perhaps 50 percent efficiency for charm tagging and ultrapure tagging for the relatively rare beauty hadrons that will appear at the higher energies.

The target in figure 1 is surrounded by a time projection chamber (TPC) tracker in a vertical dipole magnetic field. The characteristic dE/dx signatures from the tracks would identify each charged particle. Further particle ID is provided by the Cherenkov photons that are produced in the TPC gas then

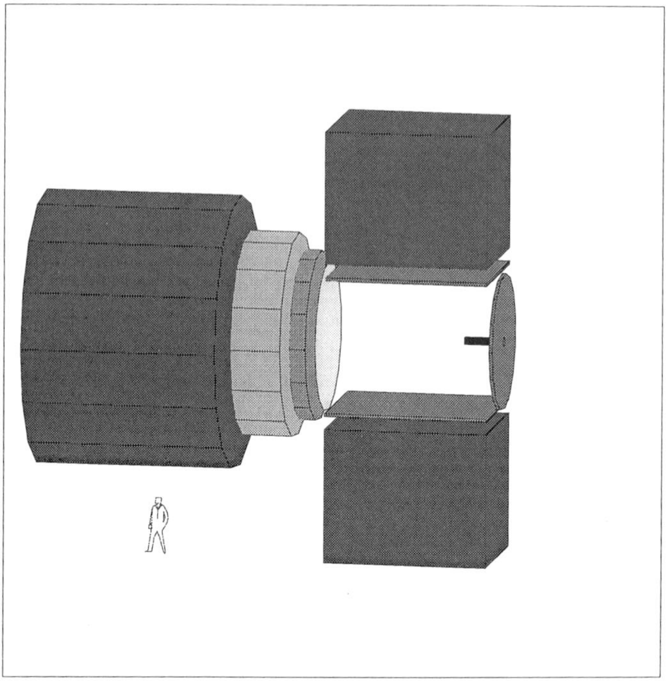

Figure 1: Example of a general purpose neutrino detector. A human figure in the lower left corner illustrates its size. The neutrino target is the small horizontal cylinder at mid-height on the right hand side of the detector. Its radial extent corresponds roughly to the radial spread of the neutrino pencil beam, which is incident from the right hand side. Further details are given in the text.

reflected by a spherical mirror at the downstream end of the tracker and focused onto a read-out plane at the upstream end of the target. The mirror is backed by electromagnetic and hadronic calorimeters and, lastly, by iron-core toroidal magnets for muon ID.

The relativistically invariant quantities that are routinely extracted in DIS experiments, along with their interpretations in the naive quark-parton model, are 1) Feynman x, the fraction of the nucleon 4-momentum carried by the struck quark, 2) the inelasticity, $y = E_{\text{hadronic}}/E_\nu$, which is related to the scattering angle of the neutrino in the neutrino-quark CoM frame (where E_{hadronic} is the energy contained in the hadron shower), and 3) the momentum-transfer-squared, $Q^2 = 2M_pE_\nu xy$ for M_p the mass of the proton. The high performance detectors at nufacts will have the further capability of reconstructing the hadronic 4-vector, resulting in a much better characterization of each interaction, particularly for NC interactions.

2.4 Cross Sections and Event Rates

This section derives approximate estimates of the event sample sizes expected in HR targets at nufacts.

The DIS cross sections for multi-GeV neutrinos are approximately proportional to the neutrino energy, E_ν, and the charged current (CC) and neutral current (NC) interaction cross sections for neutrinos and antineutrinos have numerical values of[11]:

$$\sigma_{\nu \mathrm{N}} \text{ for } \begin{pmatrix} \nu - CC \\ \nu - NC \\ \overline{\nu} - CC \\ \overline{\nu} - NC \end{pmatrix} \simeq \begin{pmatrix} 0.72 \\ 0.23 \\ 0.38 \\ 0.13 \end{pmatrix} \times \mathrm{E}_\nu[\mathrm{GeV}] \times 10^{-38}\,\mathrm{cm}^2. \tag{10}$$

This energy dependence allows us to express the number of DIS events in a target as the simple product of the average beam energy, $< E_\nu >$, and the integrated luminosity:

$$\text{No. DIS events} = \sigma^{\mathrm{R}}_{\nu \mathrm{N}}[\mathrm{cm}^2.\mathrm{GeV}^{-1}] \times < E_\nu > [\mathrm{GeV}] \times \int L\mathrm{dt}\ [\mathrm{cm}^{-2}] \tag{11}$$

where the units are given in square brackets and the constant of proportionality is the reduced cross section defined by:

$$\sigma^{\mathrm{R}}_{\nu \mathrm{N}} \equiv \sigma_{\nu \mathrm{N}} / E_\nu. \tag{12}$$

Summing equation 10 over NC and CC interactions and averaging over neutrinos and anti-neutrinos gives the numerical value of

$$\sigma^{\mathrm{R}}_{\nu \mathrm{N}} = 0.73 \times 10^{-38}\,\mathrm{cm}^2.\mathrm{GeV}^{-1}. \tag{13}$$

For a target subtending $\theta_\nu = 1/\gamma$ from the production straight section, the number of neutrinos passing through the target is seen to be exactly equal to the number of muon decays, N^{ss}_μ, since each muon decay produces two neutrinos but only half of them will be in the forward hemisphere in the muon rest frame. In this case, the total integrated luminosity into the detector is clearly

$$\int L\mathrm{dt}\ [\mathrm{cm}^{-2}] = 6.022 \times 10^{23} \times N^{ss}_\mu \times l[\mathrm{g.cm}^{-2}], \tag{14}$$

Table 1: Specifications, integrated luminosities and event rates for the HR targets discussed in this paper and for 50 GeV (500 GeV) muon storage rings. The approximation is made that the target is situated 100 m (1 km) downstream from a straight section that has $N_\mu^{ss} = 10^{20}$ decays of 50 GeV (500 GeV) muons. This corresponds to average neutrino energies of 32.5 GeV (325 GeV) and to approximately 1 (2) years running for storage ring parameters given previously in the literature.

target purpose	general	polarized	$\nu - e$ scatt.
material	Si CCD's	solid HD	liquid CH_4
ave. density	0.5 g.cm^{-3}	0.267 g.cm^{-3}	0.717 g.cm^{-3}
length	2 m	0.5 m	20 m
mass/area, l	100 g.cm^{-2}	13.4 g.cm^{-2}	1434 g.cm^{-2}
radius	0.2 m	0.2 m	0.2 m
mass	126 kg	16.8 kg	1800 kg
$\int L dt$	6.0×10^{45} cm^{-2}	8.1×10^{44} cm^{-2}	8.6×10^{46} cm^{-2}
no. DIS events:			
at 50 GeV	1.4×10^9	1.9×10^8	2.0×10^{10}
at 500 GeV	1.4×10^{10}	1.9×10^9	2.0×10^{11}
no. ν-e events:			
at 50 GeV	3.5×10^5	NA	7×10^6
at 500 GeV	3.5×10^6	NA	7×10^7

where 6.022×10^{23} is Avogadro's number and l[g.cm^{-2}] is the detector mass per unit area in appropriate units.

Table 1 gives a summary of some characteristics for examples of the three types of targets discussed in this paper, and also gives realistic but very approximate integrated luminosities and event sample sizes for 2 illustrative nufact energies: 50 GeV and 500 GeV. A muon beam energy of about 50 GeV is a likely choice for a dedicated muon storage ring [12], with default specifications of 10^{20} muon decays per year in the straight section providing the neutrino beams. Five hundred GeV muons corresponds to a 1 TeV center-of-mass muon collider and this parameter set is discussed in reference [9]¿

The event samples in table 1 are truly impressive. It is seen, for example, that high performance detectors with fully-active tracking neutrino targets might collect and precisely reconstruct data samples with many billions of neutrino-nucleon DIS interactions – more than three orders of magnitude larger than any of the data samples collected using today's much larger and cruder neutrino targets.

3 Nucleon Structure Functions and QCD Studies

3.1 Structure Function Measurements with Unpolarized Targets

It can be shown quite generally that the differential cross sections for neutrino-nucleon and anti-neutrino interactions can be written in terms of nucleon *structure functions* (SF), F_1, F_2 and F_3, as:

$$\frac{d^2\sigma^{\nu(\bar{\nu})}}{dxdy} = \frac{G_F^2 M_p E_\nu}{\pi}\left[xy^2 F_1^{\nu(\bar{\nu})} + (1-y)\, F_2^{\nu(\bar{\nu})} \mp xy\,(1-y/2)\, F_3^{\nu(\bar{\nu})}\right], \quad (15)$$

with G_F the Fermi couplng constant, M_p the proton mass and x and y in 2.3 and where small correction factors have been neglected for simplicity. The SF are found to exhibit approximate scaling behavior for $Q^2 \to \infty$, i.e. $F_i(x,Q^2) \to F_i(x)$, which is an encouraging approximate verification of the exact scaling behaviour that is predicted in the naive quark-parton model.

The SF must be experimentally determined by measuring the differential cross sections as functions of x, y and Q^2 and extracting the SF using binned fits that exploit their differing y dependences. The parity-violating structure function, xF_3, can only be measured in neutrino-nucleon scattering (and at HERA, although much less precisely and in a different kinematic regime). The F_1 and F_2 SF's that are defined for neutrino-nucleon scattering also probe different combinations of quarks to the analagous SF's defined for charged lepton DIS experiments.

Superscripts on the SF in equation 15 distinguish the potentially different structure functions for ν and $\bar{\nu}$ scattering. However, neutrino experiments to date on isoscalar targets have struggled to resolve the differences, due to insufficient statistics and insufficient control of experimental systematics. Nufacts will therefore provide the first clean extractions of all 6 SF's and this will provide an opportunity that is unique, for any physics process, to lay bare the quark-by-quark content of the nucleon. This will now be made plausible using a simplified discussion in the context of the naive quark-parton model.

In the framework of the naive quark-parton model, the SF can be expressed in terms of quark densities as:

$$\begin{aligned}
2F_1^{\nu N}(x,Q^2) &= d(x) + s(x) + \bar{u}(x) + \bar{c}(x)\\
2F_1^{\bar{\nu} N}(x,Q^2) &= u(x) + c(x) + \bar{d}(x) + \bar{s}(x)\\
F_3^{\nu N}(x,Q^2) &= d(x) + s(x) - \bar{u}(x) - \bar{c}(x)\\
F_3^{\bar{\nu} N}(x,Q^2) &= u(x) + c(x) - \bar{d}(x) - \bar{s}(x)
\end{aligned} \quad (16)$$

and

$$F_2(x,Q^2) \equiv 2xF_1(x,Q^2). \quad (17)$$

An isoscalar target (i.e. equal numbers of neutrons and protons, as is the case for the silicon tracking target of subsection 2.3) is expected to provide four further approximate equalities:

$$\begin{aligned} u(x) &= d(x) \\ \bar{u}(x) &= \bar{d}(x) \\ s(x) &= \bar{s}(x) \\ c(x) &= \bar{c}(x). \end{aligned} \tag{18}$$

The 8 equations in 16 and 18 can therefore be solved in this naive model to derive, from the measured SF's, the quark-by-quark density distributions, at each x, for isoscalar targets. (In detail, the first two equations of 16 are actually identical in this simplified model, so additional information from, e.g., charm tagging is needed to completely solve for all of the quark densities.)

As an extension of this argument, SF measurements from both hydrogen and deuterium targets would, using isospin symmetry, give measurements of the quark-by-quark densities in both protons and neutrons. Thus, even the polarization-averaged measurements from the polarized hydrogen targets in subsection 3.2 will clearly be extremely valuable.

In real life, there will clearly be complications beyond the simple picture from the naive quark model, including nuclear effects, so-called "higher twist" processes and charm thresholds, to name just a few. Despite the complications, the above discussion amounts to at least a plausibility argument that nufacts will, for the first time in any experimental process, have the potential to disentangle the quark-by-quark structure of isoscalar nuclear targets, protons and neutrons.

The uniquely precise and detailed quark-by-quark characterization of nucleon structure will provide an invaluable reference source for many diverse analyses in collider and fixed target physics including, of course, the other precision analyses at nufacts. Precise measurements at high x, $x \rightarrow 1$, are particularly relevant to the modeling of rates for interesting physics processes and backgrounds at hadron colliders.

The universality and applicability of the quark-parton model will be able to be further tested by comparing the nufact SF's with those from charged lepton DIS measurements, where the quarks couple to the electromagnetic current in proportion to their charge squared. As an example, the quark model relationship between the F_2 structure functions is

$$F_2^{\mu} = \frac{5}{18} F_2^{\nu} + \frac{1}{6} x(s(x) + \bar{s}(x)). \tag{19}$$

Current global SF analyses find puzzling deviations from this relationship at low x which, it is speculated, might be related to differing nuclear shadowing effects and would vanish for simple nucleon targets. A first high-statistics measurement of F_2^ν on deuterium at a nufact that can be compared with existing F_2^μ measurements on deuterium might well resolve the discrepancy.

As the second broad area for SF studies at nufacts, DIS has long been a dominant process in testing and understanding perturbative quantum chromodynamics (pQCD). For example, the scaling behavior of the xF_3 SF is broken by a predictable pQCD evolution that is logarithmic in Q^2 and is independent, to first order in perturbation theory, of the nucleon's gluon distribution. The observed evolution of xF_3 in νN scattering already provides one of the most precise measurements of the strong coupling constant, α_s, as does its integral over x:

$$\int_0^1 F_3(x, Q^2)dx = 3\left[1 - \frac{\alpha_s}{\pi} + O\left(\frac{\alpha_s}{\pi}\right)^2\right], \tag{20}$$

a measurement known as the Gross Llewellyn-Smith (GLS) Sum Rule. Both of these measurements of α_s are currently limited by experimental systematic uncertainties and great improvements can be expected with the superior experimental conditions at nufacts.

Heavy quark production in neutrino interactions will provide both additional complications and oppportunities for pQCD studies, due to the additional mass scale, m_Q, for $Q = c, b$. Nufacts will provide a unique facility to test and extend theoretical treatments on this topic. Improved understanding will be valuable for pQCD analyses both within and beyond neutrino physics and, for example, correct modeling of the heavy quark threshold suppression is vital for the CKM studies of section 5.

In addition to these inclusive charm production studies, semi-inclusive measurements involving, for example, the production of Λ_C^+ or J/ψ's will provide opportunities for further insights into QCD, as is addressed elsewhere in these proceedings [13].

3.2 Measurements with Polarized Targets

As well as vastly improving on today's νN SF measurements with unpolarized targets, nufacts will provide the first neutrino beams with sufficient intensity to allow the use of polarized targets. Neutrino scattering experiments using polarized targets have considerable potential for further resolution of the structure of the nucleon and for additional tests of QCD, even at the lower energy beams of dedicated nufacts.

Polarized lepton-nucleon DIS studies have so far been the domain of charged lepton experiments, where a rich program includes SLAC E155x, the recently approved COMPASS experiment at CERN, HERMES at DESY, ELSA in Bonn, MAMI in Mainz and experiments at the Thomas Jefferson National Accelerator Facility. Related studies at collider energies will soon become available in polarized proton-proton collisions at BNL's RHIC collider, and HERA may also eventually polarize their protons beam to provide polarized positron-proton collisions at center-of-momentum energies of approximately 300 GeV.

Polarized neutrino-nucleon scattering retains the experimental advantages over charged lepton DIS experiments that were discussed in subsection 3.1 for non-polarized targets. In addition, the absence of significant target heating from the beam will allow the use of polarized solid protium-deuterium (HD) targets that cannot survive in charged lepton experiments and have so far only been used in experiments with low intensity neutron or photon [14] beams. The preparation of such targets is a detailed craft [15] involving doping the targets with ortho-hydrogen and holding them for long periods of time at very low temperatures and high magnetic fields, e.g. 30-40 days at 17 T and 15 mK. In order to avoid building an entire new detector around the polarized target, it would be ecomonical to place the polarized target directly upstream from another detector, such as the tracking detector described in subsection 2.3.

The polarized SF that might be measured at a nufact are the transversely-polarized and longitudinally-polarized parity-conserving spin-SF's, g_1 and g_2, respectively, as well as the parity-violating spin-SF g_5.

As was discussed above for the unpolarized parity-violating SF, xF_3, polarized target experiments at nufacts should provide easily the most precise measurements of the parity-violating spin structure functions, g_5, that can only be measured in CC weak interactions. The only other future opportunity to measure g_5 that has been widely discussed is the possibility of eventually polarizing the proton beams in the HERA e-p collider. Because of kinematic constraints on reconstructing events, a polarized HERA would be able to make less precise measurements [16] for protons in the complementary high Q^2 region, $Q^2 > 225\,\text{GeV}^2$, that will not be accessible to nufacts. It would not provide measurements for neutrons, of course.

The quark content for the polarized SF's in the naive quark model is as follows:

$$
\begin{aligned}
g_1^{\nu N} &= \Delta d + \Delta s + \Delta\overline{u} + \Delta\overline{c} \\
g_1^{\overline{\nu} N} &= \Delta u + \Delta c + \Delta\overline{d} + \Delta\overline{s}, \\
g_2^{\nu N} &= 0
\end{aligned}
$$

$$
\begin{aligned}
g_2^{\bar{\nu}N} &= 0 \\
g_5^{\nu N} &= \Delta d + \Delta s - \Delta\overline{u} - \Delta\overline{c} \\
g_5^{\bar{\nu}N} &= \Delta u + \Delta c - \Delta\overline{d} - \Delta\overline{s},
\end{aligned}
\tag{21}
$$

where each $\Delta q \equiv q^{\uparrow\uparrow} - q^{\uparrow\downarrow}$ is the difference between quarks polarized parallel to the nucleon spin and those polarized anti-parallel. By similar arguments to that presented in subsection 3.1 for unpolarized SF's it can be seen that the measurement of the polarized SF's for both protons and neutrons should provide much information about the quark-by-quark spin content of the nucleon. This should give nufacts a central role in resolving the current so-called "spin crisis" that has been a dominant topic for DIS spin experiments over the last decade, as follows.

The spin crisis refers to the experimental observation [17] that only a small fraction of the nucleon spin is contributed by the quarks, which was considered to be counter-intuitive and has lead to efforts to investigate the other possible contributions to the spin. The contributions are summarized in the helicity sum rule for the nucleon's longitudinal spin, S_z^N:

$$S_z^N = \frac{1}{2} = \frac{1}{2}(\Delta u + \Delta d + \Delta s) + L_q + \Delta G + L_G, \tag{22}$$

where the quark contribution is $\Delta\Sigma = \Delta u + \Delta d + \Delta s$, ΔG is the gluon spin and L_q and L_G are the possible angular momentum contributions from the quarks and gluons circulating in the nucleon.

The above method for extracting the various quark spin distributions at nufacts, from inclusive SF, should be much cleaner theoretically than the semi-inclusive measurements needed in charged lepton experiments, which rely on semi-inclusive measurements plus assumptions about fragmentation functions. Even so, nufacts should also provide novel and extended capabilities for such semi-inclusive measurements. They can use, for example, the semi-muonic tagging of charm production. This can be calibrated by vertex-tagging experiments in other detectors and is sensitive to the spin of the strange quarks in the nucleon and, perhaps in some kinematic regions, to the spin contribution of the gluon. Such a capability, if realized, would be very valuable in solving the spin crisis, particularly since the gluon contribution is extremely hard to measure and yet it is the leading suspect for providing the bulk of the nucleon's spin.

4 Precision Electroweak Studies

Nufacts will improve enormously on current neutrino experiments in allowing access to aspects of the electroweak interaction that cannot be readily probed at colliders. As the most important single electroweak topic, they might well restore the historical role of neutrino scattering experiments in providing some of the most precise measurements of the weak mixing angle, $\sin^2\theta_W$. The electroweak theory predicts that $\sin^2\theta_W$ is related to the mass ratio of the W and Z intermediate vector bosons:

$$\sin^2\theta_W = 1 - \left(\frac{M_W}{M_Z}\right)^2 \tag{23}$$

to first order in perturbation theory. Precise measurements in various processes can probe higher order diagrams to test the consistency of the electroweak theory and provide differing sensitivities to new physics processes occurring, for example, through loop diagram contributions to the scattering amplitudes.

Nufacts should provide vastly improved determinations of the weak mixing angle, $\sin^2\theta_W$, in measurements from (i) the ratio of neutral current (NC) to charged current (CC) DIS events and (ii) measuring absolute cross sections for neutrino-electron scattering, as discussed in the following subsections.

4.1 Neutrino-Electron Scattering

Neutrino-electron elastic scattering,

$$\nu e^- \to \nu e^-, \tag{24}$$

is an interaction between point elementary particles with a precise theoretical prediction for its cross section as a function of $\sin^2\theta_W$. It therefore provides measurements of $\sin^2\theta_W$ that will be essentially limited only by statistics (3 orders of magnitude down from DIS) and by ingenuity in minimizing the experimental uncertainties.

Four different ν-e elastic scattering processes occur in total in the $(\nu_\mu\overline{\nu}_e)$ and $(\overline{\nu}_\mu\nu_e)$ beams:

$$\nu_\mu e^- \to \nu_\mu e^- \tag{25}$$

$$\overline{\nu_e} e^- \to \overline{\nu_e} e^- \tag{26}$$

$$\overline{\nu_\mu} e^- \to \overline{\nu_\mu} e^- \tag{27}$$

$$\nu_e e^- \to \nu_e e^-, \tag{28}$$

where the first two processes occur in the $(\nu_\mu\overline{\nu}_e)$ beam and the final two in the $(\overline{\nu}_\mu\nu_e)$ beam. The two scattering processes in a given beam cannot be experimentally separated so the experimental measurement involves counting the sum of events from the two processes. The $(\nu_\mu\overline{\nu}_e)$ and $(\overline{\nu}_\mu\nu_e)$ beams will

Reaction	g_L	g_R	$g_L^2 + \frac{1}{3}g_R^2$
$\nu_\mu e^- \to \nu_\mu e^-$	$-\frac{1}{2} + \sin^2\theta_W$	$\sin^2\theta_W$	0.0925
$\overline{\nu_e} e^- \to \overline{\nu_e} e^-$	$\sin^2\theta_W$	$\frac{1}{2} + \sin^2\theta_W$	0.2258
$\overline{\nu_\mu} e^- \to \overline{\nu_\mu} e^-$	$\sin^2\theta_W$	$-\frac{1}{2} + \sin^2\theta_W$	0.0758
$\nu_e e^- \to \nu_e e^-$	$\frac{1}{2} + \sin^2\theta_W$	$\sin^2\theta_W$	0.5425

Table 2: g_L and g_R by $\nu - e$ scattering process listed for the cross section formula in equation 29. The value $\sin^2\theta_W = 0.225$ has been used for the final column.

provide two physically distinct measurements because the scattering involves different diagrams for the neutrino species in the two beams: an s-channel (annihilation) diagram contributes to equation 26 while equation 28 includes a charged current ($W^\pm$ exchange) t-channel diagram.

Because of the small ratio of the electron to proton mass, the cross section for neutrino-electron scattering is much smaller than that for DIS. The numerical values for the cross sections after integrating over y are:

$$\sigma(\nu e^- \to \nu e^-) = 1.6 \times 10^{-41} \times E_\nu[GeV] \times \left[g_L^2 + \frac{1}{3}g_R^2\right], \tag{29}$$

where the left-handed and right-handed coupling constants, g_L and g_R, are different for each of the processes in equations 27 through 26. Their values and the values for the term in square brackets are given in table 4.1.

The experimental signature for $\nu - e$ scattering is a single negatively charged electron with very low transverse momentum, $p_t \lesssim \sqrt{m_e E_\nu}$. A tracking detector with very good p_t resolution is needed to resolve the signal peak from the much broader background distributions from quasi-elastic $\nu - N$ scattering and other low-multiplicity $\nu - N$ scattering events. An attractive target/detector option is a low-Z liquid that can form tracks of ionization electrons and drift them to an electronic read-out. Liquids under consideration [9] include argon and methane or other saturated alkanes, and the readout goemetry might be a TPC or, to reduce pile-up backgrounds, a printed-circuit kaptan strip geometry [18] with more channels and shorter drift distances.

Besides background rejection, the other big experimental challenge for the measurement will be the determination of the absolute neutrino flux. For the $(\nu_\mu\overline{\nu}_e)$ beam, signal processes can probably [9] be precisely normalized to the theoretically predictable processes involving muon production off electrons: $\nu_\mu e^- \to \nu_e \mu^-$ and $\overline{\nu_e} e^- \to \overline{\nu_\mu}\mu^-$. The $(\overline{\nu}_\mu \nu_e)$ beam requires an additional stage of relative flux normalization, which might be accomplished [9] using the relative sizes of the event samples for quasi-elastic neutrino-nucleon scattering, $\nu N \to l^\pm N'$, in the $(\overline{\nu}_\mu \nu_e)$ and $(\nu_\mu \overline{\nu}_e)$ runs.

Specific example parameters for the target/detector and for the event sample sizes are given in table 1. Predicted event samples are in the range of millions to tens-of-millions of events and if the experimental uncertainties can be controlled then these sample sizes will correspond [9] to limiting statistical uncertainties of $\Delta\sin^2\theta_W = 0.000\,3$ and $0.000\,1$ for the $(\nu_\mu\overline{\nu}_e)$ and $(\overline{\nu}_\mu\nu_e)$ beams, respectively, at the 50 GeV nufact and $\Delta\sin^2\theta_W = 0.000\,1$ and $0.000\,03$ for the corresponding measurements at 500 GeV. It remains, of course, to be demonstrated that the experimental uncertainties will ever allow these measurement accuracies, but the potential of this measurement is anyway impressive.

4.2 Measurement of the WMA in DIS

The most precise current measurement of $\sin^2\theta_W$ from ν-N DIS, from NuTeV [19],

$$\sin^2\theta_W = 0.2253 \pm 0.0019(\text{stat.}) \pm 0.0010(\text{syst.}) \qquad \text{(preliminary)}, \qquad (30)$$

gives an equivalent uncertainty on the W mass, $\Delta M_W \simeq 100\ \text{MeV}/c^2$, that is competitive with direct measurements at colliders. Measurements at nufacts may well extend the historical tradition of neutrino-nucleon DIS experiments in providing some of the most precise measurements of $\sin^2\theta_W$.

The complexity of nucleon targets makes it necessary, as in previous neutrino experiments, to consider NC-to-CC cross section ratios in order to make theoretical sense of the results. Since both the $(\nu_\mu\overline{\nu}_e)$ and $(\overline{\nu}_\mu\nu_e)$ beams at nufacts are mixtures of a neutrino and and antineutrino flavor, the appropriate experimental NC-to-CC ratios, $R_\nu^{\mu^-}$ and $R_\nu^{\mu^+}$, respectively, are each linear combinations of the traditional NC-to-CC cross section ratios for neutrinos, R^ν, and antineutrinos, $R^{\overline{\nu}}$:

$$R_\nu^{\mu^-} \simeq 0.70R^\nu + 0.30R^{\overline{\nu}},$$

$$R_{\overline{\nu}}^{\mu^+} \simeq 0.63R^\nu + 0.37R^{\overline{\nu}}.$$

Because the linear combinations are so similar, the two measurements will have almost equal numerical values, $R_\nu^{\mu^-} \simeq 0.330$ and $R_{\overline{\nu}}^{\mu^+} \simeq 0.332$ for $\sin^2\theta_W = 0.225$, and their physics content will clearly be nearly identical.

Traditional heavy target neutrino detectors could not distinguish ν_e-induced CC interactions from NC interactions and this separation will probably be the most demanding experimental requirement for analyses at nufacts. However, with huge statistics and high performance tracking detectors such as those described in section 2, the DIS measurement will presumably eventually be systematically limited by theoretical hadronic uncertainties rather than statistical or experimental uncertainties.

The vastly improved statistics and experimental conditions at nufacts makes it difficult to extrapolate the measurement accuracy from that at today's neutrino experiments. Reference [5] estimates that the predicted uncertainty in M_W from a nufact analysis might be of order 10 MeV, which improves by an order of magnitude on today's neutrino experiments [19,20] and is approximately equal to the projected best direct measurements from future collider experiments. Thus, the $\sin^2\theta_W$ measurement from DIS could be a useful complement to the ν-e scattering measurements described in the previous subsection.

5 Measurements of CKM Matrix Elements

With huge samples of flavor tagged events, nufacts have the potential to make impressive measurements of the absolute squares of several of the elements in the fundamental Cabbibo-Kobayashi-Maskawa (CKM) mixing matrix that characterises the charged current (CC) weak interactions of quarks. Lower energy nufacts will provide an opportunity for unique and precise measurements of the elements $|V_{cd}|$ and $|V_{cs}|$ and further measurements of the more theoretically interesting elements $|V_{ub}|$ and $|V_{cb}|$ will become available at the higher energies required for B production. Each of the analyses [9] would involve vertex tagging of heavy final state quarks and will be somewhat analagous to, but vastly superior to, current neutrino measurements of $|V_{cd}|^2$ that use dimuon events for final state tagging of charm quarks.

Today's measurements of $|V_{cd}|$ in νN scattering are already the most precise in any process and a fundamental advantage of νN DIS over all other types of CKM measurements is that the scattering process involves the interaction of an external W boson probing the quarks inside a nucleon rather than an internal W interaction inside a hadron. (In principle, the HERA ep collider could also do such measurements, but they turn out not to be feasible in practice.) Such external W probes allow measurements that are theoretically cleaner than other processes because the asymptotic freedom property of QCD predicts quasi-free quarks with reduced influence from their hadronic environment for Q significantly above the GeV-scale.

Another difference relative to, say, the CKM measurements at B factories is that the measurements are of the magnitudes of individual CKM matrix elements rather than of interference terms. This is understood easily from the naive quark-parton model approximation, where the differential cross sections for quark transitions, $\frac{d\sigma}{dx}$, are given as products of quark densities with the absolute squares of matrix elements:

$$\frac{d\sigma}{dx}(d \to c) \propto x\, d(x)|V_{cd}|^2 \tag{31}$$

$$\frac{d\sigma}{dx}(s \to c) \propto x\, s(x)|V_{cs}|^2 \tag{32}$$

$$\frac{d\sigma}{dx}(u \to b) \propto x\, u(x)|V_{ub}|^2 \tag{33}$$

$$\frac{d\sigma}{dx}(c \to b) \propto x\, c(x)|V_{cb}|^2, \tag{34}$$

with $d(x)$, $s(x)$, $u(x)$ and $c(x)$ the respective initial-state quark densities and these expressions are very approximate only because important threshold correction factors have been neglected.

The $|V_{cd}|$ analysis at a nufact, with hundreds of millions of vertex-tagged charm events in a high-performance detector, should have profound statistical, experimental and theoretical advantages over today's measurement and may well reach [9] the parts-per-mil level of accuracy. As is evident from equation 32, the measurement of $|V_{cs}|$ is intrinsically more difficult than $|V_{cd}|$ because it requires a knowledge of the strangeness content of the nucleon. Reference [9] speculates that $|V_{cs}|$ could nevertheless be measured at the percent level, which would also become the best direct measurement of the element and would provide a much improved unitarity test on its value.

The B-production analyses at higher energy nufacts should be experimentally rather similar to the charm analyses but would have vastly greater theoretical interest. Both $|V_{ub}|$ and $|V_{cb}|$ determine the lengths of sides of the "unitarity triangle" that is predicted to exist if the CKM matrix is indeed unitary. The main goal of today's B factories is to measure the interior angles of this triangle to confirm that it is indeed a triangle, and the complementary input from a nufact will be an enormous help in this verification process. In particular, the predicted [5,9] 1-2 % accuracy in $|V_{ub}|^2$ is several times better than predicted accuracies in any future measurements of other processes, and will obviously provide a very strong constraint on the unitarity triangle.

Table 3 summarizes the predicted [9] nufact contributions to determining the CKM matrix elements, giving the current, experimentally determined values for the 9 mixing probabilities along with their current percentage uncertainties and speculative projections [5] for how 4 of the 9 uncertainties might be reduced at a 500 GeV nufact.

6 Rare and Exotic Processes

The potential for studying rare and exotic processes at nufacts is limited relative to collider experiments due to the more modest center-of-mass energies. Nevertheless there will still be some opportunities to both further our knowledge of the SM by studying rare processes and to search for processes not

Table 3: Absolute squares of the elements in the Cabbibo-Kobayashi-Maskawa (CKM) quark mixing matrix. The second row for each quark gives current percentage uncertainties in the absolute squares and speculative projections of the uncertainties after analyses from a 500 GeV nufact. The measurements of $|V_{cd}|^2$ and $|V_{cs}|^2$ might be comparably good for a 50 GeV nufact but $|V_{ub}|^2$ and $|V_{cb}|^2$ would not be measured. The uncertainties assume that no unitarity constraints have been used.

	d	s	b
u	**0.95** $\pm 0.1\%$	**0.05** $\pm 1.6\%$	**0.00001** $\pm 50\% \rightarrow$ 1-2%
c	**0.05** $\pm 15\% \rightarrow$ 0.2-0.5%	**0.95** $\pm 35\% \rightarrow \sim 1\%$	**0.002** $\pm 15\% \rightarrow$ 3-5%
t	**0.0001** $\pm 25\%$	**0.001** $\pm 40\%$	**1.0** $\pm 30\%$

predicted by the SM. Examples of a few potential processes involving exotic physics that might be observable at nufacts include:

1. a non-standard flavor-changing neutral current (FCNC) interaction converting valence u quarks to charm quarks, $u \rightarrow c$, would appear as an excess of charm events over anti-charm events at high x. In contrast to FCNC searches in decays, this is a unique opportunity to observe fundamental flavor-changing neutral currents at the quark level [9]
2. some types of unstable exotic neutral leptons [9]
3. R-parity violating SUSY searches [21]
4. hypothetical new neutrino interactions involving microscopic extra dimensions that are much larger than the Planck scale [21].

7 Charm Physics Studies

It should be clear from the discussions in sections 2.3 and 5 that nufact's will be rather impressive factories for the study of charm – with a clean, well reconstructed sample of several times 10^8 charmed hadrons produced in 10^{10} neutrino interactions.

There are several interesting physics motivations for charm studies at a nufact [9]. Measurement of charm decay branching ratios and lifetimes are useful for both QCD studies and for the theoretical calibration of the physics analyses on B hadrons. Charm decays also provide a "clean laboratory" to search for exotic physics contributions since the SM predicts 1) tiny branching ratios for rare decays, 2) small CP asymmetries and 3) slow $D^0 \to \overline{D^0}$ oscillations, with only of order 1 in 10^4 oscillating before decay.

It is a unique advantage of CC-induced charm production in neutrinos that the production sign of the charm quark is tagged with very high efficiency and purity by the charge of the final state lepton:

$$\begin{aligned} \nu q &\to l^- c \\ \overline{\nu}\,\overline{q} &\to l^+ \overline{c}, \end{aligned} \tag{35}$$

where $q = d$ or s and $\overline{q} = \overline{d}$ or $\overline{s}$. This is of particular benefit to oscillation and CP studies, as is the expected precise vertexing reconstruction of the proper lifetime of decays. As an example of the advances in charm studies that might result, particle-antiparticle mixing has yet to be observed in the charm sector and it is quite plausible [5] that a nufact would provide the first observation of $D^0 - \overline{D^0}$ mixing.

8 Summary

It has been shown that nufacts will have unique capabilities for HR neutrino physics, with huge event samples collected in high-performance detectors. The physics reach should extend well beyond traditional neutrino physics topics and should complement or improve upon many analyses in diverse areas of collider and fixed target physics, including:

- the only realistic opportunity, in any physics process, to determine the detailed quark-by-quark structure of the nucleon
- with polarized targets, additionally map out the quark-by-quark spin structure of the nucleon and, perhaps, determine the gluon contribution to the nucleon's spin
- some of the most precise measurements and tests of perturbative QCD
- some of the most precise tests of the electroweak theory through measurements of $\sin^2\theta_W$ with fractional uncertainties approaching the 10^{-4} level
- measurements of the CKM quark mixing matrix that will be interesting for lower energy nufacts ($|V_{cd}|$ and $|V_{cs}|$) and will become extremely important ($|V_{ub}|$ and $|V_{cb}|$) at higher energies

- a new realm to search for exotic physics processes
- as a bonus outside neutrino physics, a charm factory with unique capabilities.

The expected experimental conditions at nufacts are so novel and impressive that the topics presented in this paper must surely represent no more than a limited first attempt to understand their full potential for HR neutrino physics.

Acknowledgments

This work was performed under the auspices of the U.S. Department of Energy under contract no. DE-AC02-98CH10886.

1. An unpublished note by A.C. Melissinos (1960) is the first reference found by K.T. McDonald in a compilation of muon collider accelerator physics prior to 1995 that can be found at *http://www.hep.princeton.edu/mumu/physics/index.html#accel.*
2. The Muon Collider Collaboration, $\mu^+\mu^-$ *Collider: A Feasibility Study*, BNL-52503, Fermilab-Conf-96/092, LBNL-38946, July 1996, unpublished.
3. The Muon Collider Collaboration, *Status of Muon Collider Research and Development and Future Plans*, Phys. Rev. ST Accel. Beams, 3 August, 1999.
4. B.J. King, *Assessment of the prospects for muon colliders*, paper submitted in partial fulfillment of requirements for Ph.D., Columbia University, New York (1994), available from LANL preprint archive as *physics/9907026.*
5. B.J. King, *Neutrino Physics at a Muon Collider*, Proc. Workshop on Physics at the First Muon Collider and Front End of a Muon Collider, Fermilab, November 6-9, 1997, *hep-ex/9907033.*
6. S. Geer, *Neutrino Beams from Muon Storage Rings: Characteristics and Physics Potential*, Phys. Rev. **D57**, 6989 (1998).
7. A compilation of papers with physics topics relevant to neutrino factories, compiled by K.T. McDonald, can be found at *http://www.hep.princeton.edu/mumu/nuphys/index.html* .
8. M. Goodman, these proceedings.
9. I. Bigi *et al.*, *The Potential for High Rate Neutrino Physics at Muon Colliders and Other Muon Storage Rings*, contact author B.J. King, to be published.

10. K. Abe *et al.*, *Design and performance of the SLD vertex detector: a 307 Mpixel tracking system.*, NIM A 400 (1997) 287-343.
11. See, for example, Chris Quigg, *Neutrino Interaction Cross Sections*, FERMILAB-Conf-97/158-T.
12. NuFact'99, Lyon, 5-9 July, 1999, *http://lyoinfo.in2p3.fr/nufact99/general.html.*
13. A. Petrov, these proceedings.
14. D. Babusci *et al.*, Proc. 11th Int. Symp. High-Energy Spin Phys., AIP Conf. Proc. 343, p.523 (1995).
15. D.G. Crabb and W. Meyer, *Solid Polarized Targets for Nuclear and Particle Physics Experiments*, Ann. Rev. Nucl. Part. Sci. 1997, 47:67-109.
16. See, for example, A. Deshpande, *The Physics Case for Polarized Protons at HERA*, hep-ex/9908051.
17. J. Ashman *et al.*, the EMC Collaboration, Phys. Lett. B **206** (1988) 364.
18. Private communication with P. Rehak.
19. K.S.McFarland, NuTeV collaboration, to be published in the proceedings of the XXXIIIrd Rencontres de Moriond (1998).
20. C.Arroyo *et. al.*, Phys. Rev. Lett. **72**, 3452 (1994).
21. Private communication with F. deJongh.

NEUTRINO MASS: WHAT HAS BEEN ACCOMPLISHED, AND WHAT DO WE NEED TO DO?

BORIS KAYSER

Physics Division, National Science Foundation
Arlington, VA 22230, USA
E-mail: bkayser@nsf.gov

The present evidence for neutrino oscillation is reviewed, and future measurements that would help confirm or refute this evidence are mentioned. Illustrative neutrino mass and mixing scenarios suggested by the present evidence are described, and some interesting questions raised by this evidence are briefly explored. Finally, the possibility of CP violation in neutrino oscillation is discussed. Most of the material in this talk is covered in detail in a recent review article.[1]

References

1. P. Fisher, B. Kayser, and K. McFarland, eprint hep-ph/9906244, to appear in *Ann. Rev. Nucl. Part. Sci.*

THE LINK BETWEEN NEUTRINO MASSES AND PROTON DECAY IN SUPERSYMMETRIC UNIFICATION*

JOGESH C. PATI
Physics Department
University of Maryland
College Park, MD 20742, USA

* *Based in part on talks delivered at (i) the International Conf. on Non-Accelerator Physics, Dubna, Russia (July, 1999); (ii) The Johns Hopkins Symposium on Neutrino Physics (June, 1999); (iii) the International Workshop on Next Generation Nucleon Decay and Neutrino Detector, Stony Brook, NY (Sept, 1999) and (iv) The Symposium on Frontiers of Particle Physics, Birla Science Center, Hyderabad, India (January, 1999).*

Following recent joint works with K. Babu and F. Wilczek, I stress here that supersymmetric unification, based on symmetries like SO(10) or a string-derived $G(224) = SU(2)_L \times SU(2)_R \times SU(4)^C$ possesses some crucial features that are intimately linked to each other. They are: (a) gauge-coupling unification, (b) the masses and mixings of all fermions, including especially the neutrinos, and last but not least (c) proton decay. In this context, it is noted that the value of $m(\nu_L^\tau) \sim 1/20$ eV, suggested by the SuperK result, goes extremely well with the unification-hypothesis, based on the ideas of (i) SU(4)-color, (ii) left-right symmetry and (iii) supersymmetry.

A concrete proposal is presented within an economical SO(10)-framework that makes five successful predictions for the masses and mixings of the quarks and the charged leptons. The same framework explains why the $\nu_\mu - \nu_\tau$ oscillation angle is so large ($\sin^2 2\theta^{osc}_{\nu_\mu\nu_\tau} \approx 0.82 - 0.96$) and yet V_{bc} is so small (≈ 0.04), both in accord with observation. The influence of the masses of the neutrinos and of the charged fermions on proton decay, is discussed concretely, within this framework. The $\bar{\nu}K^+$ mode is expected to be dominant for SUSY SO(10) as well as SU(5). A distinctive feature of the SO(10) model, however, is the likely prominence of the $\mu^+ K^o$ mode, which, for SU(5), is highly suppressed. Our study shows that while current limits on the rate of proton decaying into $\bar{\nu}K^+$ is compatible with theoretical expectations, improvements in these limits by a factor of 5 - 10 should either turn up events, or else the SO(10)-framework described here, which is otherwise so successful, will be in jeopardy. Prominence of the $\mu^+ K^o$ mode, if observed, will be most significant in that it will reveal the intriguing link that exists between neutrino masses and proton decay in the context of supersymmetric unification.

1 Introduction

The SuperKamiokande (SK) result, convincingly showing the oscillation of ν_μ to ν_τ (or ν_X) with a value of $\delta m^2 \approx 10^{-2} - 10^{-3} eV^2$ and an almost maximal oscillation angle $sin^2 2\theta > 0.83$ [1], clearly seems to require new physics beyond that of the standard model [2]. [a] This as well as the other relatively firm result of solar neutrino-deficit [4] serve as important clues to physics at a deeper level. Understanding these neutrino-anomalies as well as the bizarre pattern of masses and mixings of the quarks and the charged leptons is a major challenge that ought to be met within a fundamental unified theory.

It is of course known that the ideas of grand unification [5,6,7,8], as well as those of superstrings [9] - call for gauge coupling unification at a high scale and for nucleon-instability. Furthermore, both of these features are known to acquire a new perspective [10,11] in the context of supersymmetry [12]. (For recent reviews on this topic and relevant references see e.g. Ref. 13 and 14.) While proton decay is yet to show, the clearest empirical support in favor of grand unification and supersymmetry has so far come from the dramatic meeting of the three gauge couplings of the standard model, that is found to occur at a scale of $M_X \approx 2 \times 10^{16}$ GeV, when these couplings are extrapolated from their measured values at LEP to high energies, in the context of supersymmetry [10].

One major goal of this talk will be to stress that supersymmetric unification based on symmetries like SO(10) [15], or (for most purposes) a string-derived $G(224) = SU(2)_L \times SU(2)_R \times SU(4)^C$ [6,16], has implications not only for (i) gauge coupling unification and (ii) proton decay, but also for (iii) the masses and mixings of the charged fermions, as well as for (iv) those of the neutrinos. In fact, *within a unified theory, all four features (i) - (iv) get intimately linked to each other, much more so, than commonly thought.* Each of these, including even charged fermion and neutrino-masses, provides some essential clue to the nature of higher unification. As regards the link between the four features, even neutrino-masses turn out to have direct influence on proton decay. This is because the latter receives important contributions through *a new set of d=5 operators* that depend directly on the Majorana masses of the right-handed neutrinos [17]. These new d=5 operators, which were missed in the literature, contibute significantly to proton decay amplitudes, in addition of course to the "standard" d=5 operators [11], which arise through the exchange of the color-triplet Higgsinos related to the electroweak doublets. The standard and the new d=5 operators, related to the charged fermion as well as the neutrino-masses, *together* raise our expectation that

[a] This is because the standard model, having ν_L (but no ν_R) and a Higgs doublet ϕ_H, in the presence of quantum gravity, can possibly lead to an effective Majorana mass-term violating B-L for ν_L of the form: $\frac{\nu_L \nu_L <\phi_H>^2}{M_{Pl}}$ [3], yielding m(ν_L) $\sim \frac{(250 GeV)^2}{2\times 10^{18}} GeV \sim 3 \times 10^{-5}$ eV. This is, however, too small to account for the SuperK result. [see e.g. Ref. 2 for discussion of this point and estimate of m(ν_τ) in the context of left-right symmetric unified theories].

proton decay should be observed in the near future [18].

I elucidate these remarks in the next four sections, covering the following topics:

1.) In Sec. 2, I first recall briefly the motivations for left-right symmetric unified theories, utilizing neutrino-masses, suggested by the SuperKamiokande result, as a guide. The support for supersymmetric unification in the light of the LEP data is noted. Further, the origin of such a unification in the context of superstrings as well as the potential problem of rapid proton decay that arises within supersymmetric theories are briefly reviewed. These discussions provide the background needed to cover the materials in the remaining sections.

2.)I then present arguments [2] to show that the SuperK result, especially the observed δm^2, interpreted as $m(\nu_\tau)^2$, receives a simple and natural explanation within the ideas of higher unification based on the symmetry group G(224) [6], and thus SO(10) or E_6. Such an explanation would not be possible within SU(5).

3.) In Sec. 4, I present the first part of a recent work by Babu, Wilczek and myself [18], in which we attempt to understand, in the context of supersymmetric SO(10), the masses and mixings of the neutrinos, suggested by the atmospheric and the solar neutrino anomalies, *in conjunction with* those of the quarks and the charged leptons. Adopting familiar ideas of generating hierarchical eigenvalues through off-diagonal mixings, and correspondingly cabibbo-like mixing angles, we find that the bizarre pattern of masses and mixings observed in the charged fermion-sector, remarkably enough, can be adequately described (with $\sim$ 10% accuracy) within an economical and thus predictive SO(10)-framework. A concrete proposal is presented involving a minimal Higgs system that provides *five successful predictions* for the masses and mixings of the quarks and the charged leptons in the three families. The same description goes extremely well with a value of $m(\nu_\tau) \sim \frac{1}{20}$ eV as well as with a large $\nu_\mu - \nu_\tau$ oscillation-angle ($sin^2 2\theta^{osc}_{\nu_\mu \nu_\tau} \approx 0.82 - 0.96$), despite highly non- degenerate masses of the light neutrinos. Both these features are in good agreement with the SuperK result. Furthermore, this framework generically seems to support the small-angle MSW-explanation for the solar neutrino deficit [19].

4.) I next present the second part of the work by Babu, Wilczek and myself [Ref. 18] in which we link the rather successful supersymmetric SO(10)-framework describing fermion masses (noted above), with expectations for proton decay. We find that, given the SuperK result that suggests $m(\nu_\tau) \sim 1/20$ eV and a large oscillation angle, the contribution from the new d=5 operators mentioned above, and to some extent that from the standard operators as well, are significantly enhanced. As a result, in spite of generous allowance for uncertainties in the matrix elements and the SUSY-spectrum, the inverse decay rate for the dominant $\bar{\nu}K^+$-mode is found to be bounded from above by about 7×10^{33} years. Typically the lifetime should of course be lower than this bound. Furthermore, the $\mu^+ K^o$ mode is found to be prominent, with a branching ratio typically in the range of 10 - 50%, entirely because of contribution from the new operators. For comparison, minimal SUSY SU(5), which has only the standard operators, typically leads to branching ratios $\leq 10^{-3}$ for this mode. Thus our study of proton decay, correlated with fermion masses, strongly suggest that at least candidate-events for proton decay should be observed in the very near future, already at SuperK. The $\mu^+ K^o$ mode, if observed, would be specially important in exhibiting the link between neutrino masses and proton decay, that exists within the G(224)/SO(10)-route to supersymmetric unification [18].

2 Learning From Neutrino Masses About Higher Unification

2.1 Motivations for SU(4)-Color Left-Right Symmetric Theories

If one assumes a hierarchical pattern of masses for the light neutrinos (with $m_{\nu_e} \ll m_{\nu_\mu} \ll m_{\nu_\tau}$), which goes well within a quark-lepton unified theory, the SuperK-result interpreted as $\nu_\mu - \nu_\tau$ oscillation, suggests a value for the ν_τ-mass: $m_{\nu_\tau} \approx 1/20$ eV ($\frac{1}{2}$ to 2). One can argue, as shown later in this section (see also Ref. 2 and footnote a), that a ν_τ-mass of this order can be understood simply within supersymmetric unified theories which are forced to introduce the existence of right-handed (RH) neutrino, accompanying the observed left-handed ones. Postponing an estimate of the ν_τ-mass for a moment if one asks the question: What symmetry on the one hand dictates the existence of the RH neutrinos, and on the other hand also ensures quantization of electric charge, together with quark-lepton unification, one is led to two very beautiful conclusions: (i) quarks and leptons must be unified minimally within the symmetry SU(4)-color, and that, (ii) deep down, the fundamental theory should possess a left-right symmetric gauge structure: $SU(2)_L \times SU(2)_R$. In short, the standard model symmetry must be extended minimally to the gauge symmetry [5,6].

$$G(224) = SU(2)_L \times SU(2)_R \times SU(4)^C \tag{1}$$

With respect to G(224), all members of the electron-family fall into the neat pattern:

$$\mathrm{F}^{\mathrm{e}}_{\mathrm{L,R}} = \begin{bmatrix} u_r & u_y & u_b & \nu_e \\ d_r & d_y & d_b & e^- \end{bmatrix}_{\mathrm{L,R}} . \tag{2}$$

The left-right-conjugate multiplets F^e_L and F^e_R transform as (2,1,4) and (1,2,4) respectively, with respect to G(224); likewise for the mu and the tau families.

Viewed against the background of the standard model, the symmetry structure G(224) brought some attractive features to particle physics which include:

(i) Organization of all members of a family $(8_L + 8_R)$ within one left-right self-conjugate multiplet, with their peculiar hypercharges fully explained.

(ii) Quantization of electric charge, explaining why $Q_{\text{electron}} = -Q_{\text{proton}}$.

(iii) Quark-lepton unification through SU(4)-color.

(iv) Left-Right (i.e. Parity) and Particle-Antiparticle symmetries in the fundamental laws which are violated only spontaneously [6,20]. Thus, within the symmetry-structure G(224), quark-lepton distinction and parity violation may be viewed as low-energy phenomena which should disappear at sufficiently high energies.

(v) Existence of Right-Handed Neutrinos: Within G(224), there must exist a right-handed (RH) neutrino (ν_R), accompanying the left-handed one (ν_L), for each family, because ν_R is the fourth color - partner of the corresponding RH up- quarks. It is also the $SU(2)_R$ -doublet partner of the associated RH charged lepton (see eq. (2)). The RH neutrinos seem to be essential now (see later discussions) for understanding the non- vanishing light masses of the neutrinos, as suggested by the recent observations of neutrino-oscillations.

(vi) B-L as a local Gauge Symmetry: SU(4)-color introduces B-L as a local gauge symmetry. Thus, following the limits from Eötvos experiments, one can argue that B-L must be violated spontaneously. It has been realized, in the light of recent works, that to implement baryogenesis in spite of electroweak sphaleron effects, such spontaneous violation of B-L at high temperatures may well be needed [21].

2.2 Route to Higher Unification: SU(5) versus G(224)/SO(10)

To realize the idea of a single gauge coupling governing the three forces [5,6], one must embed the standard model symmetry, or G(224), into a simple (or effectively simple, like SU(N) × SU(N)) gauge group. The smallest such group is SU(5) [7] which contains the standard model symmetry, but not G(224). As a result, SU(5) does not possess some of the main advantages of G(224) listed above. In particular, SU(5) splits members of a family into two multiplets: $\bar{5}+10$, where as G(224), subject to L-R symmetry, groups them into just one multiplet. SU(5) violates parity explicitly. It does not possess SU(4)-color and therefore does not gauge B-L as a local symmetry. Further, SU(5) does not contain the RH neutrinos as an integral feature. As I will discuss below, these distinctions between SU(5) versus G(224), or its extensions (see below), turn out to be especially relevant to considerations of neutrino as well as charged fermion masses, and thereby to those of proton decay.

Since G(224) is isomorphic to SO(4) $\times$ SO(6), the smallest simple group to which it can be embedded is SO(10) [15]. Historically, by the time SO(10) was proposed, all the advantages of G(224) [(i) to (vi), listed above] and the ideas of higher unification were in place. Since SO(10) contains G(224), the features (i) to (vi) are of course retained by SO(10). In addition, the 16-fold left-right conjugate set $(F_L^e + \bar{F}_R^e)$ of G(224) corresponds to the spinorial 16 of SO(10). *Thus, SO(10) preserves even the 16-plet family-structure of G(224), without a need for any extension.* If one extends G(224) to the still higher symmetry E_6 [22], the advantages (i) to (vi) are retained, as in SO(10), but in this case, one must extend the family-structure from a 16 to a 27-plet.

Comparing G(224) with SO(10), as mentioned above, SO(10) possesses *all* the features (i) to (vi) of G(224); in addition it offers gauge coupling unification. I should, however, mention at this point that the perspective on coupling unification and proton decay has changed considerably in the context of supersymmetry and superstrings. In balance, a string-derived G(224) offers some advantages over a string-derived SO(10), while the reverse is true as well. Thus, it seems that a definite choice of one over the other, as an effective theory below the string-scale, is hard to make at this point. I will return to this point shortly.

2.3 Gauge Coupling Unification: Need for Supersymmetry

It has been known for some time that the precision measurements of the standard model coupling constants (in particular $\sin^2\theta_W$) at LEP put severe constraints on the idea of grand unification. Owing to these constraints, the non-supersymmetric minimal $SU(5)$, and for similar reasons, the one-step breaking minimal non-supersymmetric $SO(10)$-model as well, are now excluded [23].

But the situation changes radically if one assumes that the standard model is replaced by the minimal supersymmetric standard model (MSSM), above a threshold of about $1 TeV$. In this case, the three gauge couplings are found to meet [10], at least approximately, provided $\alpha_3(m_Z)$ is not too low (see figs. in e.g. Refs. [23,13]). Their scale of meeting is given by

$$M_X \approx 2 \times 10^{16} GeV \quad \text{(MSSM or SUSYSU(5))} \tag{3}$$

M_X may be interpreted as the scale where a supersymmetric grand unification symmetry (GUT) (like minimal SUSY SU(5) or SO(10)) — breaks spontaneously into the supersymmetric standard model symmetry $SU(2)_L \times U(1) \times SU(3)^c$.

The dramatic meeting of the three gauge couplings thus provides a strong support for both grand unification and supersymmetry.

2.4 Compatibility Between MSSM and String-Unifications

The superstring theory [9], and now the M theory [24] provide the only known framework that seems capable of providing a good quantum theory of gravity as well as a unity of all forces, including gravity. It thus becomes imperative that the meeting of the gauge couplings of the three non-gravitational forces, which occur by the extrapolation of the LEP data in the context of MSSM, be compatible with string-unification.

Now, string theory does provide gauge coupling unification for the effective gauge symmetry, below the compactification-scale. The new feature is that even if the effective symmety is not simple, like SU(5) or SO(10), but instead is of the form G(213) or G(224) (say), the gauge couplings of G(213) or G(224) should still exhibit familiar unification at the string-scale, for compactification involving appropriate Kac-Moody levels (i.e. $k_2 = k_3 = 1$, $k_Y = \frac{5}{3}$ for G(213)), barring of course string-threshold corrections [25]. And even more, the gauge couplings unify with the gravitational coupling ($\frac{8\pi G_N}{\alpha'}$) at the string scale, where G_N is the Newton's constant and α' is the Regge slope.

Thus one can realize coupling unification without having a GUT-like symmetry below the compactification scale. This is the *new perspective* brought forth by string theory. There is, however, an issue to be resolved. Whereas the MSSM-unification scale, obtained by extrapolation of low energy data is given by $M_X \approx 2 \times 10^{16}$ GeV, the expected one-loop level string-unification scale [25] of $M_{st} \approx g_{st} \times (5.2 \times 10^{17} GeV) \approx 3.6 \times 10^{17}$ GeV is about twenty times higher. Here, one has used $\alpha_{st} \approx \alpha_{GUT}$ (MSSM) ≈ 0.04.

Possible resolutions of this mismatch between M_X and M_{st} by about a factor of 20 have been proposed (for a comprehensive review see e.g. Refs. [13] and [14]). These include: (i) utilizing the idea of *string-duality* that allows a lowering of M_{st} [26] compared to the value suggested by Ref. [25]; alternatively

(ii) the idea of a *semi-perturbative unification*, that assumes the existence of two vector-like families at the TeV-scale, $(16 + \bar{16})$ which raise α_{GUT} to about 0.25 - 0.3, and therebt also M_X to a few $\times 10^{17}$ GeV [27]; or (iii) the alternative of a *string-GUT solution*, which would arise if superstrings yield an intact grand unification symmetry like SU(5) of SO(10), together with supersymmetry *and* the right spectrum - i.e. three chiral families and a suitable Higgs system - at M_{st}, and if the symmetry would break spontaneously at $M_X \sim 1/20 M_{st}$ to the standard model symmetry. In this last case, the gauge couplings would run together between M_X and M_{st} and thus the question of a mismatch between the two scales would not even arise. However, as yet, there does not seem to be even a semi-realistic string-derived GUT model [28]. Further, to- date, no string-GUT solution exists with a resolution of the well-known doublet-splitting problem, without which one would face the problem of rapid proton decay through the d=5 operators [11] (see discussions below). This does not necessarily mean that a realistic GUT- solution exhibiting doublet-triplet splitting can not ultimately emerge from the string or the M-theory.

While each of the solutions mentioned above possesses a certain degree of plausibility (see Ref. [13]) for some additional possibilities), it is not clear, which, if any is utilized by the true string vacuum. This is related to the fact that, as yet, there is unfortunately no insight as to how the true vacuum is selected in the string or in the M-theory.

2.5 A GUT or a Non-GUT String-Solution?

Comparing string-derived GUT solutions with non-GUT solutions, where the former yield symmetries like SU(5) or SO(10), while the latter lead to symmetries like G(2113) or G(224) at the string scale, we see from the discussions above that each class has a certain advantage, and possible disadvantages as well, compared to the other. In particular, a string- GUT solution has the positive feature, explained above, that the issue of a mismatch between M_{st} and M_X does not arise for such a solution. For a non-GUT solution, however, although plausible mechanisms of the type mentioned above could remove the mismatch, a priori it is not clear whether any such mechanism is realized.

On the other hand, for a string-dervied GUT-solution [28], achieving doublet-triplet splitting so as to avoid rapid proton decay, is still a major burden. In this regard, the non-GUT solutions possess a distinct advantage, because the dangerous color triplets are often naturally projected out [29,30]. Furthermore, these solutions invariably possess new "flavor" gauge symmetries, which are not available in GUTS. The flavor symmetries turn out to be immensely helpful in (a) providing the desired protection against gravity-induced rapid proton decay [31], (b) resolving certain naturalness problems of supersymmetry such as those pertaining to the issues of squark-degeneracy, neutrino-Higgsino mixing and CP violation (see e.g. Refs. [32,33,34]), and (c) explaining qualitatively the observed fermion mass-hierarchy [see e.g. Ref. 29].

Weighing the advantages and possible disadvantages of both, it seems hard at present to make a clear choice between a GUT verus a non-GUT string solution. We will therefore keep our options open and look for other means, for example certain features of proton decay and neutrino- masses, to provide a distinction. We will thus proceed by assuming that for a GUT-solution, string theory will somehow provide a resolution of the problem of the doublet-triplet splitting, while for a non-GUT string- solution, we will assume that one of the mechanisms mentioned above (for instance, that based on string-duality [26]), does materialize, removing the mismatch between M_X and M_{st}. In general, a combination of the two mechanisms [26,27] may also play a role.

It turns out that there are many similarities between the predictions of SO(10) and of a string-derived G(224), especially as regards neutrino and charged fermion masses, primarily because both contain SU(4)-color.

With these discussions on higher unification, including the ideas of super-symmetry and superstrings, to serve as a background. I proceed to discuss more concretely, in the next three sections, the masses and mixings of all fermions (in Secs. 3 and 4) and finally their link to proton decay (in Sec. 5). First, an estimate of m_{ν_τ} is presented in the next section.

3 Mass of ν_τ: An Evidence In Favor of the G(224) Route

One can now obtain an estimate for the mass of ν_L^τ in the context of G(224) or SO(10) by using the following three steps [2]:

(i) First, assume that B-L and I_{3R}, contained in a string-derived G(224) or SO(10), break near the unification-scale:

$$\mathrm{M_X} \sim 2 \times 10^{16}\mathrm{GeV}, \tag{4}$$

through VEVs of Higgs multiplets of the type suggested by string-solutions [35] — i.e. $< (1,2,4)_H >$ for G(224) or $< \overline{16}_H >$ for SO(10), as opposed to 126_H. In the process, the RH neutrinos (ν_R^i), which are singlets of the standard model, can and generically will acquire superheavy Majorana masses of the type $M_R^{ij}\nu_R^{iT}C^{-1}\nu_R^j$, by utilizing the VEV of $< \overline{16}_H >$ and effective couplings of the form:

$$\mathcal{L}_{\rm M}({\rm SO(10)}) = {\rm f_R^{ij} 16_i \cdot 16_j \overline{16}_H \cdot \overline{16}_H/M + hc} \tag{5}$$

A similar expression holds for G(224). Here i,j=1,2,3, correspond respectively to e, μ and τ families. Such gauge-invariant non-renormalizable couplings might be expected to be induced by Planck-scale physics, involving quantum gravity or stringy effects and/or tree-level exchange of superheavy states, such as those in the string tower. With f_{ij} (at least the largest among them) being of order unity, we would thus expect M to lie between $M_{Planck} \approx 2 \times 10^{18}$ GeV and $M_{string} \approx 4 \times 10^{7}$ GeV. Ignoring for the present off-diagonal mixings (for simplicity), one thus obtains [b]:

$$M_{3R} \approx \frac{f_{33} < \bar{16}_H >^2}{M} \approx f_{33}(2 \times 10^{14} GeV)\eta^2 (M_{Planck}/M) \tag{6}$$

This is the Majorana mass of the RH tau neturino. Guided by the value of M_X we have substituted $< \bar{16}_H >= (2 \times 10^{16} GeV)\eta$ where $\eta \approx 1/2$ to 2, for this estimate.

(ii) Second, assume that the effective gauge symmetry below the string-scale contains SU(4)-color. Now using SU(4)-color and the Higgs multiplet $(2, 2, 1)_H$ of G(224) or equivalently 10_H of SO(10), one obtains the relation $m_\tau(M_X) = m_b(M_X)$, which is known to be successful. Thus, there is a good reason to believe that the third family gets its masses primarily from the 10_H or equivalently $(2, 2, 1)_H$. In turn, this implies:

$$\rm m(\nu^\tau_{Dirac}) \approx m_{top}(M_X) \approx (100 - 120) GeV \tag{7}$$

Note that this relationship between the Dirac mass of the tau-neutrino and the top-mass is special to SU(4)-color. It does not emerge in SU(5).

(iii) Given the superheavy Majorana masses of the RH neutrinos as well as the Dirac masses, as above, the see-saw mechanism [36] yields naturally light masses for the LH neutrinos. For ν^τ_L (ignoring mixing), one thus obtains, using eqs. (6) and (7),

$$\rm m(\nu^\tau_L) \approx \frac{m(\nu^\tau_{Dirac})^2}{M_{3R}} \approx [(1/20) eV (1 to 1.44)/f_{33}\eta^2](M/M_{Planck}) \tag{8}$$

Considering that on the basis of the see-saw mechanism, we naturally expect that $m(\nu^e_L) \ll m(\nu^\mu_L) \ll m(\nu^\tau_L)$, and assuming that the SuperK observation represents $\nu^\mu_L - \nu^\tau_L$ (rather than $\nu^\mu_L - \nu_X$) oscillation, so that the

[b] The effects of neutrino-mixing and of possible choice of $M = M_{string} \approx 4 \times 10^{17}$ GeV (instead of $M = M_{Planck}$) on M_{3R} are considered in Ref. 18 (secs. 5 and 6) and here, briefly, in sec. 4

observed $\delta m^2 \approx 1/2(10^{-2} - 10^{-3})eV^2$ corresponds to $m(\nu_L^\tau)_{obs} \approx$ (1/15 to 1/40) eV, it seems *truly remarkable* that the expected magnitude of $m(\nu_L^\tau)$, given by eq.(8), is just about what is observed, if $f_{33}\eta^2(M_{Planck}/M) \approx 1.3$ to 1/2. Such a range for the value of $f_{33}\eta^2(M_{Planck}/M)$ seems most plausible and natural (see discussion in Ref. [2]). It should be stressed that the estimate (10) utilizes the ideas of both supersymmetric unification, which yields the scale of M_{3R} (eq.(6)), and of SU(4)-color that yields $m(\nu_{Dirac}^\tau)$ (eq.(7)). The agreement between the expected and the SuperK result thus suggests that, *at a deeper level, near the string or the coupling unification scale M_X, the symmetry group G(224) and thus the ideas of SU(4)-color and left-right symmetry are likely to be relevant to nature.*

By providing clear support for G(224), the SK result selects out SO(10) or E_6 as the underlying grand unification symmetry, rather than SU(5). Either SO(10) or E_6 or both of these symmetries ought to be relevant at some scale, and in the string context, as discussed in Sec. 2, that may well be in higher dimensions, above the compactification-scale, below which there need be no more than just the G(224)-symmetry. If, on the other hand, SU(5) were regarded as a fundamental symmetry, first, there would be no compelling reason, based on symmetry alone, to introduce a ν_R, because it is a singlet of SU(5). Second, even if one did introduce ν_R^i by hand, their Dirac masses, arising from the coupling $h^i\bar{5}_i < 5_H > \nu_R^i$, would be unrelated to the up-flavor masses and thus rather arbitrary (contrast with eq. (7)). So also would be the Majorana masses of the ν_R^i's, which are SU(5)-invariant and thus can even be of order Planck scale (contrast with Eq. (6)). This would give $m(\nu_L^\tau)$ in gross conflict with the observed value. In this sense, the SK result appears to disfavor SU(5) as a fundamental symmetry, with or without supersymmetry.

4 Fermion Masses and Neutrino Oscillations in SO(10)

4.1 Preliminaries

I now discuss the masses and mixings of the quarks and charged leptons in conjunction with those of the neutrinos, to see first of all how well they can be understood *together* within the ideas of higher unification.

The most striking regularity in the masses of the fermions belonging to the three families (at least of the charged ones) is their inter-family hierarchy. This is reflected by the uniform pattern: $m_t \gg m_c \gg m_u$; $m_b \gg m_s \gg m_d$; and $m_\tau \gg m_\mu \gg m_e$. Apart from this gross feature, however, if one examines the pattern in more detail, it looks rather bizarre, especially when one compares intra- family mass-splittings of the three families. For instance, while

$m_t^\circ/m_b^\circ \sim 60$, one finds that $m_c^\circ/m_s^\circ \sim 10$ and $m_u^\circ/m_d^\circ \sim 1/2$. Here, the superscript zero denotes that the respective mass is evaluated at the unification scale. Note that the ratio of the up and down-flavor masses within a family varies widely in going from the third to the second to the first family. Further, comparing quark versus lepton masses of the down-flavor within a family in contrast to $m_b^\circ \approx m_\tau^\circ$, that suggests $b-\tau$ unification for the third family, one finds: $m_s^\circ \sim m_\mu^\circ/3$ and $m_d^\circ \sim 3m_e^\circ$ [37]. In short, there does not seeem to be any obvious regularity in the intra-family mass-splittings. The question is: do these apparent irregularities may still have a simple origin.

The pattern seems to be equally bizarre when one examines the mixing angles. While the parameter $V_{us} = \theta_c$, representing the mixing between the electron and the muon-families in the quark-sector, is moderately large (≈ 0.21), the parameter V_{cb}, representing $\mu-\tau$ family mixing, also in the quark-sector, is small (≈ 0.04). This feature seems even more strange, when one compares V_{cb} with the $\nu_\mu - \nu_\tau$ oscillation angle, which also represents $\mu - \tau$ family mixing, although in the leptonic sector. This angle seems to be almost maximal: $\sin^2 2\theta^{osc}_{\nu_\mu\nu_\tau} > 0.83$ (see Sec. 3). One might have been tempted to associate such a large mixing angle with near degeneracy of ν_μ and ν_τ, as has been attempted by several authors. But, then, such degeneracy does not go well with the see-saw formula, especially within a unified scheme in which the Dirac masses of the neutrinos are related to those of the quarks, which exhibit a large inter-family hierarchy. Thus one major puzzle is: *Why V_{bc} is so small and yet $\theta^{osc}_{\nu_\mu\nu_\tau}$ so large?* could the smallness of one imply the largeness of the other within a quark-lepton unified theory? Further, are these peculiarities of the mixing angles related to the irregularities in the intra-family mass-splittings mentioned above.

From a theoretical viewpoint, the goal is to resolve some of these puzzles within a unified predictive theory, in particular to understand the masses and mixings of the neutrinos *in conjunction with* those of the quarks and the charged leptons, rather than in isolation. It is, however, known that there is no obvious way to address any of these puzzles in the context of the standard model (SM), because, apriori, the SM allows for all the masses and mixings to be arbitrary parameters. Even ignoring CP violation for the present discussion, there are 12 such observables: $m_t, m_b, m_\tau, m_c, m_s, m_\mu, m_u, m_d, m_\ell, V_{us}, V_{cb}$ and V_{ub}. The 3×3 mass-matrices of the 3 sectors (up, down and charged lepton) would in general have as many as $9 \times 3 = 27$ real parameters, which represent, however, only 12 observables. The parameters would even increase if one introduces RH neutrinos and considers both the Dirac *and* the Majorana mass-matrices of the three neutrinos.

To reduce the number of parameters, it thus seems that one may have to appeal to *symmetries of two kinds*: first like those in G(224) or SO(10), which relate quark versus lepton as well as up-versus-down Yukawa couplings, and second "flavor" symmetries which distinguish between the three families (e, μ, and τ) and could account for inter-family mass hierarchy. Interestingly enough, these latter symmetries do seem to arise in string-solutions (see e.g. [29], [30]), though not in GUT's. To proceed further, we will use the following guidelines.

1) Hierarchy Through Off-Diagonal Mixings: Recall earlier attempts [38] that attribute hierarchy in the quark mass-matrices of the first two families to matrices of the type:

$$M = \begin{pmatrix} 0 & \epsilon \\ \epsilon & 1 \end{pmatrix} m_s^{(o)} , \qquad (9)$$

for the (d,s) quarks, and likewise for the (u,c)-quarks. Here $\epsilon \sim 1/10$. Note the symmetric form of (9) (i.e. $M_{12} = M_{21}$) and especially the hierarchical pattern: $(1,1) \ll (1,2) \ll (2,2)$, where $(1,1) \leq O(\epsilon^3)$. The symmetric nature of (9) is guaranteed by group theory if the relevant Higgs field is a **10** of SO(10). The hierarchical entries in (9) can be ensured by imposing a suitable flavor symmetry, that distinguishes between the two families (Origin of such symmetries must ultimately be attributed to, for example, string theory). The pattern (9) has the virtues that (a) it generates a hierarchy larger than the input parameter ϵ : $|\frac{m_d}{m_s}| \approx \epsilon^2 \ll \epsilon$, and (b) it leads to the rather successful expression for the cabibbo angle:

$$\theta_C \simeq \left| \sqrt{\frac{m_d}{m_s}} - e^{i\phi} \sqrt{\frac{m_u}{m_c}} \right| . \qquad (10)$$

Using $\sqrt{m_d/m_s} \simeq 0.22$ and $\sqrt{m_u/m_c} \simeq 0.06$, we see that Eq. (7) works within 30% for any value of the phase ϕ, and perfectly for a value of the phase parameter ϕ around $\pi/2$.

A generalization of the pattern (9) to the case of three families would suggest that the first and the second families (i.e. the e and the μ-families) receive their masses primarily through their mixings with the third family (τ); the (3,3)-element in this case is then the leading one in each sector. One must also rely on flavor symmetries, that distinguish between the e, μ and τ-families, so as to ensure that the (1,3) and (1,2) mixing elements are smaller than the (2,3)-element. We will follow this guideline, except, however, for the modification noted below.

2) The Need for an Antisymmetric Component: Although the symmetric hierarchical mass-matrix (9) works well for the first two families, a matrix of the same form fails altogether to reproduce V_{cb}, for which it would yield:

$$|V_{cb}| \simeq \left| \sqrt{\frac{m_c}{m_t}} - e^{i\chi} \sqrt{\frac{m_s}{m_b}} \right| , \tag{11}$$

Given that $\sqrt{m_s/m_b} \simeq 0.17$ and $\sqrt{m_c/m_t} \simeq 0.06$, we see that eq. (11) would yield $|V_{cb}|$ varying between 0.11 and 0.23, depending upon the value of the phase χ. This is however, too big, compared to the observed value of $V_{cb} \approx 0.04 \pm 0.003$, by at least a factor of 3. We thus see that the simple square root formula for the mixing angle in each sector ($\sin\theta_{ij} \approx \tan\theta_{ij} = \sqrt{\frac{m_i}{m_j}}$, see eq. (10) or (11)), arising from a symmetric matrix of the form (9), fails for V_{cb}. We would interpret this failure as a *clue* to the presence of *antisymmetric contribution* to off-diagonal mixing in the mass-matrix together with a symmetric one, (thus $m_{ij} \neq m_{ji}$) which would modify the square-root formula for the mixing angle to $\sqrt{\frac{m_i}{m_j}}\sqrt{\frac{m_{ij}}{m_{ji}}}$, where m_i and m_j denote the respective eigenvalues. We will note below a simple group theoretical origin of such an antisymmetric component in SO(10), even for a minimal Higgs system, and point out its crucial role in resolvinng some of the puzzles, alluded to above. The resolution would depend, however, on an additional feature, noted below.

3) The Need for a Contribution Proportional to B-L: The success of the relations $m_b^\circ \approx m_t^\circ$, and also $m_\tau^\circ \approx m(\nu_\tau)^\circ_{Dirac}$ (see sec. 3) suggests that members of the third family receive their masses primarily from the VEV of a Higgs field, which is a singlet of SU(4)-color, and thus independent of B-L. That is in fact the case for the Higgs transforming as (2,2,1) of G(2,2,4) or **10** of SO(10). However, the empirical observations of $m_s^\circ \sim m_\mu^\circ/3$ and $m_d^\circ \sim 3m_e^\circ$, as well as the suppression of V_{bc} (noted above) together with the enhancement of $\theta^{osc}_{\nu_\mu\nu_\tau}$ (SuperK-result) clearly calls for a contribution proportional to B-L as well. This would be the case for contributions from the VEV of a Higgs transforming as 15 of SU(4)-color. We note below how such a contribution can arise simply, for a minimal Higgs system in SO(10). The amusing thing is that such a contribution, while it is proportional to B-L, turns out to be anti-symmetric as well, in the family-space, fulfilling the need (2).

I now present, following Ref. [18], a simple and predictive mass-matrix, based on SO(10), which is constructed by using the guidelines (1) - (3). For simplicity, I first consider only the μ and the τ-families. The discussion is extended later to include the electron family.

4.2 The Minimal Higgs System for SO(10)-Breaking and Fermion Masses

The minimal Higgs system, capable of breaking SO(10) at the unification scale M_X into the SM symmetry G(213), consists of a $\mathbf{45_H}$, a $\mathbf{16_H}$ and (for supersymmetry) a $\overline{\mathbf{16}}_\mathbf{H}$. Of these $< 45_H > \sim M_X$ breaks SO(10) into $G(2213) = SU(2)_L \times SU(2)_R \times U(1)_{B-L} \times SU(3)^C$, while $< 16_H > = < \overline{16}_H > \sim M_X$ breaks $SU(2)_R$ and B-L and thus G(2213) into G(213). To break G(213) into $U(1)_{em} \times SU(3)^C$ at the electroweak scale, one minimally needs in addition the VEV of a $\mathbf{10_H}$. Thus the minimal Higgs system, that is needed for appropriate SO(10)-breaking, consists of the set:

$$H_{minimal} = \{45_H, 16_H, \overline{16}_H, 10_H\} \tag{12}$$

Of these, only 10_H can have Yukawa coupling with the fermions at the cubic level of the form $h_{ij} 16_i 16_j 10_H$, which could be the dominant source of mases, especially for fermions belonging to the third family. But, the first two families must have additional sources for their masses, because a $< 10_H >$, by itself, would lead to three undesirable results: (a) $V_{CKM} = 1$, (b) purely symmetric mass-matrices, and (c) (B-L)- independent masses. We have on the other hand argued above that antisymmetric and (B-L) dependent contributions to mass-matrices are needed.

Now, there exist large-dimensional tensorial multiplets of SO(10), that is 126_H and 120_H, which can have cubic-level Yukawa couplings with the fermions and can give (B-L)-dependent contributions. Further $< 120_H >$, gives purely family-antisymmetric contributions, as needed. There are, however, two apriori reasons why we prefer not to use these large-dimensional multiplets: (a) They seem to be hard, if not impossible, to emerge from string solutions [35], and (b) generically, such large-dimensional multiplets tend to give large threshold corrections (typically exceeding 20 %) to $\alpha_3(m_Z)$, thereby rendering observed coupling unification fortuitous. By contrast, the multiplets in the minimal set can arise in string soutions leading to SO(10) ((45_H) arises at kac-Moody level ≥ 2, while 16_H, $\overline{16}_H$ and 10_H arise at level 1), and their threshold corrections have been computed. They were found not only to be smaller in magnitude, but also to have the right sign to go well with observed coupling unification [18].

Given these advantages of the minimal Higgs system (compared to those containing large multiplets like 126_H and/or 120_H) for SO(10)- breaking, the question arises: can this minimal system meet the requirements arising from fermion masses and mixings – that is (a) $V_{CKM} \neq 1$, (b) presence of antisymmetric, and (c) that of (B-L)- dependent contributions? It was noted in Ref. [18] that the minimal Higgs system can indeed meet all three requirements quite simply, if one allows for not just cubic, but also (seemingly) non-renormalizable effective quartic couplings of this minimal set with the 16-plets of fermions. Such quartic couplings could well arise through exchanges of superheavy particles (for example those in the string-tower) involving renormalizable couplings, and/or through quantum gravity.

Allowing for such cubic and quartic couplings of the minimal Higgs sytem and adopting the guideline (1) of family hierarchical couplings, we are led to suggest the following effective Lagrangian for generating masses and mixings of the μ and the τ-families [18]. (The same consideration is extended later to include the electron family). (For a related but different pattern, see Ref. [39]).

$$\mathcal{L}_{\text{Yukawa}} = h_{33}\mathbf{16_3 16_3 10_H} + \frac{a_{23}}{M}\mathbf{16_2 16_3 10_H 45_H} + \frac{g_{23}}{M}\mathbf{16_2 16_3 16_H 16_H} + h_{23}\mathbf{16_2 16_3 10_H} \,. \tag{13}$$

Note that a mass-matrix of the type shown in eq. (9) (barring its symmetric form) results if the first term $h_{33} < 10_H >$ is dominant. This ensures $m_b^\circ \approx m_\tau^\circ$ and $m_t^\circ \approx m(\nu^\tau_{Dirac})^\circ$. The smallness of the remaining terms responsible for off-diagonal mixings, by about an order of magnitude compared to the h_{33}-term, may come about as follows. First, as mentioned before, the smallness of the SO(10)-invariant coupling $h_{22}16_2 16_2 10_H$ (not shown) compared to the h_{23}-coupling and that of h_{23} compared to h_{33} (i.e. $h_{22} \ll h_{23} \ll h_{33}$) may well have its origin in a flavor symmetry (or symmetries), which assigns *different* charges to the three different families, and also to the Higgs-like fields. In this case, assuming that the h_{33}-term is allowed by the flavor symmetries and that the second and the third families have different flavor-charges, the h_{23}-term will not be allowed as a genuine cubic coupling. It can still arise effectively by utilizing an effective non-renormalizable coupling $\hat{h}_{23}16_2 16_3 10_H < S > /M$ where S is an SO(10)-singlet carrying appropriate flavor charge(s), and acquires a VEV $\sim M_U$. In this case, $h_{23}(= \hat{h}_{23} < S > /M)$ can naturally be $O(1/10)h_{33}$, if $\hat{h}_{23} \sim h_{33}$ and $< S > /M \sim M_U/M_{st} \sim 1/10$. The h_{22} term would then be suppressed by $(< S > /M)^2 \sim 10^{-2}$, compared to h_{33}, as desired [c]. Now, as regards the effective non-renormalizable terms in eq. (13), assuming that

[c]Although, no explicit string solution with the hierarchy in h_{ij} mentioned above, together with the a_{23} and g_{23} couplings of eq. (13), exists as yet, flavor symmetries of the type alluded to as well as SM-singlets carrying flavor- charge and acquiring VEVs of order M_U do emerge generically in string solutons. And, there exist solutions with the top Yukawa coupling being leading (see e.g. Refs. [29] and [16]).

they are generated by quantum gravity or stringy effects and/or by tree-level exchanges of superheavy states (see e.g. those in the string-tower), the scale M is naturally expected to be of order $M_{st} \sim$ few $\times 10^{17}$ GeV, while $< 45_H > \sim < 16_H > \sim M_U \sim$ few $\times 10^{16}$ GeV . Thus effectively $a_{23} < 45_H > /M$ and $g_{23} < 16_H > /M$ could quite plausibly be of order $h_{33}/10$.

It is interesting to observe the symmetry properties of the a_{23} and g_{23}-terms. Although $10_H \times 45_H = 10 + 120 + 320$, given that $< 45_H >$ is along B-L, which is needed to implement doublet-triplet splitting (see sec. 5), only 120 in the decomposition contributes to the mass-matrices. This contribution is, however, antisymmetric in the family-index and, at the same time, proportional to B-L. *Thus the a_{23} term fulfills the requirements of both (2) and (3) simultaneously.* With only h_{ij} and a_{ij}-terms, however, the up and down quark mass-matrices will be proportional to each other, which would yield $V_{CKM} = 1$. This is remedied by the g_{ij} coupling as follows. While the 16_H has a VEV, primarily along its SM singlet component (transforming as $\tilde{\bar{\nu}}_R$) which is of order M_U, it also has an electroweak down-type Higgs doublet (but not up- type) – call it 16_d – which can acquire a VEV of the electroweak scale by its mixing with the corresponding doublet – call it 10_d in the 10_H. The MSSM doublet H_d, which is light, is then a mixture of 10_d and 16_d, while the orthogonal combination is superheavy. This feature is discussed in more detail in conjunction with the mechanism for doublet-triplet splitting in Ref. [18] and also briefly in the following section. With $< 16_d >$ contributing only to the down- flavor mass matrices, but not to the up-flavor, the g_{23} coupling generates non-trivial CKM-mixing. *We thus see that the minimal Higgs system satisfies apriori all the qualitative requirements (1) - (3), and also the condition $V_{CKM} \neq 1$.* I now discuss that this system works well even quantitatively.

With these four effective Yukawa couplings, the Dirac mass matrices of quarks and leptons of the second and the third families at the unification scale take the form:

$$U = \begin{pmatrix} 0 & \epsilon + \sigma \\ -\epsilon + \sigma & 1 \end{pmatrix} m_U, \quad D = \begin{pmatrix} 0 & \epsilon + \eta \\ -\epsilon + \eta & 1 \end{pmatrix} m_D,$$
$$N = \begin{pmatrix} 0 & -3\epsilon + \sigma \\ 3\epsilon + \sigma & 1 \end{pmatrix} m_U, \quad L = \begin{pmatrix} 0 & -3\epsilon + \eta \\ 3\epsilon + \eta & 1 \end{pmatrix} m_D \,. \quad (14)$$

Here the matrices are multiplied by left–handed fermion fields from the left and by anti–fermion fields from the right. (U, D) stand for the mass matrices of up and down quarks, while (N, L) are the Dirac mass matrices of the neutrinos and the charged leptons.

The entries $(1, \epsilon, \sigma)$ arise respectively from the h_{33}, a_{23} and h_{23} terms in Eq. (13), while η entering into D and L receives contributions from both g_{23} and h_{23}; thus $\eta \neq \sigma$. Note the quark–lepton correlations between (U, N) as well as (D, L), and the up–down correlation between (U, D) as well as (N, L). These correlations arise because of the symmetry structure of G(224). The relative factor of -3 between quarks and leptons involving the ϵ entry reflects the fact that $\langle \mathbf{45_H} \rangle \propto (B - L)$, while the antisymmetry in this entry arises from the $SO(10)$ structure as explained above.

Assuming $\epsilon, \eta, \sigma \ll 1$, we obtain at the unification scale:

$$\left|\frac{m_c}{m_t}\right| \simeq |\epsilon^2 - \sigma^2|, |\frac{m_s}{m_b}| \simeq |\epsilon^2 - \eta^2| \, ,$$

$$\left|\frac{m_\mu}{m_\tau}\right| \simeq |9\epsilon^2 - \eta^2|, \quad |m_b| \simeq |m_\tau||1 - 8\epsilon^2|, \tag{15}$$

$$|V_{cb}| \simeq |\sigma - \eta| \approx \left|\sqrt{m_s/m_b}\left(\frac{\eta + \epsilon}{\eta - \epsilon}\right)^{1/2} - \sqrt{m_c/m_t}\left(\frac{\sigma + \epsilon}{\sigma - \epsilon}\right)^{1/2}\right| \tag{16}$$

$$\theta^{\ell}_{\mu\tau} \approx -3\epsilon + \eta \approx \sqrt{m_\mu/m_\tau}\left(\frac{-3\epsilon + \eta}{3\epsilon + \eta}\right)^{1/2} . \tag{17}$$

The relations in eqs.(15) and (16) lead to two sum rules:

$$\left|\frac{m_b}{m_\tau}\right| \simeq \left|1 - 8\{\left|\frac{m_\mu}{m_\tau}\right| - \left|\frac{m_s}{m_b}\right|\}\right|$$

$$\frac{m_s}{m_b} \simeq \frac{m_c}{m_t} - \frac{5}{4}V_{cb}^2 \pm V_{cb}\left[\frac{9}{16}V_{cb}^2 + \frac{1}{2}\frac{m_\mu}{m_\tau} - \frac{9}{2}\frac{m_c}{m_t}\right]^{1/2} \tag{18}$$

The superscript zero, meaning unification-scale values, is not exhibited, but should be understood in all the relations in eqs. (15) - (18).

The mass matrices in eq. (14) contain 5 parameters [d]: ϵ, σ, η, $m_D = h_{33} < 10_d >$ and $m_U = h_{33} < 10_U >$. These may be determined by using, for example, the following input values: $m_t^{phys} = 174 GeV$, $m_c(m_c) = 1.37$ GeV, $m_s(1GeV) = 110 - 116 MeV$ and the observed masses of μ and τ. While the input value of m_s is somewhat lower than that advocated in Ref. [40], it is in good agreement with recent lattic calculations [41]. With these input values, the parameters are found to be:

$$\sigma \simeq -0.110 \; \eta_{cb} \; , \; \eta \simeq -0.151 \; \eta_{cb} \; , \; \epsilon \simeq 0.095 \; \eta_\epsilon \; ,$$

[d]Of these, $m_U^0 \approx m_t^0$ can in fact be estimated to within 20% accuracy by either using the argument of radiative electroweak symmetry breaking or in the context of a class of string solutions (see e.g. Ref. [29]).

$$m_U \simeq m_t(M_U) \simeq (100 - 120)\ \text{GeV}\ ,\ m_D \simeq m_b(M_U) \simeq 1.5\ \text{GeV}\ . \quad (19)$$

Here η_ϵ and η_{cb} denote the phases of ϵ and V_{cb} respectively (i.e. $\epsilon = \eta_\epsilon|\epsilon|$ etc.). We assume for simplicity that they are real (barring phase angles of $\pm 10°$). Thus, $\eta_\epsilon = \pm 1$ and $\eta_{cb} = \pm 1$. The relative signs of σ, η and ϵ get fixed by ensuring that the results are optimized as regards their agreeement with observation. This yields $\eta_{cb} = \eta_\epsilon$. Note that in accord with our general expectations discussed above, each of these parameters are found to be of order 1/10, as opposed to being [e] O(1) or O(10^{-2}), compared to the leading (3,3)- element. Having determined these parameters, one can now obtain the following predictions:

$$\begin{aligned} m_b\,(m_b) &\simeq (4.6 - 4.9)\,GeV; V_{cb} \simeq 0.045 && (20)\\ m^D_{\nu_\tau}(M_U) &\simeq m_t(M_U) \simeq 100 - 120\ \text{GeV}\ , \\ m^D_{\nu_\mu}(M_U) &\simeq (9\epsilon^2 - \sigma^2) m_U \simeq 8\ \text{GeV}\ , \\ \theta^\ell_{\mu\tau} &\simeq -3\epsilon + \eta \simeq -0.437\eta_\epsilon\ (\text{for}\ \eta_{\text{cb}}/\eta_\epsilon = +1)\ . && (21) \end{aligned}$$

In quoting the numbers in Eq. (20), we have extrapolated the GUT scale values down to low energies using the beta functions of the minimal supersymmetric extension of the Standard Model (MSSM), assuming $\alpha_s(M_Z) = 0.118$, an effective SUSY threshold of 500 GeV and tan$\beta = 5$. Our results depend only weakly on these input choices, so long as $\tan\beta$ is neither too large (≥ 30) nor too small (≤ 2). The first two of the predictions listed above (eq. (20)) correspond to directly observed entities. The last three (eq. (21)) cannot be observed directly, but they are important because they need to be combined with the Majorana masses of the RH neutrinos to yield observable entities (see below).

Given the bizarre pattern of quark and lepton masses and mixings, it seems remarkable that the simple pattern of fermion mass-matrices, motivated by the group theory of G(224)/SO(10), gives an overall fit to all of them which is good to within 10%. This includes the two successful predictions on m_b and V_{cb} (eq.(20)). It is worth noting that, in supersymmetric unified theories, the "observed" value of $m_b(m_b)$ and renormalization-group studies suggest that, for a wide range of the parameter $\tan\beta$, m_b° should in fact be about 10 - 20 % *lower* than m_τ° [42]. This is neatly explained by the relation: $m_b^\circ \approx m_\tau^\circ(1 - 8\epsilon^2)$ (eq. (15)), where exact equality holds in the limit $\epsilon \to 0$ (due to SU(4)-color), while the decrease by $8\epsilon^2 \sim 10\%$ is precisely because the off-diagonal ϵ-entry is proportional to B-L (see eq. (14)).

[e]This is one characteristic difference between our work and that of Ref. [38], where the (2,3)-element is even bigger than the (3,3).

Specially intriguing is the result on $V_{cb} \approx 0.045$ which compares well with the observed value of $\simeq 0.04$. The suppression of V_{cb}, compared to the value of 0.17 ± 0.06 obtained from eq. (6), is now possible because the mass matrices (eq. (14)) contain an *antisymmetric component* $\propto\epsilon$. Such a component corrects the square-root mixing angle formula $\theta_{sb} = \sqrt{m_s/m_b}$ (appropriate for symmetric matrices of the type given by eq. (9)) by the asymmetry factor $(\frac{\eta+\epsilon}{\eta-\epsilon})^{1/2}$ (see eq. (15)), and similarly for the angle θ_{ct}. This factor suppresses V_{cb} if η and ϵ have opposite signs. The interesting point is that, *the same feature necessarily enhances the corresponding mixing angle* $\theta^{\ell}_{\mu\tau}$ *in the leptonic sector*, since the asymmetry factor in this case is given by $[(-3\epsilon+\eta)/(3\epsilon+\eta)]^{1/2}$ (see eq. (15)). This enhancement of $\theta^{\ell}_{\mu\tau}$ also seems to be borne out by observation in the sense that that is a key factor in accounting for the nearly maximal oscillation angle observed at SuperK (see discussion below). Note that this intriguing correlation between the mixing angles in the quark versus leptonic sectors – *that is suppression of one implying enhancement of the other* – has become possible because the ϵ-contribution is simultaneously antisymmetric and is proportional to B-L. As a result, *it changes sign* as one goes from the quarks to the leptons.

Taking stock, we see an overwhelming set of evidences in favor of B-L and in fact for the full SU(4)-color-symmetry. These include: (i) the suppression of V_{cb}, together with the enhancement of $\theta^{\ell}_{\mu\tau}$, just mentioned above, (ii) the successful relation $m_b^\circ \approx m_\tau^\circ(1-8\epsilon^2)$, where the near equality follows from SU(4)-color, while the decrease of m_b° relative to m_τ° by $8\epsilon^2 \sim 10\%$ is a consequence of the (B-L)-dependence of the off-diagonal ϵ-entry, (iii) the usefulness again of the SU(4)-color-relation $m(\nu^\tau_{Dirac})^\circ \approx m_t^\circ$ in accounting for $m(\nu_L^\tau)$, as discussed in Sec. 3, and (iv) the agreement of the relation $|m_s^\circ/m_\mu^\circ| = |(\epsilon^2-\eta^2)/(9\epsilon^2-\eta^2)|$ with the data, in that the ratio is naturally *less than* 1, if $\eta \sim \epsilon$. The presence of $9\epsilon^2$ in the denominator as opposed to ϵ^2 in the numerator is again a consequence of the *off-diagonal entry being proportional to B-L.* Finally, as mentioned in Sec. 2, a spontaneously broken (B-L)-local symmetry may well be needed to ensure preservation of baryon excess in the presence of electroweak sphaleron effects [21].

Although all the entries for the Dirac mass matrix are now fixed, to obtain the parameters for the light neutrinos one needs to specify the Majorana mass matrix of the RH neutrinos (ν_R^μ and ν_R^τ). For concreteness, we assume that this too has the hierarchical form of eq. (9):

$$M_\nu^R = \begin{pmatrix} 0 & y \\ y & 1 \end{pmatrix} M_R \,, \tag{22}$$

In the spirit of our discussion that flavor symmetries are the origin of hierarchical masses, we will assume that $10^{-2} \ll |y| \leq 1/10$ as opposed to $|y|$ being ≥ 0.3 (say). Apriori, $y = \eta_y |y|$ can have either sign, i.e., $\eta_y = \pm 1$. Note that Majorana mass matrices are constrained to be symmetric by Lorentz invariance. The seesaw mass matrix $(-N(M_\nu^R)^{-1}N^T)$ for the light $(\nu_\mu - \nu_\tau)$ system is then

$$M_\nu^{\text{light}} = \begin{pmatrix} 0 & A \\ A & B \end{pmatrix} \frac{m_U^2}{M_R} \,, \tag{23}$$

where $A \simeq (\sigma^2 - 9\epsilon^2)/y$ and $B \simeq -(\sigma + 3\epsilon)(\sigma + 3\epsilon - 2y)/y^2$. With $A \ll B$, this yields

$$m_{\nu_3} \simeq B\frac{m_U^2}{M_R}; \quad \frac{m_{\nu_2}}{m_{\nu_3}} \simeq -\frac{A^2}{B^2}; \quad \tan\theta^\nu_{\mu\tau} = \sqrt{\frac{m_{\nu_2}}{m_{\nu_3}}} \,. \tag{24}$$

For a given choice of the sign of y relative to that of ϵ, and for a given mass ratio m_{ν_2}/m_{ν_3}, we can now determine y using Eqs. (23) and (24), and the values of ϵ and σ obtained in Eq. (19). Taking $m_{\nu_2}/m_{\nu_3} = (1/10, 1/15, 1/20, 1/25, 1/30)$ [f], the requirement of hierarchy mentioned above - i.e. *10*$^{-2} \ll |y| \leq$ *0.2* (say) - can be satisfied only provided y *is positive relative to* ϵ - i.e. $\eta_y = \eta_\epsilon$; the corresponding values for y are: $y = (0.0543, 0.0500, 0.0468, 0.0444, 0.0424)\eta_\epsilon$. With $\eta_y = \eta_\epsilon = \pm 1$, we obtain for the neutrino oscillation angle:

$$\theta^{osc}_{\nu_\mu\nu_\tau} \simeq \theta^\ell_{\mu\tau} - \theta^\nu_{\mu\tau} \simeq (0.437 + \sqrt{\frac{m_{\nu_2}}{m_{\nu_3}}})(-\eta_\epsilon) \,. \tag{25}$$

$$\begin{aligned} \sin^2 2\theta^{osc}_{\nu_\mu\nu_\tau} &= (0.96, 0.91, 0.86, 0.83, 0.81) \\ \text{for } \; m_{\nu_2}/m_{\nu_3} &= (1/10, 1/15, 1/20, 1/25, 1/30) \,. \end{aligned} \tag{26}$$

Note the interesting point that just the requirement, that $|y|$ should have a natural hierarchical value, leads to $\eta_y = \eta_\epsilon$ and that in turn implies that the two contributions in eq.(25) must *add* rather than subtract, leading to an *almost maximal oscillation angle.* The other factor contributing to the enhancement of $\theta^{osc}_{\nu_\mu\nu_\tau}$ is, of course, also the asymmetry-ratio which increased $|\theta^\ell_{\mu\tau}|$ from 0.25 to 0.437 (see eqs. (17) and (21)). We see that one can derive rather plausibly a large $\nu_\mu - \nu_\tau$ oscillation angle $\sin^2 2\theta^{osc}_{\nu_\mu\nu_\tau} \geq 0.8$, together with an understanding of hierarchical masses and mixings of the quarks and the charged leptons, while maintaining a large hierarchy in the seesaw derived masses $(m_{\nu_2}/m_{\nu_3} = 1/10 - 1/30)$ of ν_μ and ν_τ, all within a unified framework

[f] With $m_{\nu_3} \sim (1/10-1/40)$ eV (see Sec. 3), these values of the ratio m_{ν_2}/m_{ν_3} are suggested by the MSW solution for the solar neutrino puzzle. Aposteriori, they also go well with a hierarchical value of y - i.e. $10^{-2} \ll |y| \leq 1/10$ (see text).

including both quarks and leptons. In the example exhibited here, the mixing angles for the mass eigenstates of neither the neutrinos nor the charged leptons are really large, $\theta^{\ell}_{\mu\tau} \simeq 0.437 \simeq 23^0$ and $\theta^{\nu}_{\mu\tau} \simeq (0.18-0.31) \approx (10-18)^0$, *yet the oscillation angle obtained by combining the two is near-maximal.* This contrasts with most previous work, in which a large oscillation angle is obtained either entirely from the neutrino sector (with nearly degenerate neutrinos) or almost entirely from the charged lepton sector.

It is worth noting that the interplay due to the mixing in the Dirac and the Majorana mass matrices via the see-saw mechanism has the net effect of enhancing $M_R \approx B(m_{\nu_2}/m_{\nu_3})$, for a given m_{ν_3}, precisely by a factor of $|B| \approx 5$ (see eq. (23)), compared to what it would be without mixing. Using $m_U \approx 100$ GeV (see eq. (7) or (19)), $m_{\nu_3} \approx (1/10 - 1/30)$ eV (SuperK-result) and $|B| \approx 5$, one gets:

$$M_R \approx (5 - 15) \times 10^{14} GeV \tag{27}$$

compare this with its counterpart, estimated in eq. (6), which yields $M_{3R} \approx (few) \times 10^{14}$ GeV, for $f_{33}\eta^2 \approx 1$, if $M \approx M_{Planck}$. It is interesting that the larger value of $M_R \approx 10^{15}$ GeV goes well with the theoretical estimate of eq. (6) if the characteristic mass M is chosen (perhaps more appropriately) to be $M_{string} \approx 4 \times 10^{17}$ GeV rather than M_{Planck}. Further, this larger value of M_R also goes well with the observed m_{ν_3}, once one includes the effect of mixing (which was dropped in sec. 3).

4.3 Inclusion of the First Family

The first family may now be included following the spirit of the hierarchical structure shown in eqs. (9) and (14). As mentioned before, this may have its origin in flavor symmetries of a deeper theory. In the absence of such a deeper understanding, however, the theoretical uncertainties in dealing with the masses and mixings of the first family are much greater than for the heavier families, simply because the masses of the first family are so small, that relatively small perturbations can significantly affect their values.

Assuming that flavor symmetries and SO(10) permit the (3,3)-coupling at a genuine cubic level, but the (2,3)-couplings only at the quartic level, which are thus effectively suppressed by about an order of magnitude compared to the (3,3)-element (see discussion following eq. (13)), we would naturally expect that the (1,2) and (1,3)- couplings (e.g. a_{12} and g_{12}, see below) would be suppressed compared to the corresponding (23)-couplings. This in turn would account for the observed interfamily mass-hierarchy.

Following this as a guide, and in the interest of economy, we add only two effective quartic couplings to eq. (13), to include the first family: $a_{12}\mathbf{16}_1\mathbf{16}_2\mathbf{45_H}\mathbf{10_H}/M$ and $g_{12}\mathbf{16}_1\mathbf{16}_2\mathbf{16_H}\mathbf{16_H}/M$. The first coupling introduces an ϵ'-term in the (1,2)-entry, which is antisymmetric and proportional to B-L (analog of ϵ); the second introduces an η'-term in the (1,2)-entry of only D and L, which is symmetric. The resulting 3×3 Dirac mass-matrices are:

$$U = \begin{pmatrix} 0 & \epsilon' & 0 \\ -\epsilon' & 0 & \epsilon+\sigma \\ 0 & -\epsilon+\sigma & 1 \end{pmatrix} m_U, \quad D = \begin{pmatrix} 0 & \epsilon'+\eta' & 0 \\ -\epsilon'+\eta' & 0 & \epsilon+\eta \\ 0 & -\epsilon+\eta & 1 \end{pmatrix} m_D,$$

$$N = \begin{pmatrix} 0 & -3\epsilon' & 0 \\ 3\epsilon' & 0 & -3\epsilon+\sigma \\ 0 & 3\epsilon+\sigma & 1 \end{pmatrix} m_U, \quad L = \begin{pmatrix} 0 & -3\epsilon'+\eta' & 0 \\ 3\epsilon'+\eta' & 0 & -3\epsilon+\eta \\ 0 & 3\epsilon+\eta & 1 \end{pmatrix} m_D. \tag{28}$$

With $\epsilon, \sigma, \eta, m_U$ and m_D determined essentially by considerations of the second and the third families (Eq. (19)), we now have just two new parameters in Eq. (28) – i.e. ϵ' and η' – which describe five *new* observables in the quark and charged lepton sector: m_u, m_d, m_e, θ_C and V_{ub}. Thus with $m_u \approx 1.5$ MeV (at $M_{\rm U}$) and m_e/m_μ taken as inputs one obtains: $\epsilon' \simeq \sqrt{m_u/m_c}(m_c/m_t) \approx 2 \times 10^{-4}$ and $|\eta'| \simeq \sqrt{m_e/m_\mu}(m_\mu/m_\tau) \simeq 4.4 \times 10^{-3}$. We can now calculate m_d, θ_c and V_{ub}. Combining the two predictions for the second and the third families, obtained before (see Eq. (19)), we are thus led to a total of five predictions for the observable parameters of the quarks and charged leptons belonging to the three families.

$$\begin{aligned} m_b(m_b) &\simeq (4.6 - 4.9)\ \text{GeV} \\ V_{cb} &\simeq 0.045 \\ m_d(1\ \text{GeV}) &\simeq 8\ \text{MeV} \\ \theta_C &\simeq \left|\sqrt{m_d/m_s} - e^{i\phi}\sqrt{m_u/m_c}\right| \\ |V_{ub}/V_{cb}| &\simeq \sqrt{m_u/m_c} \simeq 0.07\,. \end{aligned} \tag{29}$$

Further, the Dirac masses and mixings of the neutrinos and the mixings of the charged leptons also get determined. Including those for the $\mu - \tau$ families listed in Eq. (21) we obtain:

$$\begin{aligned} m^D_{\nu_\tau}(M_{\rm U}) &\approx 100 - 120\ \text{GeV};\ m^D_{\nu_\mu}(M_{\rm U}) \simeq 8\ \text{GeV},\ \theta^\ell_{\mu\tau} \simeq -.0.437\eta_\epsilon \\ m^D_{\nu_e} &\simeq [9\epsilon'^2/(9\epsilon^2 - \sigma^2)]m_U \simeq 0.4\ \text{MeV} \\ \theta^\ell_{e\mu} &\simeq \left[\frac{\eta' - 3\epsilon'}{\eta' + 3\epsilon'}\right]^{1/2} \sqrt{m_e/m_\mu} \simeq 0.85\sqrt{m_e/m_\mu} \simeq 0.06 \end{aligned}$$

$$\theta^{\ell}_{e\tau} \simeq \frac{1}{.85}\sqrt{m_e/m_\tau}(m_\mu/m_\tau) \simeq 0.0012 \, . \tag{30}$$

In evaluating $\theta^{\ell}_{e\mu}$, we have assumed ϵ' and η' to be relatively positive.

Note that the first five predictions in Eq. (29) pertaining to observed parameters in the quark system are fairly successful. Considering the bizarre pattern of the masses and mixings of the fermions in the three families (recall comments on V_{cb}, m_b/m_τ, m_s/m_μ and m_d/m_e), we feel that the success of the mass pattern exhibited by Eq. (28) is rather remarkable. This is one reason for taking patterns like Eq. (28) seriously as a guide for considerations on proton decay. A particularly interesting variant is obtained in the limit $\epsilon' \to 0$, as I will mention at the end of this section.

To obtain some guidelines for the neutrino system involving ν_e, we need to extend the Majorana mass matrix of eq. (22), by including entries for ν^e_R. Guided by economy and the assumption of hierarchy, as in eq. (9), we consider the following pattern:

$$M^R_\nu = \begin{pmatrix} x & 0 & z \\ 0 & 0 & y \\ z & y & 1 \end{pmatrix} M_R \, . \tag{31}$$

Eq. (30) introduces four effective parameters: x, y, z and M_R. The magnitude of $M_R \approx (5to15) \times 10^{14}$ GeV can quite plausibly be justified in the context of supersymmetric unificaton (see estimate give in eq. (6) and discussion following eq. (27)). And, to the same extent, the magnitude of $m(\nu_\tau) \approx (1/10 - 1/30)$ eV, which is consistent with the SuperK value, can also be anticipated. Since all the Dirac parameters are determined, there are, effectively, three new parameters: x, y, and z. However, there are six observables in the light three neutrino system: the three masses and the three oscillation angles. Thus one can expect three predictions for the light neutrinos. These may be taken to be $\theta^{osc}_{\nu_\mu\nu_\tau}$ (eq. (25)), m_{ν_τ} (see eqs. (8) and (24)), and for example $\theta^{osc}_{\nu_e\nu_\mu}$.

Recall that the parameter y was determined above by assuming that the MSW (small or large angle) solution for the solar neutrino-deficit corresponds to $\nu_e - \nu_\mu$ oscillation, with $(\delta m^2)_{MSW} \approx m(\nu_\mu)^2 \sim 10^{-5} eV^2$. This gave a value of $|y| \approx 1/20$, in full accord with our general expectation of a hierarchy of order (1/10) for the (2,3)-entry compared to the (3,3). We do not, however, have much experimental information at present, to determine the other two parameters x and y, reliably, because very little is known about the observale parameters involving ν_e. To have a feel, consistent with our presumption that the inter-family hierarchical masses arise through successively smaller off-diagonal mixing elements, we will assume that $y \approx 1/20$ (as above), $z \leq$

$y/10$ and $x \sim z^2$. Thus, in addition to $M_R \approx (5-15) \times 10^{14}$ GeV and $y \approx 1/20$, which (as mentioned above) are better determined, we take as a guide: $z \sim (1-5) \times 10^{-3}$ and $x \sim (1tofew)(10^{-6}-10^{-5})$. Including the three predictions mentioned above, the mass eigenvalues and the oscillation angles are then:

$$\begin{aligned} m_{\nu_\tau} &\approx (1/10 - 1/30)\ eV \\ m_{\nu_\mu} &\simeq 10^{-3}(5to1)\ eV \\ m_{\nu_e} &\simeq (10^{-5} - 10^{-4}(1tofew)\ eV \\ \theta^{\text{osc}}_{\mu\tau} &\simeq 0.437 + \sqrt{m_{\nu_2}/m_{\nu_3}} \\ \theta^{\text{osc}}_{e\mu} &\simeq \theta^{\ell}_{e\mu} - \theta^{\nu}_{e\mu} \simeq 0.06 \pm 0.015 \\ \theta^{\text{osc}}_{e\tau} &\simeq \theta^{\ell}_{e\tau} - \theta^{\nu}_{e\tau} \simeq 10^{-3} \pm 0.03\ . \end{aligned} \tag{32}$$

We see that the masses of ν_e and ν_μ and the oscillation angle $\theta^{osc}_{e\mu}$ goes well with the small angle MSW explanation of the solar neutrino-deficit. [g]

Although the superheavy Majorana masses of the RH neutrinos cannot be observed directly, they can be of cosmological significance. The pattern given above and the arguments given in Sec. 3 and in this section suggests that $M(\nu^\tau_R) \approx (5-15) \times 10^{14}$ GeV, $M(\nu^\mu_R) \approx (1-4) \times 10^{12}$ GeV (for $y \approx 1/20$); and $M(\nu^e_R) \sim (1/2-10) \times 10^9 GeV (forx \sim (1/2-10)10^{-6} > z^2)$. A mass of $\nu^e_R \sim 10^9$ GeV is of the right magnitude for producing ν^e_R following reheating and inducing lepton asymmetry in ν^e_R decay into $H^\circ + \nu^i_L$, that is subsequently converted into baryon asymmetry by the electroweak sphalerons [21].

We have demonstrated that a rather simple pattern for the four Dirac mass matrices, motivated and constrained by the group structure of $SO(10)$, is consistent within 10% with the observed masses and mixing of all the quarks and the charged leptons. This fit is significantly overconstrained, leading to five predictions, which are successful. The same pattern, supplemented with a similar structure for the Majorana mass matrix, quite plausibly accounts for the SuperKamiokande result with the large $\nu_\mu - \nu_\tau$ oscillation angle required for the atmospheric neutrinos and accomodates a small $\nu_e - \nu_\mu$ oscillation angle relevant for theories of the solar neutrino deficit.

Before turning to proton decay, it is worth noting that much of our discussion of fermion masses and mixings, including those of the neutrinos, is essentially unaltered if we go to the limit $\epsilon' \to 0$ of Eq. (28). This limit

[g] Although the small angle MSW solution appears to be more generic within the approach outlined above [18], the large angle solution can still plausibly emerge within this approach, in some limited region of the parameter space. This will be analyzed.

clearly involves:

$$m_u = 0, \ \theta_C \simeq \sqrt{m_d/m_s}$$
$$|V_{ub}| \simeq \sqrt{\frac{\eta - \epsilon}{\eta + \epsilon}} \sqrt{m_d/m_b}(m_s/m_b) \simeq (2.1)(0.039)(0.023) \simeq 0.0019$$
$$m_{\nu_e} = 0, \ \theta^{\nu}_{e\mu} = \theta^{\nu}_{e\tau} = 0 \ .$$

All other predictions will remain unaltered. Now, among the observed quantities in the list above, $\theta_C \simeq \sqrt{m_d/m_s}$ is indeed a good result. Considering that $m_u/m_t \approx 10^{-5}$, $m_u = 0$ is also a pretty good result. There are of course, plausible small corrections arising from higher dimensional operators (for example) involving Planck scale physics which could induce a small value for m_u through the (1,1)-entry $\delta \approx 10^{-5}$. For considerations of proton decay, it is worth distinguishing between these two variants, which we will refer to as cases I and II respectively.

$$\begin{aligned} &\text{Case I}: \ \epsilon' \approx 2 \times 10^{-4}, \ \delta = 0 \\ &\text{Case II}: \delta \approx 10^{-5}, \ \epsilon' = 0 \end{aligned} \tag{33}$$

5 Link Between Fermion Masses and Proton Decay in Supersymmetric SO(10)

5.1 Preliminaries

Following paper II [18], I present now the results of a recent study of proton decay in SUSY SO(10), which was carried out by paying attention specially to the link that exists in SUSY SO(10) between proton decay and the masses and mixings of all fermions, including especially the neutrinos.

It is well known that in supersymmetric unified theories (GUTs), with $M_X \sim 2 \times 10^{16}$ GeV, the gauge-boson mediated d=6 proton decay operators, for which $e^+\pi^0$ would have been the dominant mode, are strongly suppressed. The dominant mechanism for proton decay in these theories is given by effective d=5 operators of the form $Q_iQ_jQ_kL_\ell/M$ in the superpotential, which arise through the exchange of color triplet Higgsinos, that are the GUT partners of the standard Higgs doublets, such as those in $5_H + \overline{5}_H$ of SU(5) or the 10_H of SO(10). Subject to a doublet-triplet splitting mechanism which makes these color triplets acquire heavy GUT-scale masses, while the doublets remain light, these *standard* d=5 operators, suppressed by just one power of the heavy mass and the small Yukawa couplings, lead to proton decay, with a lifetime $\tau_p \sim 10^{30} - 10^{34}$ yrs. [43,44,45,46]. Note that these standard d=5 operators are proportional to the product of two Yukawa couplings, which are related to the masses and mixings of the *charged* fermions. Further, for these operators to induce proton decay, they must be dressed by wino (or gluino)-exchange so as to convert a pair of squarks to quarks. Owing to a) Bose symmetry of the superfields in QQQL/M, b) color antisymmetry, and especially c) the hierarchical Yukawa couplings of the standard Higgs doublets, it turns out that these operators exhibit a strong preference for the decay of a proton into channels involving $\overline{\nu}$ rather than e^+ or (even) μ^+ and those involving an $\overline{s}$ rather than a $\overline{d}$. Thus the standard operators lead to dominant $\overline{\nu}K^+$ and comparable $\overline{\nu}K^+$ modes, but in all cases to highly suppressed $e^+\pi^o, e^+K^o$ and even μ^+K^o modes. For instance, for SUSY SU(5) one obtains (for $tan\beta \leq 15$, say): $[\Gamma(\mu^+K^o)/\Gamma(\overline{\nu}_\mu K^+)]_{std} \sim [m_u/m_c sin^2\theta_C]^2 R \approx 10^{-3}$ where $R \approx 0.1$ is the ratio of the products of the relevant $|\text{matrixelement}|^2 \times$ (phase space) for the two modes.

Now, it was recently realized that in left-right symmetric unified theories possessing supersymmetry, such as those based on G(224) or SO(10), there is very likely a new source of d=5 proton decay operators, which are related to the Majorana masses of the right-handed neutrinos [17]. For instance, in the context of the minimal set of Higgs mulitplets [h] $\{45_H, 16_H, \overline{16}_H$ and $10_H\}$, which have been utilized in secs. 3 and 4 to break SO(10) and generate fermion masses, these new d=5 operators arise by combining three effective couplings: - i.e. (a) the couplings $f_{ij}16_i16_j\overline{16}_H\overline{16}_H/M$ (see eq. (5)) which are essential to assign Majorana masses to the right-handed neutrinos, (b) the couplings $g_{ij}16_i16_j16_H16_H/M$, which are needed to generate non-trivial CKM mixing and (c) the mass-term $M_{16}16_H\overline{16}_H$. In the presence of these three (unavoidable) effective couplings and the VEVs $< 16_H >=< \overline{16}_H >\sim M_X$ (see secs. 3 and 4), the color triplet Higgsinos in 16_H and $\overline{16}_H$ of mass M_{16} can be exchanged between $\tilde{q}_i q_j$ and $\tilde{q}_k \ell_\ell$-pairs. This exchange gives rise to a new set of effective d=5 couplings of the form:

$$L^{d=5}_{new} = [f_{ij}g_{kl}(16_i16_j)(16_k16_l)/M_{16}]\frac{< \overline{16}_H < 16_H >}{M^2} \qquad (34)$$

which induce proton decay, just as the standard operators do. Note that these new d=5 operators depend, through the couplings f_{ij} and g_{kl}, both on the Majorana and on the Dirac masses of the respective fermions. *This is why within SUSY G(224) or SO(10), proton decay gets intimately linked to the masses and mixings of all fermions, including neutrinos.*

[h] The orignin of these new d=5 operators in the context of other Higgs muliplets, in particular in the cases where 126_H and $\overline{126}_H$ are used to break B-L, has been discussed in Ref. [17].

Specifically, it is found that the Super-K result on atmospheric neutrinos, that suggests $m(\nu_L^\tau) \sim 1/20$ eV and a large $\nu_\mu - \nu_\tau$ oscillation angle, leads to a *significant enhancement* especially in the new d=5 operators, compared to previous estimate, which were based on guesses of much larger values of $m(\nu_L^\tau) \sim (2-4)$ eV [17]. Curiously enough, the net effect of including the enhancement of f_{33} (due to a lowering of $m(\nu_L^\tau)$) and the suppression of the relevant CKM mixings is such that *the strength of the new d=5 operators is found to be comparable to that of the standard ones* [18]. The flavor-structure of the new operators are, however, very different from those of the standard ones, in part because the former depend on the Majorana masses of the RH neutrinos, and the latter do not. As a result, the new operators lead to some characteristic differences in the proton-decay pattern (that is branching ratios of different decay modes) compared to the standard ones (see below).

5.2 Framework for Calculating Proton Decay Rate

To establish notations, consider the case of minimal SUSY SU(5) and, as an example, the process $\tilde{c}\tilde{d} \to \bar{s}\bar{\nu}_\mu$, which induces $p \to \overline{\nu}_\mu K^+$. Let the strength of the corresponding d=5 operator, multiplied by the product of the CKM mixing elements entering into wino-exchange vertices, (which in this case is $\sin\theta_C \cos\theta_C$) be denoted by $\hat{A}$. Thus (putting $\cos\theta_C = 1$), one obtains:

$$\begin{aligned}\hat{A}_{\tilde{c}\tilde{d}}(SU(5)) &= (h_{22}^u h_{12}^d / M_{H_C}) \sin\theta_c \simeq (m_c m_s \sin^2\theta_C / v_u^2)(\tan\beta / M_{H_C}) \\ &\simeq (1.9 \times 10^{-8})(\tan\beta / M_{H_C}),\end{aligned} \quad (35)$$

where $\tan\beta \equiv v_u/v_d$, and we have put v_u =174 GeV and the fermion masses extrapolated to the unification-scale - i.e. $m_c \simeq 300$ MeV and $m_s \simeq 40$ MeV. The amplitude for the associated four- fermion process $dus \to \overline{\nu_\mu}$ is given by:

$$A_5(dus \to \overline{\nu}_\mu) = (\hat{A}_{\tilde{c}\tilde{d})} \times (2f) \quad (36)$$

where f is the loop-factor associated with wino-dressing. Assuming $m_{\tilde{w}} \ll m_{\tilde{q}} \sim m_{\tilde{l}}$ one gets: $f \simeq (m_{\tilde{w}}/m_{\tilde{q}}^2)(\alpha_2/4\pi)$. Using the amplitude for $(\mathrm{du})(\mathrm{s}\nu_\ell)$, as in eq. (36), ($\ell = \mu$ or τ), one then obtains [44-46, 18]:

$$\Gamma^{-1}(p \to \overline{\nu}_\tau K^+) \approx (2.2 \times 10^{31}) \text{ yrs} \times$$
$$\left(\frac{.67}{A_S}\right)^2 \left[\frac{0.006\text{GeV}^3}{\beta_H}\right]^2 \left[\frac{(1/6)}{(m_{\tilde{W}}/m_{\tilde{q}})}\right]^2 \left[\frac{m_{\tilde{q}}}{1\text{ TeV}}\right]^2 \left[\frac{2 \times 10^{-24}\text{ GeV}^{-1}}{\hat{A}(\overline{\nu})}\right]^2. \quad (37)$$

Here β_H denotes the hadronic matrix element defined by $\beta_H u_L(\vec{k}) \equiv \epsilon_{\alpha\beta\gamma} \left\langle 0|(d_L^\alpha u_L^\beta) u_L^\gamma|p, \vec{k}\right\rangle$. While the range $\beta_H = (0.003 - 0.03)$ GeV^3 has been used in the past [45], given that one lattice calculations yield $\beta_H = (5.6 \pm 0.5) \times 10^{-3}$ GeV^3 [50], we will take as a plausible range: $\beta_H = (0.006\ \text{GeV}^3)(1/2$ to $2)$. $A_S \approx 0.67$ stands for the short distance renormalization factor of the $d = 5$ operator. Note that the familiar factors that appear in the expression for proton lifetime – i.e., M_{H_C}, $(1 + y_{tK})$ representing the interference between the $\tilde{t}$ and $\tilde{c}$ contributions and $\tan\beta$ – are all effectively contained in $\hat{A}(\overline{\nu})$. Allowing for plausible and rather generous uncertainties in the matrix element and the spectrum we take:

$$\beta_H = (0.006GeV^3)(1/2\text{to}2)$$
$$(m_{\tilde{w}}/m_{\tilde{q}}) = 1/6(1/2to2), \text{ and } m_{\tilde{q}} \approx m_{\tilde{\ell}} \approx 1TeV(1/\sqrt{2}\text{to}\sqrt{2}) \tag{38}$$

Using eqs. (37) - (38), we get:

$$\Gamma^{-1}(p \to \overline{\nu}_\tau K^+) \approx (2.2 \times 10^{31})\ \text{yrs}[2.2 \times 10^{-24}GeV^{-1}/\hat{A}(\overline{\nu}_\ell)]^2[32\text{to}1/32]. \tag{39}$$

This relation is general, depending only on $\hat{A}(\overline{\nu}_\ell)$ and on the range of parameters given in eq. (38). It can thus be used for both SU(5) and SO(10). The experimental lower limit on the inverse rate for the $\bar{\nu}K^+$ modes is given by [47],

$$[\sum_\ell \Gamma(p \to \overline{\nu_\ell}K^+)]^{-1}_{expt} > 7 \times 10^{32}\text{yrs}. \tag{40}$$

Allowing for all the uncertainties to stretch in the same direction (in this case, the square bracket = 32), and assuming that just one neutrino flavor (e.g. ν_μ for SU(5)) dominates, the observed limit eq. (40) provides an upper bound on the amplitude [i]:

$$\hat{A}(\overline{\nu_\ell}) \leq 2 \times 10^{-24}GeV^{-1} \tag{41}$$

which holds for both SU(5) and SO(10). For minimal SU(5), using eq. (35) and $\tan\beta \geq 2$ (which is suggested on several grounds), one obtains a lower limit on M_{HC} given by:

$$M_{HC} \geq 2 \times 10^{16}GeV(SU(5)) \tag{42}$$

At the same time, higher values of $M_{HC} > 3 \times 10^{16}$ GeV do not go very well with gauge coupling unification [48]. Thus keeping $M_{HC} \leq 3 \times 10^{16}$ and $\tan\beta \geq 2$, we obtain from eq. (35): $\hat{A}(SU(5)) \geq (4/3) \times 10^{-24}GeV^{-1}$. Using eq. (39), this in turn implies that

$$\Gamma^{-1}(p \to \bar{\nu}K^+) \leq 1.5 \times 10^{33}yrs(SU(5)) \tag{43}$$

[i] If there are sub- dominant $\overline{\nu_i}K^+$ modes with branching ratio R, the right side of eq. (41) should be divided by $\sqrt{1+R}$.

This a conservative upper limit. In practise, it is unlikely that all the uncertainties, including that in M_{HC}, would stretch in the same direction to nearly extreme values so as to prolong proton lifetime. A more reasonable upper limit, for minimal SU(5), thus seems to be: $\Gamma^{-1}(p \to \bar{\nu}K^+)(SU(5)) \leq (0.7) \times 10^{33}$ yrs. Given the experimental lower limit (eq. (40)), we see that minimal SUSY SU(5) is almost on the verge of being excluded by proton decay-searches. We have of course noted in Sec. 3 that SUSY SU(5) does not go well with the neutrino oscillations observed at SuperK.

Now, to discuss proton decay in the context of supersymmetric SO(10), it is necessary to discuss first the mechanism for doublet-triplet splitting. Details of this discussion may be found in Ref. [18]. Here, I present only a synopsis.

5.3 A Natural Doublet-Triplet Splitting Mechanism in SO(10)

In supersymmetric $SO(10)$, a natural doublet–triplet splitting can be achieved by coupling the adjoint Higgs $\mathbf{45_H}$ to a $\mathbf{10_H}$ and a $\mathbf{10'_H}$, with $\mathbf{45_H}$ acquiring a unification–scale VEV in the $B-L$ direction [49]: $\langle\mathbf{45_H}\rangle = (a,a,a,0,0)\times\tau_2$ with $a \sim M_U$. As discussed in Section 2, to generate CKM mixing for fermions we require an $\langle\mathbf{16_H}\rangle_d$ to acquire an electroweak scale vacuum expectation value. To insure accurate gauge coupling unification, the effective low energy theory should not contain split multiplets beyond those of MSSM. Thus the MSSM Higgs doublets must be linear combinations of the $SU(2)_L$ doublets in $\mathbf{10_H}$ and $\mathbf{16_H}$. A simple set of superpotential terms that ensures this and incorporates doublet–triplet splitting is:

$$W_H = \lambda \mathbf{10_H 45_H 10'_H} + M_{10}\mathbf{10'_H}^2 + \lambda' \overline{\mathbf{16_H}}\,\overline{\mathbf{16_H}}\mathbf{10_H} + M_{16}\mathbf{16_H}\overline{\mathbf{16_H}}\ . \quad (44)$$

A complete superpotential for $\mathbf{45_H}, \mathbf{16_H}, \overline{\mathbf{16_H}}, \mathbf{10}_H, \mathbf{10}'_H$ and possibly other fields, which ensure that $\mathbf{45_H}$, $\mathbf{16_H}$ and $\overline{\mathbf{16_H}}$ acquire unification scale VEVs with $\langle\mathbf{45_H}\rangle$ being along the $(B-L)$ direction, that exactly two Higgs doublets (H_u, H_d) remain light, with H_d being a linear combination of $(\mathbf{10_H})_d$ and $(\mathbf{16_H})_d$, and that there are no unwanted pseudoGoldstone bosons, can be constructed With the vacuum expectation value $\langle\mathbf{45_H}\rangle$ in the $B-L$ direction, it does not contribute to the doublet matrix, so one pair of Higgs doublet remains light, while all triplets acquire unification scale masses. The light MSSM Higgs doublets are

$$H_u = \mathbf{10}_u,\ H_d = \cos\gamma\mathbf{10}_d + \sin\gamma\mathbf{16}_d\ , \quad (45)$$

with $\tan\gamma \equiv \lambda'\langle\overline{\mathbf{16_H}}\rangle/M_{16}$. Consequently, $\langle\mathbf{10}\rangle_d = \cos\gamma\ v_d, \langle\mathbf{16}_d\rangle = \sin\gamma\ v_d$, with $\langle H_d\rangle = v_d$ and $\langle\mathbf{16}_d\rangle$ and $\langle\mathbf{10_d}\rangle$ denoting the electroweak VEVs of those multiplets. Note that the H_u is purely in $\mathbf{10_H}$ and that $\langle\mathbf{10}_d\rangle^2 + \langle\mathbf{16}_d\rangle^2 = v_d^2$.

This mechanism of doublet-triplet (DT) splitting is rather unique for the minimal Higgs systems in that it meets the requirements of both D-T splitting and CKM-mixing. In turn, it has three important consequences:

(i) It modifies the familiar SO(10)-relation $\tan\beta \equiv v_u/v_d = m_t/m_b \approx 60$ to:

$$\tan\beta/\cos\gamma \approx m_t/m_b \approx 60 \tag{46}$$

As a result, even low to moderate values of $\tan\beta \approx 3$ to 10 (say) are perfectly allowed in SO(10) (corresponding to $\cos\gamma \approx 1/20$ to $1/6$).

(ii) In contrast to SU(5), for which the strengths of the standard d=5 operators are proportional to $(M_{H_c})^{-1}$, where $M_{H_C} \sim M_U \sim few \times 10^{16}$ GeV (see eq. (35)), for the SO(10)-model, with DT-splitting given as above, they become proportional to M_{eff}^{-1}, where $M_{eff} = (\lambda a)^2/M_{10'} \sim M_U^2/M_{10'}$. $M_{10'}$ can be naturally smaller than M_U and thus M_{eff} correspondingly larger (than M_U) by one to two orders of magnitude (see Ref. [18]). Now, the proton decay amplitudes for SO(10) in fact possess an intrinsic enhancement compared to those for SU(5), owing primarily due to differences in their Yukawa couplings for the up sector (see Ref. [18], Appendix C). As a result, these larger values of $M_{eff} \sim 10^{18}$ GeV are found to lead to expected proton decay lifetimes that are on the one hand compatible with observed limits, but on the other hand allow optimism as regards future observation of proton decay (see below).

(iii) M_{eff} gets bounded above by considerations of coupling unification and GUT-scale threshold effects. Owing to mixing between 10_d and 16_d (see eq. (45)), the correction to $\alpha_3(m_z)$ due to doublet-triplet splitting becomes proportional to $ln(M_{eff}/\cos\gamma)$. Inclusion of this correction and those due to splittings within the gauge and the Higgs multiplets (i.e. $45_H, 16_H$, and $\overline{16}_H$) [j], together with the observed degree of coupling unification allows us to obtain a conservative upper limit on $M_{eff} \leq 3 \times 10^{18}$ GeV [see Ref. 18]. This in turn helps provide an upper limit on the expected proton decay lifetime (see below).

The calculation of the amplitudes $\hat{A}_{std}$ and $\hat{A}_{new}$ for the standard and the new operators for the SO(10) model, are given in detail in Ref. [18]. Here, I will present only the results. It is found that the four amplitudes $\hat{A}_{std}(\bar{\nu}_\tau K^+), \hat{A}_{std}(\bar{\nu}_\mu K^+)$, $\hat{A}_{new}(\bar{\nu}_\tau K^+)$ and $\hat{A}_{new}(\bar{\nu}_\nu K^+)$ are in fact very comparable to each other, within about a factor of two, either way. Since there is no reason to expect a near cancellation between the standard and the new operators, especially for both $\bar{\nu}_\tau K^+$ and $\bar{\nu}_\mu K^+$ modes, we expect the net amplitude (standard + new) to be in the range exhibited by either one.

[j] The correction to $\alpha_3(m_z)$ due to Planck scale physics through the effective operator $F_{\mu\nu}F^{\nu\mu}45_H/M$ vanishes due to antisymmetry in the SO(10)-contraction.

Following Ref. [18], I therefore present the contributions from the standard and the new operators separately. Using the upper limit on $M_{eff} \geq 3 \times 10^{18}$ GeV, we obtain a lower limit for the standard proton decay amplitude given by

$$\hat{A}(\overline{\nu}_\tau K^+)_{\rm std} \geq \left[\begin{matrix} (7 \times 10^{-24} {\rm GeV}^{-1})(1/6 \text{ to } 1/4) \\ (3 \times 10^{-24} \ {\rm GeV}^{-1})(1/6 \text{ to } 1/2) \end{matrix} \right] \tag{47}$$

Substituting into Eq. (39) and adding the contribution form the second competing mode $\overline{\nu}_\mu K^+$ with a typical branching ratio $R \approx 0.3$, we obtain

$$\Gamma^{-1}(\overline{\nu} K^+)_{\rm std} \leq \begin{bmatrix} (3 \times 10^{31} {\rm yrs.})(1.6 \text{ to } 0.7) \\ (6.8 \times 10^{31} {\rm yrs.})(4 \text{ to } 0.44) \end{bmatrix} (32 \text{ to } 1/32) \tag{48}$$

The upper and lower entries in eqs.(47) and (48) and henceforth correspond to the cases I and II of the fermion mass-matrix (i.e. $\epsilon' \neq 0$ and $\epsilon' = 0$, respectively, see eq. (33)). The uncertainty shown inside the square brackets correspond to that in the relative phases of the different contributions. The uncertainty (32 to 1/32) corresponds to the uncertainty in β_H, $(m_{\tilde{W}}/m_{\tilde{q}})$ and $m_{\tilde{q}}$, by factors of 2,2, and $\sqrt{2}$ respectively, either way, around the "central" values reflected in Eq. (38). Thus we find that for MSSM embedded in SO(10), the inverse partial proton decay rate should satisfy:

$$\begin{aligned} \Gamma^{-1}(p \to \overline{\nu} K^+)_{\rm std} &\leq \begin{bmatrix} 3 \times 10^{31 \pm 1.7} \ {\rm yrs.} \\ 6.8 \times 10^{31^{+2.1}_{-1.5}} \ {\rm yrs} \end{bmatrix} \\ &\leq \begin{bmatrix} 1.5 \times 10^{33} \ {\rm yrs.} \\ 7 \times 10^{33} {\rm yrs.} \end{bmatrix} \quad ({\rm SO}(10)) \end{aligned} \tag{49}$$

The central value of the upper limit in Eq. (49) essentially reflects the upper limit on $M_{\rm eff}$, while the remaining uncertainties of matrix elements and spectrum are reflected in the exponents.

Evaluating similarly the contribution from the new operator, we obtain:

$$\hat{A}(\overline{\nu}_\mu K^+)_{\rm new} \approx (1.5 \times 10^{-24} \ {\rm GeV}^{-1})(1/4 \text{ to } 1.3) \tag{50}$$

$$\Gamma^{-1}(\overline{\nu} K^+)_{\rm new} \approx (3 \times 10^{31} \ {\rm yrs})[16 \text{ to } 1/1.7]\{32 \text{ to } 1/32\} \tag{51}$$

In this estimate we have included the contribution of the $\overline{\nu}_\tau K^+$ mode with a typical branching ratio $R \approx 0.4$. Here the second factor, inside the square bracket, reflects the uncertainties in the amplitude, while the last factor corresponds to varying β_H, $(m_{\tilde{W}}/m_{\tilde{q}})$ and $m_{\tilde{q}}$ around the central values reflected in Eq. (38). With a net factor of even 20 to 100 arising jointly from the square and the curly brackets, *i.e.* without going to extreme ends of all parameters, the new operators related to neutrino masses lead by themselves to proton decay lifetimes

$$\Gamma^{-1}(\bar{\nu}K^+)^{expected}_{new} \approx (0.6 - 3) \times 10^{33} yrs.(\mathrm{SO(10)}) \tag{52}$$

5.4 The Charged Lepton Decay Mode ($p \to \mu^+ K^o$)

I now discuss a special feature of the SO(10) model pertaining to the possible prominence of the charged lepton decay mode: $p \to \mu^+ K^o$, which is not permissible in SUSY SU(5). Allowing for uncertainties in the way the standard and the new operators can combine with each other for the three leading modes i.e. $\bar{\nu}_\tau K^+, \bar{\nu}_\mu K^+$ and $\mu^+ K^o$, we obtain (see Ref. [18] for details):

$$B(\mu^+ K^0)_{\mathrm{std+new}} \approx [1 \text{ to } (50 - 60)\%]\, \rho(\mathrm{SO(10)}) \tag{53}$$

where ρ denotes the ratio of the squares of relevant matrix elements for the $\mu^+ K^0$ and $\overline{\nu} K^+$ modes.

In the absence – presumably temporary – of a reliable lattice calculation, which is presently missing for the $\bar{\nu}K^+$ mode [50], one should remain open to the possibility of $\rho \approx 1/2$ to 1 (say). Using Eq.(53), we find that for a large range of parameters, the branching ratio $B(\mu^+ K^0)$ can lie in the range of 20 to 30% (if $\rho \approx 1$). Thus we see that the $\mu^+ K^0$ mode is likely to be prominent in the $SO(10)$ model presented here, and if $\rho \approx 1$, it can even become a dominant mode. This contrasts sharply with the minimal $SU(5)$ model, in which the $\mu^+ K^0$ is expected to have a branching ratio of only about 10^{-3}. In the $SO(10)$ model, the standard operator by itself gives a branching ratio for this mode of (1 to 10) % while *the potential prominence of the $\mu^+ K^0$ mode arises only through the new operator related to neutrino masses.*

6 Some Crucial Observations Pertaining to Unification: A Summary

The discussion in the preceding sections can be best summarized by listing the implications of some crucial findings which bear on unification.

A. The Family Multiplet Structure: The observed multiplet- structure in each family consisting of either sixteen members (including the ν_R) or fifteen members (without ν_R) is the *first empirical hint* in favor of an underlying gauge symmetry like G(224), SO(10) or SU(5). As mentioned in Sec. 2, while the standard model organizes the 15 members of a family into five multiplets, SU(5) groups them into two, and G(224) with L-R discrete symmetry, as well as SO(10), place all sixteen members within just one multiplet. Further, each of these higher symmetries (G(224), SO(10) or SU(5)) explain precisely the $SU(3)^C \times SU(2)_L$-representations and the weak hypercharge (Y_W)-quantum numbers of all the members in a family. This feature, as well as the need to explain observed quantization of electric charge, have been two of the primary motivations for proposing the idea of grand unification [5 - 7].

B. Meeting of the Gauge Couplings: The meeting of the gauge couplings, which is found to occur when their measured values at LEP are extrapolated to higher energies, in the context of supersymmetry, clearly supports the ideas of:

- An underlying unity of forces, as well as of supersymmetry,
- the relevance of effective gauge symmetries like SU(5), or SO(10), or a string-derived G(224), or $[SU(3)]^3$ at the underlying level, and
- unification at a scale $M_X \sim 2 \times 10^{16}$ GeV (assuming MSSM spectrum below M_X).

C. Neutrino Masses, especially $m(\nu_\tau) \sim 1/20$ eV: This single piece of information, suggested by the SuperK-result, brings to light the existence of the RH neutrinos, accompanying the left-handed ones, and reinforces the ideas of (see secs. 2 and 3):

- SU(4)-color
- Left-Right symmetry
- supersymmetric unification, and
- see-saw.

In short, the SuperK-result, suggesting $m(\nu_\tau) \sim \frac{1}{20}$ eV, selects out the route to higher unification based on a string-derived G(224) or SO(10), as opposed to SU(5). Further, it suggests that B-L breaking occurs at the unification scale, $M_{B-L} \sim M_X \sim 2 \times 10^{16}$ GeV rather than at an intermediate scale.

D. Masses and Mixings of All Fermions: (q, ℓ, ν): Adopting familiar ideas of generating lighter eigenvalues through off- diagonal mixings and using the group-theory of SO(10) for the effective Yukawa couplings of the *minimal Higgs system*, it was found in [18], that, remarkably enough, the bizarre pattern of the masses and mixings of the charged fermions as well as of the neutrinos can be adequately described (with $\sim 10\%$ accuracy) within an economical and predictive SO(10)-framework. In particular, the framework provides five successful predictions for the masses and mixings of the quarks and the charged leptons. The same description goes extremely well with a value of $m_{\nu_\tau} \sim 1/20$ eV as well as with a large $\nu_\mu - \nu_\tau$ oscillation angle ($sin^2 2\theta_{\nu_\mu \nu_\tau} \approx 0.82 - 0.96$), despite highly non-degenerate masses for the light neutrinos. Both these features are in good agreement with the SuperK results on atmospheric neutrinos. The same framework also typically leads to the small angle MSW solution for the solar neutrino puzzle, with $m_{\nu_e} \sim 3 \times 10^{-3} eV \gg m_{\nu_e}$. These results are listed in Sec. 4.

One intriguing feature of the SO(10)-framework, presented in Sec. 4, is that the largeness of the $\nu_\mu - \nu_\tau$ oscillation angle emerges naturally *together with* the smallness of the analogous mixing parameter in the quark-sector: $V_{bc} \approx 0.04$. This remarkable correlation between the leptonic versus the quark mixing angles clearly points to the presence of a contribution to the mass- matrices, which is proportional to B-L, and is antisymmetric in the family-space. The minimal Higgs system together with the group theory of SO(10) precisely yields such a contribution.

E. Proton Decay: The Hall-Mark of Quark-Lepton Unification: Proton decay, if seen, would directly verify the idea of quark- lepton unification. Note that this crucial aspect of grand unification is not probed directly by the other three observations listed above: B, C, and D.

We have argued that three different sets of observations - i.e. (a) the observed meeting of the three gauge couplings, (b) the SuperK result on atmospheric neutrino oscillations, and (c) fermion masses and mixings - go extremely well with the idea of supersymmetric unification, based on symmetry structures such as SO(10). Babu, Wilczek and I have studied proton decay in this context, paying attention to its correlation with fermion masses and mixings [18]. We found that the proton decay amplitudes receive a major contribution from a set of *new* d=5 operators which are directly related to the Majorana masses of the RH neutrinos and to the CKM mixings [17,18]. This is in addition to the contribution from the standard d=5 operators, which are related to the Dirac masses of the charged fermions. The study shows that the mass of $m_{\nu_\tau} \sim 1/20$ eV (as opposed to previously considered values of a few eV) and the large oscillation angle, suggested by the SuperK result, in fact imply a net enhancement in the rates of proton decay into the $\bar{\nu} K^+$ and especially the $\mu^+ K^o$-modes [18], relative to previous estimates.

There are of course uncertainties in the prediction for proton decay rates owing to those in the SUSY-spectrum, the hadronic matrix elements and the relative phases of the different contributions. Allowing for rather generous uncertainties in this regard (see Sec. 5), we expect proton to decay dominantly into the $\bar{\nu}K^+$ and very likely to the μ^+K^0-mode as well, with a lifetime:

$$\tau_{proton} \leq 7 \times 10^{33} yrs(\mathrm{SO}(10)) \tag{54}$$

This is a conservative upper limit which is obtained only if all the uncertainties are stretched in the same direction to nearly their extreme values, so as to extend proton longevity. Since the likelihood of this happening is small, we expect that within either a string- derived G(224) or the SO(10)-model, of the sort presented here, proton should decay with a lifetime *shorter* than the limit shown above. With the current experimental lower limit already at 7×10^{32} years, we conclude that improvement in the present limit for $p \to \bar{\nu}K^+$ and $p \to \mu^+K^0$-modes by a factor of 2 to at most 10 should either turn up events, or else the remarkably successful SO(10)-framework described here will be called into question seriously. On the basis of our study, we expect that the SuperK detector should in fact see a few proton decay events in the $\overline{\nu}K^+$ and quite possibly in the μ^+K^o-channel, in the near future. To establish the reality of this important process firmly and also to study efficiently the branching ratios of some crucial modes, like the μ^+K^o, next- generation detectors with sensitivity of at least 5×10^{34} and perhaps 10^{35} years are essential.

We have stressed that observation of proton decay into μ^+K^o with a branching ratio exceding 20% (say) would provide a clear signature in favor of (a) supersymmetric unification based on symmetry structures such as a string-derived G(224) or SO(10), as well as (b) the mechanism described here of generating the masses and mixings of all fermions including especially the neutrinos [18].

To conclude, proton decay has been anticipated for quite some time as a hallmark of grand unification. With coupling unification and neutrino masses revealed, proton decay is the missing link. While its discovery, with dominance of the $\overline{\nu}K^+$mode, would confirm supersymmetric unification, prominence of the μ^+K^o mode would establish the beautiful link that exists between the neutrino masses and proton decay within the G(224)/SO(10)-route to unification.

7 Acknowledgements

I would like to thank specially Kaladi S. Babu and Frank Wilczek for a most enjoyable collaboration on the research described here. I would also like to thank Susan and Gabor Domokos for arranging a stimulating conference and for their kind invitation. The research presented here is supported in part by DOE grant NO. DE-FGO2-96ER-41015.

References

1. SuperKamiokande Collaboration, Y. Fukuda et. al., Phys. Rev. Lett. **81**, 1562 (1998).
2. J.C. Pati, "Implications of The SuperKamiokande Result on the Nature of New Physics", in Neutrino 98, Takayama, Japan, June, 98, hep-ph/9807315; Nuclear Phys. B (Proc. Suppl.) 77, 299(1999).
3. S. Weinberg, Phys. Rev. Lett. **43**, 1566 (1979);
   E. Akhmedov, Z. Berezhiani and G. Senjanovic, Phys. Rev. Lett. **69**, 3013 (1992).
4. For a recent update see: J.N. Bachall, P. Krastev and A.Yu. Smirnov, Phys. Rev. **D58**, 096016 (1998).
5. J.C. Pati and Abdus Salam; Proc. 15th High Energy Conference, Batavia, reported by J.D. Bjorken, Vol. 2, p. 301 (1972); Phys. Rev. 8 (1973) 1240.
6. J.C. Pati and Abdus Salam, Phys. Rev. Lett. **31** (1973) 661; Phys. Rev. **D10** (1974) 275.
7. H. Georgi and S. L. Glashow, Phys. Rev. Lett. **32**, 438 (1974).
8. H. Georgi, H. Quinn and S. Weinberg, Phys. Rev. Lett. **33**, 451 (1974).
9. M. Green and J.H. Schwarz, Phys. Lett. **149B**, 117 (1984); D.J. Gross, J.A. Harvey, E. Martinec and R. Rohm, Phys. Rev. Lett. **54**, 502 (1985); P. Candelas, G.T. Horowitz, A. Strominger and E. Witten, Nucl. Phys. **B 258**, 46 (1985). For introductions and reviews, see: M.B. Green, J.H. Schwarz and E. Witten, "Superstring Theory" Vols. 1 and 2 (Cambridge University Press); M. Dine, ed., "String Theory in Four Dimensions" (North Holland, 1988); J. Polchinski, "What is String Theory?", 1994 Les Houches Lectures, hep-th/9411028.
10. For recent work, see P. Langacker and M. Luo, Phys. Rev. **D 44** (1991) 817; U. Amaldi, W. de Boer and H. Furstenau, Phys. Lett. **B 260** (1991); J. Ellis, S. Kelley and D. V. Nanopoulos, Phys. Lett. **B 260** (1991) 131; F. Anselmo, L. Cifarelli, A. Peterman and A. Zichichi, Nuov. Cim. **A 104** (1991) 1817. The essential features pertaining to coupling unification in SUSY GUTS were noted earlier by S. Dimopoulos, S. Raby and F. Wilczek, Phys. Rev. **D24**, 1681 (1981); W. Marciano and G. Senjanovic, Phys. Rev. D 25, 3092 (1982); M. Einhorn and D.R.T. Jones, Nucl. Phys. B 196, 475 (1982).
11. S. Weinberg, Phys. Rev. **D26**, 287 (1982); N. Sakai and T. Yanagida, Nucl. Phys. **B197**, 533 (1982).
12. Y.A. Gelfand and E.S. Likhtman, JETP Lett. **13**, 323 (1971); J. Wess and B. Zumino, Nucl. Phys. B70, 139 (1974); D. Volkov and V.P. Akulov,

JETP Lett. **16**, 438 (1972).
13. For a recent discussion, see K. Dienes, Phys. Reports **287**, 447 (1997), (hep-th/9602045), and references therein.
14. J.C. Pati, "With Neutrino Masses Revealed, Proton Decay is the Missing Link", hep-ph/9811442; Proc. Salam Memorial Meeting (1998), publ. by World Scientific, page 98 (1998)
15. H. Georgi, in Particles and Fields, ed. by C. Carlson (AIP, NY, 1975), p. 575; H. Fritzsch and P. Minkowski, Ann. Phys. **93**, 193 (1975).
16. I. Antoniadis, G. Leontaris and J. Rizos, Phys. Lett **B245**, 161 (1990); G. Leontaris, Phys. Lett. **B372**, 212 (1996).
17. K.S. Babu, J.C. Pati and F. Wilczek, "Suggested New Modes in Supersymmetric Proton Decay", Phys. Lett. **B423**, 337 (1998).
18. K.S. Babu, J.C. Pati and F. Wilczek, "Fermion Masses, Neutrino Oscillations and Proton Decay in the Light of the SuperKamiokande", hep-ph/981538V3; Nucl. Phys. B. (To appear).
19. S. Mikheyev and A. Smirnov, Nuovo Cim. **9C**, 17 (1986), and L. Wolfenstein, Phys. Rev. **D17**, 2369 (1978).
20. J.C. Pati and A. Salam, Phys. Rev. **D10**, 275 (1974);
R.N. Mohapatra and J.C. Pati, Phys. Rev. **D11**, 566, 2558 (1975);
G. Senjanovic and R.N. Mohapatra, Phys. Rev. **D12**, 1502 (1975).
21. V. Kuzmin, Va. Rubakov and M. Shaposhnikov, Phys. Lett **BM155**, 36 (1985); M. Fukugita and T. Yanagida, Phys. Lett. **B 174**, 45 (1986); M. A. Luty, Phys. Rev. **D45**, 455 (1992); W. Buchmuller and M. Plumacher, hep-ph/9608308.
22. F. Gürsey, P. Ramond and P. Sikivie, Phys. Lett. **B 60**, 177 (1976).
23. For recent reviews see e.g., P. Langacker and N. Polonsky, Phys. Rev. **D47**, 4028 (1993) and references therein.
24. The literature on string-dualities and M-theory is large. A few pioneering papers relevant to gauge-coupling unification are: E. Witten, Nucl. Phys. **B 443**, 85 (1995), P. Horava and E. Witten, Nucl. Phys. **B 460**, 506 (1996). For reviews, see e.g. J. Polchinski, hep-th/9511157; A. Sen, hep-th/9802051, and references therein. For a layman's guide, see M. Duff (talk at this meeting), hep-ph/9805177 V3.
25. P. Ginsparg, Phys. Lett. **B 197**, 139 (1987); V.S. Kaplunovsky, Nucl. Phys. **B 307**, 145 (1988); Erratum: *ibid.* **B 382**, 436 (1992).
26. E. Witten, hep-th/9602070.
27. J.C. Pati and K.S. Babu, "*The Problems of Unification- Mismatch and Low α_3: A Solution with Light Vector-Like Matter*", hep-ph/9606215, Physics Lett. **B384**, 140 (1996).
28. See e.g. D. Lewellen, Nucl. Phys. **B 337**, 61 (1990); A. Font, L.

Ibanez and F. Quevedo, Nucl. Phys. **B 345**, 389 (1990); S. Chaudhari, G. Hockney and J. Lykken, Nucl. Phys. **B 456**, 89 (1995) and hep-th/9510241; G. Aldazabal, A. Font, L. Ibanez and A. Uranga, Nucl. Phys. **B 452**, 3 (1995); ibid. **B 465**, 34 (1996); D. Finnell, Phys. Rev. **D 53**, 5781 (1996); A.A. Maslikov, I. Naumov and G.G. Volkov, Int. J. Mod. Phys. **A 11**, 1117 (1996); J. Erler, hep-th/9602032 and G. Cleaver, hep-th/9604183; and Z. Kakushadze and S.H. Tye, hep-th/9605221, and hep-th/9609027; Z. Kakushadze et al, hep- ph/9705202.

29. A. Faraggi, Phys. Lett. **B278**, 131 (1992); Phys. Lett. **B274**, 47 (1992); Nucl. Phys. **B403**, 101 (1993); A. Faraggi and E. Halyo, Nucl. Phys. **B416**, 63 (1994).
30. See e.g. Ref. [16]
31. J.C. Pati, "*The Essential Role of String Derived Symmetries in Ensuring Proton Stability and Light Neutrino Masses*", hep-ph/9607446, Phys. Lett. **B388**, 532 (1996).
32. A. Faraggi and J.C. Pati, "*A Family Universal Anomalous U(1) in String Models as the Origin of Supersymmetry Breaking and Squark-degeneracy*", hep-ph/9712516v3, December 97, Nucl. Phys. B (to appear).
33. A. Faraggi and J.C. Pati, "*Meeting the Constraint of Neutrino-Higgsino Mixing in Gravity-Unified Theories*", Phys. Lett. **B 400**, 314 (1997).
34. K.S. Babu and J.C. Pati, "*Towards A Resolution of the Supersymmetric CP Problem Through Flavor and Left-Right Symmetries*", (to appear).
35. See e.g. K. R. Dienes and J. March-Russell, hep-th/9604112; K. R. Dienes, hep-ph/9606467.
36. M. Gell-Mann, P. Ramond and R. Slansky, in: *Supergravity*, eds. F. van Nieuwenhuizen and D. Freedman (Amsterdam, North Holland, 1979) p. 315; T. Yanagida, in: *Workshop on the Unified Theory and Baryon Number in the Universe*, eds. O. Sawada and A. Sugamoto (KEK, Tsukuba) 95 (1979); R. N. Mohapatra and G. Senjanovic, Phys. Rev. Lett. **44**, 912 (1980).
37. H. Georgi and C. Jarlskog, Phys. Lett. **B 86**, 297 (1979).
38. S. Weinberg, I.I. Rabi Festschrift (1977);
F. Wilczek and A. Zee, Phys. Lett. **70B**, 418 (1977);
H. Fritzsch, Phys. Lett. **70B**, 436 (1977).
39. For a related but different $SO(10)$ model see C. Albright, K.S. Babu ans S.M. Barr, Phys. Rev. Lett. **81**, 1167 (1998).
40. J. Gasser and H. Leutwyler, Phys. Rept. **87**, 77 (1982).
41. See eg. R. Gupta and T. Bhattacharya, Nucl. Phys. Proc. Suppl. **53**, 292 (1997); and Nucl. Phys. Proc. Suppl. **63**, 45 (1998).

42. See for e.g. V. Barger, M. Berger and P. Ohmann, Phys. Rev. **D47**, 1093 (1993);
M. Carena, S. Pokorski and C. Wagner, Nucl. Phys. **B406**, 59 (1993);
P. Langacker and N. Polonsky, Phys. Rev. **D49**, 1454 (1994);
D.M. Pierce, J.A. Bagger, K. Matchev and R. Zhang, Nucl.Phys. **B491**, 3 (1997);
K.S. Babu and C. Kolda, hep-ph/9811308.
43. S. Dimopoulos, S. Raby and F. Wilczek, Phys. Lett. **B112**, 133 (1982);
J. Ellis, D.V. Nanopoulos and S. Rudaz, Nucl. Phys. **B202**, 43 (1982).
44. P. Nath, A.H. Chemseddine and R. Arnowitt, Phys. Rev. **D32**, 2348 (1985);
P. Nath and R. Arnowitt, hep-ph/9708469.
45. J. Hisano, H. Murayama and T. Yanagida, Nucl. Phys. **B402**, 46 (1993).
46. K.S. Babu and S.M. Barr, Phys. Rev. **D50**, 3529 (1994); **D51**, 2463 (1995).
47. SuperKamiokande collaboration, Y. Hayato, talk given at the Int. Cof. High Energy Physics, Vancouver, July (1998).
48. See e.g. Ref. 45
49. S. Dimopoulos and F. Wilczek, Report No. NSF-ITP-82-07 (1981), in *The unity of fundamental interactions*, Proc. of the 19th Course of the International School on Subnuclear Physics, Erice, Italy, Erice, Italy, 1981, Plenum Press, New York (Ed. A. Zichichi);
K.S. Babu and S.M. Barr, Phys. Rev. **D48**, 5354 (1993).
50. For a recent work, comparing the results of lattice and chiral lagrangian-calculations for the $p \to \pi^o, p \to \pi^+$ and $p \to K^o$ modes, see N. Tatsui et al (JLQCD collaboration), hep-lat/9809151.

NEUTRINO MASSES
AND
LEPTON-QUARK SYMMETRIES

P. RAMOND
Institute for Fundamental Theory
Department of Physics, University of Florida
Gainesville, Fl 32611

The recent discovery by SuperKamiokande of evidence for neutrino masses requires the addition of at least seven new parameters to the Standard Model. We discuss the general theoretical schemes which require their inclusion, and point out how quark-lepton symmetries, either in the framework of Grand Unification, or of string theories, can be used to determine them

1 Neutrino Story

The Neutrino Story starts with the experiment of O. Von Bayer, O. Hahn, and L. Meitner [1,2] who measured the spectrum of electrons in β radioactivity, and found it to be discrete! In 1914, Chadwick [3], then in Geiger's laboratory in Berlin came to the correct conclusion of a continuous electron spectrum, but was interned for the duration of the Great War. After much controversy, the issue was settled in 1927 by C.D. Ellis and W. A. Wooster [4], who found the mean energy liberated in β decay accounted to be only 1/3 of the allowed energy. The stage was set for W. Pauli's famous 1930 letter.

In December of that year, in a letter that starts with typical panache, "*Dear Radioactive Ladies and Gentlemen...*", W. Pauli puts forward a "*desperate*" way out: there is a companion particle to the β electron. Undetected, it must be electrically neutral, and in order to balance the $N - Li^6$ statistics, it carries spin 1/2. He calls it the *neutron*, but sees no reason why it could not be massive.

In 1933, E. Fermi in his β decay paper gave it its final name, the little neutron or *neutrino*, as it is clearly much lighter than Chadwick's neutron which had just been discovered.

In 1945, B. Pontecorvo [5] proposes the unthinkable: neutrinos can be detected, through the following observation: an electron neutrino that hits a ^{37}Cl atom will transform it into the inert radioactive gas ^{37}Ar, which then can be stored and be detected through its radioactive decay. Pontecorvo did not publish the report, perhaps because of its secret classification, or because Fermi thought the idea ingenious but not immediately achievable. In 1954, Davis [6] follows up on Pontecorvo's original proposal, by setting a tank of cleaning fluid outside a nuclear reactor.

In 1956, using a scintillation counter experiment they had proposed three years earlier [7], Cowan and Reines [8] discover electron antineutrinos through the reaction $\overline{\nu}_e + p \to e^+ + n$. Cowan passed away before 1995, the year Fred Reines was awarded the Nobel Prize for their discovery. There emerge two lessons in neutrino physics: not only is patience required but also longevity: it took 26 years from birth to detection and then another 39 for the Nobel Committee to recognize the achievement! This should encourage future physicists to train their children at the earliest age to follow their footsteps, in order to establish dynasties of neutrino physicists.

In 1956, it was rumored that Davis had found evidence for neutrinos coming from a pile, and Pontecorvo [9], influenced by the recent work of Gell-Mann and Pais, theorized that an antineutrino produced in the Savannah reactor could oscillate into a neutrino and be detected by Davis. The rumor went away, but the idea of neutrino oscillations was born; it has remained with us ever since, and proven the most potent tool in hunting for neutrino masses.

Having detected neutrinos, there remained to determine its spin and mass. Its helicity was measured in 1958 by M. Goldhaber [10], but convincing evidence for its mass had, until SuperK's bombshell, eluded experimentalists.

After the 1957 Lee and Yang proposal of parity violation, the neutrino is again at the center of the action. Unlike the charged elementary particles which have both left- and right-handed components, weakly interacting neutrinos are purely left-handed (antineutrinos are right-handed), which means that lepton-number is chiral.

The second neutrino, the muon neutrino is detected [11] in 1962, (long anticipated by theorists Inouë and Sakata in 1943 [12]). This time things went a bit faster as it took only 19 years from theory (1943) to discovery (1962) and 26 years to Nobel recognition (1988).

That same year, Maki, Nakagawa and Sakata [13] introduce two crucial ideas; one is that these two neutrinos can mix, and the second is that this mixing can cause one type of neutrino to oscillate into the other (called today flavor oscillation). This is possible only if the two neutrino flavors have different masses.

In 1963, the Astrophysics group at Caltech, Bahcall, Fowler, Iben and Sears [14] puts forward the most accurate of neutrino fluxes from the Sun. Their calculations included the all important Boron decay spectrum, which produces neutrinos with the right energy range for the Chlorine experiment.

In 1964, using Bahcall's result [15] of an enhanced capture rate of ^8B neutrinos through an excited state of ^{37}Ar, Davis [16] proposes to search for ^8B solar neutrinos using a 100,000 gallon tank of cleaning fluid deep underground. Soon after, R. Davis starts his epochal experiment at the Homestake mine, marking the beginning of the solar neutrino watch which continues to this day. In 1968, Davis et al reported [17] a deficit in the solar neutrino flux, a result that has withstood scrutiny to this day, and stands as a truly remarkable experimental *tour de force*. Shortly after, Gribov and Pontecorvo [18] interpreted the deficit as evidence for neutrino oscillations.

In the early 1970's, with the idea of quark-lepton symmetries [19,20] comes the idea that the proton could be unstable. This brings about the construction of underground (to avoid contamination from cosmic ray by-product) detectors, large enough to monitor many protons, and instrumentalized to detect the Čerenkov light emitted by its decay products. By the middle 1980's, several such detectors are in place. They fail to detect proton decay, but in a serendipitous turn of events, 150,000 years earlier, a supernova erupted in the large Magellanic Cloud, and in 1987, its burst of neutrinos was detected in these detectors! All of a sudden, proton decay detectors turn their attention to neutrinos, and to this day still waiting for its protons to decay!

As we all know, these detectors routinely monitor neutrinos from the Sun, as well as neutrinos produced by cosmic ray collisions.

2 Standard Model Neutrinos

The standard model of electro-weak and strong interactions contains three left-handed neutrinos. The three neutrinos are represented by two-components Weyl spinors, ν_i, $i = e, \mu, \tau$, each describing a left-handed fermion (right-handed antifermion). As the upper components of weak isodoublets L_i, they have $I_{3W} = 1/2$, and a unit of the global ith lepton number.

These standard model neutrinos are strictly massless. The only Lorentz scalar made out of these neutrinos is the Majorana mass, of the form $\nu_i^t \nu_j$; it has the quantum numbers of a weak isotriplet, with third component $I_{3W} = 1$, as well as two units of total lepton number. Higgs isotriplet with two units of lepton number could generate neutrino Majorana masses, but there is no such higgs in the Standard Model: there are no tree-level neutrino masses in the standard model.

Quantum corrections, however, are not limited to renormalizable couplings, and it is easy to make a weak isotriplet out of two isodoublets, yielding the $SU(2) \times U(1)$ invariant $L_i^t \vec{\tau} L_j \cdot H^t \vec{\tau} H$, where H is the Higgs doublet. As this term is not invariant under lepton number, it is not be generated in perturbation theory. Thus the important conclusion: *The standard model neutrinos are kept massless by global chiral lepton number symmetry.* The detection of non-zero neutrino masses is therefore *a tangible indication of physics beyond the standard model.*

3 Neutrino Mass Models

Neutrinos must be extraordinarily light: experiments indicate $m_{\nu_e} < 10$ eV, $m_{\nu_\mu} < 170$ keV, $m_{\nu_\tau} < 18$ MeV [21], and any model of neutrino masses must explain this suppression.

We do not discuss generating neutrino masses without new fermions, by breaking lepton number through interaction of lepton number-carrying Higgs fields.

The natural way to generate neutrinos masses is to introduce for each one its electroweak singlet Dirac partner, $\overline{N}_i$. These appear naturally in the Grand Unified group $SO(10)$ where they complete each family into its spinor representation. Neutrino Dirac masses stem from the couplings $L_i\overline{N}_jH$ after electroweak breaking. Unfortunately, these Yukawa couplings yield masses which are too big, of the same order of magnitude as the masses of the charged elementary particles $m \sim \Delta I_w = 1/2$.

The situation is remedied by introducing Majorana mass terms $\overline{N}_i\overline{N}_j$ for the right-handed neutrinos. The masses of these new degrees of freedom are arbitrary, as they have no electroweak quantum numbers, $M \sim \Delta I_w = 0$. If they are much larger than the electroweak scale, the neutrino masses are suppressed relative to that of their charged counterparts by the ratio of the electroweak scale to that new scale: the mass matrix (in 3×3 block form) is

$$\begin{pmatrix} 0 & m \\ m & M \end{pmatrix} , \tag{1}$$

leading, for each family, to one small and one large eigenvalue

$$m_\nu \sim m \cdot \frac{m}{M} \sim \left(\Delta I_w = \frac{1}{2}\right) \cdot \left(\frac{\Delta I_w = \frac{1}{2}}{\Delta I_w = 0}\right) . \tag{2}$$

This seesaw mechanism [22] provides a natural explanation for small neutrino masses as long as lepton number is broken at a large scale M. With M around the energy at which the gauge couplings unify, this yields neutrino masses at or below tenths of eVs, consistent with the SuperK results.

The lepton flavor mixing comes from the diagonalization of the charged lepton Yukawa couplings, and of the neutrino mass matrix. From the charged lepton Yukawas, we obtain $\mathcal{U}_e$, the unitary matrix that rotates the lepton doublets L_i. From the neutrino Majorana matrix, we obtain $\mathcal{U}_\nu$, the matrix that diagonalizes the Majorana mass matrix. The 6×6 seesaw Majorana matrix can be written in 3×3 block form

$$\mathcal{M} = \mathcal{V}_\nu^t \, \mathcal{D} \mathcal{V}_\nu \sim \begin{pmatrix} \mathcal{U}_{\nu\nu} & \epsilon \mathcal{U}_{\nu N} \\ \epsilon \mathcal{U}_{N\nu}^t & \mathcal{U}_{NN} \end{pmatrix} , \tag{3}$$

where ϵ is the tiny ratio of the electroweak to lepton number violating scales, and $\mathcal{D} = \text{diag}(\epsilon^2 \mathcal{D}_\nu, \mathcal{D}_N)$, is a diagonal matrix. $\mathcal{D}_\nu$ contains the three neutrino masses, and ϵ^2 is the seesaw suppression. The weak charged current is then given by

$$j_\mu^+ = e_i^\dagger \sigma_\mu \mathcal{U}_{MNS}^{ij} \nu_j , \tag{4}$$

where

$$\mathcal{U}_{MNS} = \mathcal{U}_e \mathcal{U}_\nu^\dagger , \tag{5}$$

is the Maki-Nakagawa-Sakata[13] (MNS) flavor mixing matrix, the analog of the CKM matrix in the quark sector.

In the seesaw-augmented standard model, this mixing matrix is totally arbitrary. It contains, as does the CKM matrix, three rotation angles, and one CP-violating phase. In the seesaw scenario, it also contains two additional CP-violating phases which cannot be absorbed in a redefinition of the neutrino fields, because of their Majorana masses (these extra phases can be measured only in $\Delta \mathcal{L} = 2$ processes). These additional parameters of the seesaw-augmented standard model, need to be determined by experiment.

4 Theories

Theoretical predictions of lepton hierarchies and mixings depend very much on hitherto untested theoretical assumptions. In the quark sector, where the bulk of the experimental data resides, the theoretical origin of quark hierarchies and mixings is a mystery, although there exits many theories, but none so convincing as to offer a definitive answer to the community's satisfaction. It is therefore no surprise that there are more theories of lepton masses and mixings than there are parameters to be measured. Nevertheless, one can present the issues as questions:

- Do the right handed neutrinos have quantum numbers beyond the standard model?
- Are quarks and leptons related by grand unified theories?
- Are quarks and leptons related by anomalies?
- Are there family symmetries for quarks and leptons?

The measured numerical value of the neutrino mass difference (barring any fortuitous degeneracies), suggests through the seesaw mechanism, a mass for the right-handed neutrinos that is consistent with the scale at which the gauge couplings unify. Is this just a numerical coincidence, or should we view this as a hint for grand unification?

Grand unified Theories, originally proposed as a way to treat leptons and quarks on the same footing, imply symmetries much larger than the standard model's. Implementation of these ideas necessitates a desert and supersymmetry, but also a carefully designed contingent of Higgs particles to achieve the desired symmetry breaking. That such models can be built is perhaps more of a testimony to the cleverness of theorists rather than of Nature's. Indeed with the advent of string theory, we know that the best features of grand unified theories can be preserved, as most of the symmetry breaking is achieved by geometric compactification from higher dimensions [23].

An alternative point of view is that the vanishing of chiral anomalies is necessary for consistent theories, and their cancellation is most easily achieved by assembling matter in representations of anomaly-free groups. Perhaps anomaly cancellation is more important than group structure.

Below, we present two theoretical frameworks of our work, in which one deduces the lepton mixing parameters and masses. One is ancient [24], uses the standard techniques of grand unification, but it had the virtue of *predicting* the large $\nu_\mu - \nu_\tau$ mixing observed by SuperKamiokande. The other [25] is more recent, and uses extra Abelian family symmetries to explain both quark and lepton hierarchies. It also predicts large $\nu_\mu - \nu_\tau$ mixing. Both schemes imply small $\nu_e - \nu_\mu$ mixings.

4.1 A Grand Unified Model

The seesaw mechanism was born in the context of the grand unified group $SO(10)$, which naturally contains electroweak neutral right-handed neutrinos. Each standard model family is contained in two irreducible representations of $SU(5)$. However, the predictions of this theory for Yukawa couplings is not so clear cut, and to reproduce the known quark and charged lepton hierarchies, a special but simple set of Higgs particles had to be included. In the simple scheme proposed by Georgi and Jarlskog [26], the ratios between the charged leptons and quark masses is reproduced, albeit not naturally since two Yukawa couplings, not fixed by group theory, had to be set equal. This motivated us to generalize their scheme to $SO(10)$, where their scheme was (technically) natural, which meant that we had an automatic window into neutrino masses through the seesaw. The Yukawa couplings were of the form

$$[A\mathbf{16}_1 \cdot \mathbf{16}_2 + B\mathbf{16}_3\mathbf{16}_3] \cdot \mathbf{126}_1 + [a\mathbf{16}_1 \cdot \mathbf{16}_2 + b\mathbf{16}_3\mathbf{16}_3] \cdot (\mathbf{10}_1 + i\mathbf{10}_2) \quad (6)$$

$$+c\mathbf{16}_2\mathbf{16}_2 \cdot \overline{\mathbf{126}}_2 + d\mathbf{16}_2\mathbf{16}_3\overline{\mathbf{126}}_3 \; . \quad (7)$$

This is of course Higgs-heavy, but the attitude at the time was "damn the Higgs torpedoes, and see what happens". This assignment was "technically" natural, enforced by two discrete symmetries. A modern treatment would include non-renormalizable operators [27], rather than introducing the **126** representations, which spoil asymptotic freedom.

The Higgs vacuum values produced the resultant masses

$$m_b = m_\tau \; ; \qquad m_d m_s = m_e m_\mu \; ; \qquad m_d - m_s = 3(m_e - m_\mu) \; . \quad (8)$$

and mixing angles

$$V_{us} = \tan\theta_c = \sqrt{\frac{m_d}{m_s}} \; ; \qquad V_{cb} = \sqrt{\frac{m_c}{m_t}} \; . \quad (9)$$

While reproducing the well-known lepton and quark mass hierarchies, it predicted a long-lived b quark, contrary to the lore of the time. It also made predictions in the lepton sector, namely **maximal** $\nu_\tau - \nu_\mu$ mixing, small $\nu_e - \nu_\mu$ mixing of the order of $(m_e/m_\mu)^{1/2}$, and no $\nu_e - \nu_\tau$ mixing.

The neutral lepton masses came out to be hierarchical, but heavily dependent on the masses of the right-handed neutrinos. The electron neutrino mass came out much lighter than those of ν_μ and ν_τ. Their numerical values depended on the top quark mass, which was then supposed to be in the tens of GeVs!

Given the present knowledge, some of the features are remarkable, such as the long-lived b quark and the maximal $\nu_\tau - \nu_\mu$ mixing. On the other hand, the actual numerical value of the b lifetime was off a bit,and the $\nu_e - \nu_\mu$ mixing was too large to reproduce the small angle MSW solution of the solar neutrino problem.

The lesson should be that the simplest $SO(10)$ model that fits the observed quark and charged lepton hierarchies, reproduces, at least qualitatively, the maximal mixing found by SuperK, and predicts small mixing with the electron neutrino [28].

4.2 A Non-grand-unified Model

There is another way to generate hierarchies, based on adding extra family symmetries to the standard model, without invoking grand unification. These types of models address only the Cabibbo suppression of the Yukawa couplings, and are not as predictive as specific grand unified models. Still, they predict no Cabibbo suppression between the muon and tau neutrinos. Below, we present a pre-SuperK model [25] with those features.

The Cabibbo supression is assumed to be an indication of extra family symmetries in the standard model. The idea is that any standard model-invariant operator, such as $\mathbf{Q}_i\overline{\mathbf{d}}_jH_d$, cannot be present at tree-level if there are additional symmetries under which the operator is not invariant. Simplest is to assume an Abelian symmetry, with an electroweak singlet field θ, as its order parameter. Then the interaction

$$\mathbf{Q}_i\overline{\mathbf{d}}_jH_d\left(\frac{\theta}{M}\right)^{n_{ij}} \tag{10}$$

can appear in the potential as long as the family charges balance under the new symmetry. As θ acquires a *vev*, this leads to a suppression of the Yukawa couplings of the order of $\lambda^{n_{ij}}$ for each matrix element, with $\lambda = \theta/M$ identified with the Cabibbo angle, and M is the natural cut-off of the effective low energy theory. As a consequence of the charge balance equation

$$X_{if}^{[d]} + n_{ij}X_\theta = 0\ , \tag{11}$$

the exponents of the suppression are related to the charge of the standard model-invariant operator [29], the sum of the charges of the fields that make up the the invariant.

This simple Ansatz, together with the seesaw mechanism, implies that the family structure of the neutrino mass matrix is determined by the charges of the left-handed lepton doublet fields.

Each charged lepton Yukawa coupling $L_i\overline{N}_jH_u$, has an extra charge $X_{L_i} + X_{Nj} + X_H$, which gives the Cabibbo suppression of the ij matrix element. Hence, the orders of magnitude of these couplings can be expressed as

$$\begin{pmatrix} \lambda^{l_1} & 0 & 0 \\ 0 & \lambda^{l_2} & 0 \\ 0 & 0 & \lambda^{l_3} \end{pmatrix} \hat{Y} \begin{pmatrix} \lambda^{p_1} & 0 & 0 \\ 0 & \lambda^{p_2} & 0 \\ 0 & 0 & \lambda^{p_3} \end{pmatrix}\ , \tag{12}$$

where $\hat{Y}$ is a Yukawa matrix with no Cabibbo suppressions, $l_i = X_{L_i}/X_\theta$ are the charges of the left-handed doublets, and $p_i = X_{N_i}/X_\theta$, those of the singlets. The first matrix forms half of the MNS matrix. Similarly, the mass matrix for the right-handed neutrinos, $\overline{N}_i\overline{N}_j$ will be written in the form

$$\begin{pmatrix} \lambda^{p_1} & 0 & 0 \\ 0 & \lambda^{p_2} & 0 \\ 0 & 0 & \lambda^{p_3} \end{pmatrix} \mathcal{M} \begin{pmatrix} \lambda^{p_1} & 0 & 0 \\ 0 & \lambda^{p_2} & 0 \\ 0 & 0 & \lambda^{p_3} \end{pmatrix}\ . \tag{13}$$

The diagonalization of the seesaw matrix is of the form

$$L_i H_u \overline{N}_j \left(\frac{1}{\overline{N}\ \overline{N}} \right)_{jk} \overline{N}_k H_u L_l \ , \tag{14}$$

from which the Cabibbo suppression matrix from the $\overline{N}_i$ fields *cancels*, leaving us with

$$\begin{pmatrix} \lambda^{l_1} & 0 & 0 \\ 0 & \lambda^{l_2} & 0 \\ 0 & 0 & \lambda^{l_3} \end{pmatrix} \hat{\mathcal{M}} \begin{pmatrix} \lambda^{l_1} & 0 & 0 \\ 0 & \lambda^{l_2} & 0 \\ 0 & 0 & \lambda^{l_3} \end{pmatrix} , \tag{15}$$

where $\hat{\mathcal{M}}$ is a matrix with no Cabibbo suppressions. The Cabibbo structure of the seesaw neutrino matrix is determined solely by the charges of the lepton doublets! As a result, the Cabibbo structure of the MNS mixing matrix is also due entirely to the charges of the three lepton doublets. This general conclusion depends on the existence of at least one Abelian family symmetry, which we argue is implied by the observed structure in the quark sector.

The Wolfenstein parametrization of the CKM matrix[30],

$$\begin{pmatrix} 1 & \lambda & \lambda^3 \\ \lambda & 1 & \lambda^2 \\ \lambda^3 & \lambda^2 & 1 \end{pmatrix} , \tag{16}$$

and the Cabibbo structure of the quark mass ratios

$$\frac{m_u}{m_t} \sim \lambda^8 \ \ \frac{m_c}{m_t} \sim \lambda^4 \ \ ; \ \ \frac{m_d}{m_b} \sim \lambda^4 \ \ \frac{m_s}{m_b} \sim \lambda^2 \ , \tag{17}$$

can be reproduced[25,31] by a simple *family-traceless* charge assignment for the three quark families, namely

$$X_{\mathbf{Q},\overline{\mathbf{u}},\overline{\mathbf{d}}} = \mathcal{B}(2,-1,-1) + \eta_{\mathbf{Q},\overline{\mathbf{u}},\overline{\mathbf{d}}}(1,0,-1) \ , \tag{18}$$

where $\mathcal{B}$ is baryon number, $\eta_{\overline{\mathbf{d}}} = 0$, and $\eta_{\mathbf{Q}} = \eta_{\overline{\mathbf{u}}} = 2$. Two striking facts are evident:

- the charges of the down quarks, $\overline{\mathbf{d}}$, associated with the second and third families are the same,
- $\mathbf{Q}$ and $\overline{\mathbf{u}}$ have the same value for η.

To relate these quark charge assignments to those of the leptons, we need to inject some more theoretical prejudices. Assume these family-traceless charges are gauged, and not anomalous. Then to cancel anomalies, the leptons must themselves have family charges.

Anomaly cancellation generically implies group structure. In $SO(10)$, baryon number generalizes to $\mathcal{B}-\mathcal{L}$, where $\mathcal{L}$ is total lepton number, and in $SU(5)$ the fermion assignment is $\overline{\mathbf{5}}=\overline{\mathbf{d}}+L$, and $\mathbf{10}=\mathbf{Q}+\overline{\mathbf{u}}+\overline{e}$. Thus anomaly cancellation is easily achieved by assigning $\eta=0$ to the lepton doublet L_i, and $\eta=2$ to the electron singlet $\overline{e}_i$, and by generalizing baryon number to $\mathcal{B}-\mathcal{L}$, leading to the charges

$$X_{\mathbf{Q},\overline{\mathbf{u}},\overline{\mathbf{d}},L,\overline{e}}=(\mathcal{B}-\mathcal{L})(2,-1,-1)+\eta_{\mathbf{Q},\overline{\mathbf{u}},\overline{\mathbf{d}}}(1,0,-1)\ , \tag{19}$$

where now $\eta_{\overline{\mathbf{d}}}=\eta_L=0$, and $\eta_{\mathbf{Q}}=\eta_{\overline{\mathbf{u}}}=\eta_{\overline{e}}=2$. It is interesting to note that η is at least in E_6. The origin of such charges is not clear, as it implies in the superstring context, rather unconventional compactification.

As a result, the charges of the lepton doublets are simply $X_{L_i}=-(2,-1,-1)$. We have just argued that these charges determine the Cabibbo structure of the MNS lepton mixing matrix to be

$$\mathcal{U}_{MNS}\sim\begin{pmatrix}1 & \lambda^3 & \lambda^3\\ \lambda^3 & 1 & 1\\ \lambda^3 & 1 & 1\end{pmatrix}\ , \tag{20}$$

implying*no Cabibbo suppression in the mixing between ν_μ and ν_τ*. This is consistent with the SuperK discovery and with the small angle MSW[33] solution to the solar neutrino deficit. One also obtains a much lighter electron neutrino, and Cabibbo-comparable masses for the muon and tau neutrinos. Notice that these predictions are subtly different from those of grand unification, as they yield $\nu_e-\nu_\tau$ mixing. It also implies a much lighter electron neutrino, and Cabibbo-comparable masses for the muon and tau neutrinos.

On the other hand, the scale of the neutrino mass values depend on the family trace of the family charge(s). Here we simply quote the results our model [25]. The masses of the right-handed neutrinos are found to be of the following orders of magnitude

$$m_{\overline{N}_e}\sim M\lambda^{13}\ ;\qquad m_{\overline{N}_\mu}\sim m_{\overline{N}_\tau}\sim M\lambda^7\ , \tag{21}$$

where M is the scale of the right-handed neutrino mass terms, assumed to be the cut-off. The seesaw mass matrix for the three light neutrinos comes out to be

$$m_0\begin{pmatrix}a\lambda^6 & b\lambda^3 & c\lambda^3\\ b\lambda^3 & d & e\\ c\lambda^3 & e & f\end{pmatrix}\ , \tag{22}$$

where we have added for future reference the prefactors a, b, c, d, e, f, all of order one, and

$$m_0 = \frac{v_u^2}{M\lambda^3} , \tag{23}$$

where v_u is the *vev* of the Higgs doublet. This matrix has one light eigenvalue

$$m_{\nu_e} \sim m_0 \lambda^6 . \tag{24}$$

Without a detailed analysis of the prefactors, the masses of the other two neutrinos come out to be both of order m_0. The mass difference announced by superK [32] cannot be reproduced without going beyond the model, by taking into account the prefactors. The two heavier mass eigenstates and their mixing angle are written in terms of

$$x = \frac{df - e^2}{(d+f)^2} , \qquad y = \frac{d-f}{d+f} , \tag{25}$$

as

$$\frac{m_{\nu_2}}{m_{\nu_3}} = \frac{1 - \sqrt{1-4x}}{1 + \sqrt{1-4x}} , \qquad \sin^2 2\theta_{\mu\tau} = 1 - \frac{y^2}{1-4x} . \tag{26}$$

If $4x \sim 1$, the two heaviest neutrinos are nearly degenerate. If $4x \ll 1$, a condition easy to achieve if d and f have the same sign, we can obtain an adequate split between the two mass eigenstates. For illustrative purposes, when $0.03 < x < 0.15$, we find

$$4.4 \times 10^{-6} \leq \Delta m^2_{\nu_e - \nu_\mu} \leq 10^{-5}\ rmeV^2 , \tag{27}$$

which yields the correct non-adiabatic MSW effect, and

$$5 \times 10^{-4} \leq \Delta m^2_{\nu_\mu - \nu_\tau} \leq 5 \times 10^{-3}\ \mathrm{eV}^2 , \tag{28}$$

for the atmospheric neutrino effect. These were calculated with a cut-off, 10^{16} GeV $< M < 4 \times 10^{17}$ GeV, and a mixing angle, $0.9 < \sin^2 2\theta_{\mu-\tau} < 1$. This value of the cut-off is compatible not only with the data but also with the gauge coupling unification scale, a necessary condition for the consistency of our model, and more generally for the basic ideas of Grand Unification.

5 Outlook

Exact predictions of neutrino masses and mixings depend on developing a credible theory of flavor. In the absence of such, we have presented two schemes, which predicted not only maximal $\nu_\mu - \nu_\tau$ mixing, but also smal $\nu_e - \nu_\mu$ mixings.

Neither scheme includes sterile neutrinos. The present experimental situation is somewhat unclear: the LSND results [34] imply the presence of a sterile neutrino; at this conference we heard that superK favors $\nu_\mu - \nu_\tau$ oscillation over $\nu_\mu - \nu_{\rm sterile}$, and the origin of the solar neutrino deficit remains a puzzle, which several possible explanations. One is the non-adiabatic MSW effect in the Sun, which our theoretical ideas seem to favor. However, it is an experimental question which is soon to be answered by the continuing monitoring of the ^8B spectrum by SuperK, and the advent of the SNO detector. Neutrino physics is at an exciting stage, and experimentally vibrant, as upcoming measurements will help us determine iour basic ideas about fundamental interactions.

6 Acknowledgments

I would like to thank Professors G. Domokos and S. Kövesi-Domokos for their usual superb hospitality and the high scientific quality of this workshop. This research was supported in part by the department of energy under grant DE-FG02-97ER41029.

1. Some of the following follows the excellent article by P. Radvanyi, *From Becquerel to Pauli* in Neutrinos, Dark Matter and the Universe, T. Stolarcyk, J. Tran Thanh Van, F. Vanucci, editors (Editions Frontières, France).
2. O. Von Bayer, O. Hahn, and L. Meitner, Phys. Zeitschr. **12**, 273(1911); *ibid* **13**, 273(1911); **13** 264(1912).
3. J. Chadwick, Verh. d. D. Phys. Ges., **16**, 383(1914).
4. C. D. Ellis and W. A. Wooster, Proc. Royal Soc. **A117**, 109(1927).
5. B. Pontecorvo, Chalk river Report PD-205, November 1946, unpublished.
6. Raymond Davis Jr., Phys Rev **97**, 766(1955).
7. C. L. Cowan and F. Reines, Phys.Rev. **90**, 492(1953).
8. C.L. Cowan, F. Reines, F.B. Harrison, H.W. Kruse, A.D. McGuire , Science **124**, 103(1956).
9. B. Pontecorvo, JETP (USSR) **34**, 247(1958).
10. M. Goldhaber, L. Grodzins, A.W. Sunyar, Phys.Rev. **109**, 1015(1958).
11. G. Danby, J.M. Gaillard, K. Goulianos, L.M. Lederman, N. Mistry, M. Schwartz, J. Steinberger, Phys.Rev.Lett. **9**, 36(1962).
12. S. Sakata and T. Inouë, Prog. Theo. Physics, **1**, 143(1946).
13. Z. Maki, M. Nakagawa and S. Sakata, Prog. Theo. Physics, **28**, 247(1962). B. Pontecorvo, Zh. Eksp. Teor. Fiz. **53**, 1717(1967).
14. J. Bahcall, W. A. Fowler, I. Iben and R. L. Sears, Astrophysics. J. **137**, 344(1963).
15. J. Bahcall, Phys. Rev. Lett. **12**, 300(1964).

16. Raymond Davis Jr., Phys. Rev. Lett. **12**, 303(1964).
17. Raymond Davis Jr., D. Harmer and K. Hoffman, Phys. Rev. Lett. **20**, 1205(1968).
18. V. Gribov and B. Pontecorvo, Phys. Lett. **B28**, 493(1969).
19. J. Pati and A. Salam, Phys. Rev. Lett. **31**, 661(1973)
20. H. Georgi and S. L. Glashow, Phys. Rev. Lett. **32**, 438(1974) ; H. Fritzsch and P. Minkowski, Annals Phys. **93**, 193(1975); H. Georgi, in AIP Conference Proceedings no 23, Williamsburg, Va, 1975; F. Gürsey, P. Ramond and P. Sikivie, Phys. Lett. **60B**, 177(1976)
21. Particle Data Group, R. M. Barnett *et al.*, Phys Rev **D54**, 1(1996).
22. M. Gell-Mann, P. Ramond, and R. Slansky in Sanibel Talk, CALT-68-709, Feb 1979 (unpublished), and in *Supergravity* (North Holland, Amsterdam 1979). T. Yanagida, in *Proceedings of the Workshop on Unified Theory and Baryon Number of the Universe*, KEK, Japan, 1979.
23. P. Candelas, G. Horowitz, A. Strominger and E. Witten, Nucl. Phys. **B258**, 46(1985)
24. J. A. Harvey, P. Ramond and D. B. Reiss, Nucl. Phys. **B199**, 223(1982)
25. N. Irges, S. Lavignac and P. Ramond, Phys. Rev. **D58**, 035003(1998).
26. H. Georgi and C. Jarlskog, Phys. Lett. **B86**, 297(1979)
27. K. S. Babu, J. Pati and F. Wilczek, hep-ph/9812538.
28. The Case for Neutrino Oscillations, P. Ramond, Proceedings of the Los Alamos Neutrino Workshop, LA-9358-C, June 1981.
29. C. Froggatt and H. B. Nielsen Nucl. Phys. B147 (1979) 277; P. Ramond, R.G. Roberts and G.G. Ross, Nucl. Phys. B406 (1993)
30. L. Wolfenstein, Phys. Rev. Lett. **51**, 1945(1983).
31. J. Elwood, N., Irges, and P. Ramond, Phys. Rev. Lett. **81**, 5064(1998)
32. Super-Kamiokande Collaboration, Phys. Rev. Lett. **81**, 1562(1998)
33. L. Wolfenstein, Phys. Rev. D17, 2369 (1978); S. Mikheyev and A. Yu Smirnov, Nuovo Cim. **9C**, 17 (1986).
34. C. Athanassopoulos *et al.*, Phys. Rev. Lett. **75**, 2560(1995); **77**, 3082(1996); nucl-ex/9706006.

MODELS OF NEUTRINO MASSES FROM OSCILLATIONS WITH LARGE MIXINGS

FERRUCCIO FERUGLIO
University of Padova and I.N.F.N, Padova, Italy
E-mail: feruglio@pd.infn.it

We discuss models of neutrino masses that, in the context of the see-saw mechanism, could lead to a large mixing angle for the atmospheric neutrino oscillations without requiring too much fine–tuning between the Dirac and the Majorana sectors. These models are compatible with Abelian flavour symmetries and with the picture of flavour expected in grand unified theories.

1 Introduction

Recent data from Superkamiokande [1] have provided a more solid experimental basis for neutrino oscillations as an explanation of the atmospheric neutrino anomaly. In addition, also the solar neutrino deficit, observed by several experiments [2], is probably an indication of a different sort of neutrino oscillations. Neutrino oscillations imply neutrino masses. The extreme smallness of neutrino masses in comparison with quark and charged lepton masses indicates a different nature of the former, presumably linked to lepton-number violation and the Majorana nature of neutrinos. Thus neutrino masses provide a window on the very large energy scale where lepton number is violated and on Grand Unified Theories (GUTs). Experimental facts on neutrino masses and mixings could give an important feedback on the problem of quark and charged lepton masses, as all these masses are possibly related in GUTs. In particular the observation of a nearly maximal mixing angle for $\nu_\mu \to \nu_\tau$ is particularly interesting. Perhaps also solar neutrinos may occur with large mixing angle. At present solar neutrino mixings can be either large or very small, depending on which particular solution will eventually be established by the data. Large mixings in the neutrino sector are very interesting because a first guess was in favour of small mixings, in analogy to what is observed for quarks. If confirmed, single or double maximal mixings can provide an important hint on the mechanisms that generate neutrino masses.

The experimental status of neutrino oscillations is still very preliminary. Thus, in order to be able to proceed, the theorist has to make a number of assumptions on how the data will finally look when the experimental situation will be completely clarified.

1.1 Three Light Neutrinos

Here we assume that only two distinct oscillation frequencies exist, the largest being associated with atmospheric neutrinos and the smallest with solar neutrinos. We assume that the hint of an additional frequency from the LSND experiment [3], not confirmed by the Karmen experiment [4] (but yet far from being completely excluded), will disappear. Thus we avoid the introduction of new sterile neutrino species and can deal with only the three known species of light neutrinos [a].

We interpret the atmospheric neutrino oscillations as nearly maximal $\nu_\mu \to \nu_\tau$ oscillations, in agreement with the Chooz results [6]. The solar-neutrino oscillations correspond to the disappearance of ν_e into nearly equal fractions of ν_μ and ν_τ. A priori we are open minded about which of the three most likely solutions for solar neutrino oscillations is adopted: the two MSW solutions with small (SA) or large (LA) mixing angle, or the vacuum oscillation solution (VO).

1.2 A Useful Parametrization

Maximal atmospheric neutrino mixing and the requirement that the electron neutrino does not participate in the atmospheric oscillations, as indicated by the Superkamiokande [1] and Chooz [6] data, lead directly to the following structure of the U_{fi} ($f = e, \mu, \tau$, i=1,2,3) real orthogonal mixing matrix, apart from sign convention redefinitions (here we are not interested in CP violation effects: all matrices are taken real)

$$U_{fi} = \begin{bmatrix} c & -s & 0 \\ s/\sqrt{2} & c/\sqrt{2} & -1/\sqrt{2} \\ s/\sqrt{2} & c/\sqrt{2} & +1/\sqrt{2} \end{bmatrix} \quad . \tag{1}$$

This result is obtained by a simple generalization of the analysis of ref. [7] (also discussed in ref. [8]) to the case of arbitrary solar mixing angle ($s \equiv \sin\theta_{sun}$, $c \equiv \cos\theta_{sun}$): $c = s = 1/\sqrt{2}$ for maximal solar mixing (e.g. for vacuum oscillations $\sin^2 2\theta_{sun} \sim 0.75$) , while $\sin^2 2\theta_{sun} \sim 4s^2 \sim 5.5 \cdot 10^{-3}$ for the

[a] Can three light neutrinos accommodate solar, atmospheric and LSND oscillations? This would seem a priori possible if the solar Cl experiment were affected by a large unknown systematic error. In this case an energy-independent suppression of the solar neutrino flux by approximately a factor of two could reasonably describe the data, leaving the solar frequency undetermined in a vast range. Thus we might associate the two independent frequencies to LSND and atmospheric oscillations. It has been observed that such an attractive scenario is incompatible with the combined results of the Chooz and SuperKamiokande experiments, when also the atmospheric neutrino asymmetries are considered [5].

small angle MSW [9] solution. The vanishing of U_{e3} guarantees that ν_e does not participate in the atmospheric oscillations and the relation $|U_{\mu 3}| = |U_{\tau 3}| = 1/\sqrt{2}$ implies maximal mixing for atmospheric neutrinos. Note that we are assuming only two frequencies, given by

$$\Delta_{sun} \propto m_2^2 - m_1^2, \qquad \Delta_{atm} \propto m_3^2 - m_{1,2}^2 \tag{2}$$

The effective light neutrino mass matrix is given by $m_\nu = U m_{diag} U^T$ with $m_{diag} = \mathrm{Diag}[m_1, m_2, m_3]$. For generic s one finds

$$m_\nu = \begin{bmatrix} 2\epsilon & \delta & \delta \\ \delta & \frac{m_3}{2} + \epsilon_2 & -\frac{m_3}{2} + \epsilon_2 \\ \delta & -\frac{m_3}{2} + \epsilon_2 & \frac{m_3}{2} + \epsilon_2 \end{bmatrix} \quad , \tag{3}$$

with

$$\epsilon = (m_1 c^2 + m_2 s^2)/2 \ , \quad \delta = (m_1 - m_2) cs/\sqrt{2} \ , \quad \epsilon_2 = (m_1 s^2 + m_2 c^2)/2 \quad . \tag{4}$$

We see that the existence of one maximal mixing and $U_{e3} = 0$ are the most important input that leads to the matrix form in eq. (3,4). The value of the solar neutrino mixing angle can be left free. While the simple parametrization of the matrix U in eq. (1) is quite useful to guide the search for a realistic pattern of neutrino mass matrices, it should not be taken too literally. In particular the data do not exclude a non-vanishing U_{e3} element. In most of the Superkamiokande allowed region the bound by Chooz [6] amounts to $|U_{e3}| < 0.2$. In the region not covered by Chooz $|U_{e3}|$ can even be larger [10,5]. Thus neglecting $|U_{e3}|$ with respect to s in eq. (1) is not really justified. Also note that in presence of a large hierarchy $|m_3| \gg |m_{1,2}|$ the effect of neglected parameters in eq. (3) can be enhanced by $m_3/m_{1,2}$ and produce seizable corrections. A non vanishing U_{e3} term can lead to different $(m_\nu)_{12}$ and $(m_\nu)_{13}$ terms. Similarly a deviation from maximal mixing $U_{\mu 3} \neq U_{\tau 3}$ distorts the ϵ_2 terms in the 23 sector of m_ν. Therefore, especially for a large hierarchy, there is more freedom in the small terms in order to construct a model that fits the data than it is apparent from eq. (3).

1.3 Hierarchical Spectrum

Since neutrino oscillations only measure differences of squared masses, the observed differences $(\Delta m^2)_{atm} = |m_3^2 - m_2^2| \gg (\Delta m^2)_{sun} = |m_2^2 - m_1^2|$ could correspond to A) hierarchical eigenvalues $|m_3| \gg |m_{2,1}|$ or to partial or total near degeneracy: B) $|m_1| \sim |m_2| \gg |m_3|$ or C) $|m_1| \sim |m_2| \sim |m_3|$. The configurations B) and C) imply a very precise near degeneracy of squared masses. For example, the case C) is the only one that could in principle accommodate neutrinos as hot dark matter together with solar and atmospheric neutrino oscillations. We think that it is not at all clear at the moment that a hot dark matter component is really needed [11] but this could be a reason in favour of the fully degenerate solution. Then the common mass should be around 1-3 eV. The solar frequency could be given by a small 1-2 splitting, while the atmospheric frequency could be given by a still small but much larger 1,2-3 splitting. A strong constraint arises in this case from the non observation of neutrinoless double beta decay which requires that the ee entry of m_ν must obey $|(m_\nu)_{ee}| \leq 0.46$ eV [12]. As observed in ref. [13], this bound can only be satisfied if bimixing is realized (that is double maximal mixing, with solar neutrinos explained by the VO or MSW-LA solutions). But we would need a relative splitting $|\Delta m/m| \sim \Delta m^2_{atm}/2m^2 \sim 10^{-3} - 10^{-4}$ and a much smaller one for solar neutrinos explained by vacuum oscillations: $|\Delta m/m| \sim 10^{-10} - 10^{-11}$. Such a tiny relative mass splitting, arranged at the large energy scale where lepton number is violated, can be easily upset by the renormalization group evolution down to the electroweak scale [14], unless a suitable flavour symmetry protects it during the running [15].

1.4 See-saw Mechanism

For reasons of simplicity, we consider the simplest version of the see-saw mechanism with one Dirac, m_D, and one Majorana, M, mass matrix, related to the neutrino mass matrix m_ν, in the basis where the charged lepton mass matrix is diagonal, by

$$m_\nu = m_D^T M^{-1} m_D \quad . \tag{5}$$

As well known this is not the most general see-saw mechanism because we are not including the left-left Majorana mass block. It is unplausible that starting from hierarchical Dirac matrices we end up via the see-saw mechanism into a nearly perfect degeneracy of squared masses and the assumption of hierarchical Dirac masses and the see-saw mechanism naturally leads to a pattern of type A with $|m_3| \gg |m_2| \gg |m_1|$. Models with degenerate neutrinos (see, for example, refs. [16]) could be natural if the dominant contributions directly arise from non renormalizable operators which are apriori unrelated to other fermion masses, but we will not explore this possibility here.

2 Two Interesting Mechanisms

In general large mass splittings correspond to small mixings because normally only close-by states are strongly mixed. In a 2 by 2 matrix context, the requirement of large splitting and large mixing leads to a condition of vanishing determinant. For example the matrix

$$m \propto \begin{bmatrix} x^2 & x \\ x & 1 \end{bmatrix} \tag{6}$$

has eigenvalues 0 and $1 + x^2$ and for x of O(1) the mixing is large. Thus, in the limit of neglecting small mass terms of order $m_{1,2}$, the demands of large atmospheric neutrino mixing and dominance of m_3 translate into the condition that the subdeterminant 23 of the 3 by 3 mass matrix vanishes. The problem is to show that this vanishing can be arranged in a natural way without fine tuning.

Without loss of generality we can go to a basis where both the charged lepton Dirac mass matrix m_D^l and the Majorana matrix M for the right-handed neutrinos are diagonal. In fact, after diagonalization of the charged lepton Dirac mass matrix, we still have the freedom of a change of basis for the right-handed neutrino fields, in that the right-handed charged lepton and neutrino fields, as opposed to left-handed fields, are uncorrelated by the SU(2)×U(1) gauge symmetry. We can use this freedom to make the Majorana matrix diagonal: $M^{-1} = V^T d_M V$ with $d_M = \mathrm{Diag}[1/M_1, 1/M_2, 1/M_3]$. Then if we parametrize the matrix $V m_D$ by z_{ab} we have:

$$(m_\nu)_{ab} \equiv (m_D^T M^{-1} m_D)_{ab} = \sum_c \frac{z_{ca} z_{cb}}{M_c}. \tag{7}$$

From this expression we see that, while we can always arrange the twelve parameters z_{ab} and M_a to arbitrarily fix the six independent matrix elements of m_ν, the hierarchical case is special in that it can be approximately reproduced in two particularly simple ways, without relying on precise cancellations among different terms:

i) there are only two large entries in the z matrix, $|z_{c2}| \sim |z_{c3}|$, and the three eigenvalues M_a are of comparable magnitude (or, at least, with a less pronounced hierarchy than for the z matrix elements). Then, the subdeterminant 23 vanishes and one only needs the ratio $|z_{c2}/z_{c3}|$ to be close to 1. This possibility was discussed for instance in [17];

ii) one of the right-handed neutrinos is particularly light and, in first approximation, it is only coupled to μ and τ. Thus, $M_c \sim \eta$ (small) and

$z_{c1} \sim 0$. In this case the 23 subdeterminant vanishes, and again one only needs the ratio $|z_{c2}/z_{c3}|$ to be close to 1. This possibility has been especially emphasized in refs. [18,19,5,20].

In a 2 by 2 matrix context (in the 23 sector), a typical example of mechanism *i* is given by a Dirac matrix m_D, defined by $\bar{R} m_D L$, which takes the approximate form:

$$m_D \propto \begin{bmatrix} 0 & 0 & 0 \\ 0 & 0 & 0 \\ 0 & x & 1 \end{bmatrix} \quad . \tag{8}$$

This matrix has the property that for a generic Majorana matrix M one finds:

$$m_\nu = m_D^T M^{-1} m_D \propto \begin{bmatrix} 0 & 0 & 0 \\ 0 & x^2 & x \\ 0 & x & 1 \end{bmatrix} \quad . \tag{9}$$

The only condition on M^{-1} is that the 33 entry is non zero. It is important for the following discussion to observe that m_D given by eq. (8) under a change of basis transforms as $m_D \to V^\dagger m_D U$ where V and U rotate the right and left fields respectively. It is easy to check that in order to make m_D diagonal we need large left mixings. More precisely m_D is diagonalized by taking V=1 and U given by

$$U = \begin{bmatrix} c & -s & 0 \\ s c_\gamma & c c_\gamma & -s_\gamma \\ s s_\gamma & c s_\gamma & c_\gamma \end{bmatrix} \quad , \tag{10}$$

with

$$s_\gamma = -x/r \quad , \qquad c_\gamma = 1/r \quad , \qquad r = \sqrt{1+x^2} \quad . \tag{11}$$

The matrix U is directly the neutrino mixing matrix. The mixing angle for atmospheric neutrino oscillations is given by:

$$\sin^2 2\theta = 4 s_\gamma^2 c_\gamma^2 = \frac{4x^2}{(1+x^2)^2} \quad . \tag{12}$$

Thus the bound $\sin^2 2\theta > 0.8$ translates into $0.6 < |x| < 1.6$. As is clear, this mechanism is based on asymmetric Dirac matrices, with, in the case of the example, a large left-handed mixing already present in the Dirac matrix.

If, for some reason, one prefers symmetric or nearly so matrices, then one can use mechanism *ii*. For example, one could want to preserve left–right symmetry at the GUT scale. Then, the observed smallness of left-handed mixings for quarks would also demand small right-handed mixings. So we now assume that m_D is nearly diagonal (always in the basis where m_D^l and M are diagonal) with all its off-diagonal terms proportional to some small parameter ϵ. Working in the subsector 23 and starting from

$$m_D \propto \begin{bmatrix} \epsilon^p & x\epsilon \\ x\epsilon & 1 \end{bmatrix}, \qquad M^{-1} \propto \begin{bmatrix} r_2 & 0 \\ 0 & 1 \end{bmatrix} \quad , \tag{13}$$

where x is of O(1) and $r_2 \equiv M_3/M_2$, we obtain:

$$m_\nu \propto \begin{bmatrix} \epsilon^{2p} r_2 + x^2\epsilon^2 & x\epsilon^{p+1} r_2 + x\epsilon \\ x\epsilon^{p+1} r_2 + x\epsilon & x^2\epsilon^2 r_2 + 1 \end{bmatrix} \quad . \tag{14}$$

For sufficiently small M_2 the terms in r_2 are dominant. For $p = 1, 2$, which we consider as typical cases, it is sufficient that $\epsilon^2 r_2 \gg 1$. Assuming that this condition is satisfied, consider first the case with $p = 2$. We have

$$m_\nu \propto x^2\epsilon^2 r_2 \begin{bmatrix} \frac{\epsilon^2}{x^2} & \frac{\epsilon}{x} \\ \frac{\epsilon}{x} & 1 \end{bmatrix} \quad . \tag{15}$$

In this case the determinant is naturally vanishing (to the extent that the terms in r_2 are dominant), so that the mass eigenvalues are widely split. However, the mixing is nominally small: $\sin 2\theta$ is of $O(2\epsilon/x)$. It could be numerically large enough if $1/x \sim$ 2–3 and ϵ is of the order of the Cabibbo angle $\epsilon \sim$ 0.20–0.25. This is what we call "stretching": the large neutrino mixing is explained in terms of a small parameter; this is not so small and can give a perhaps sufficient amount of mixing if enhanced by a possibly large coefficient. This minimalistic view was endorsed in refs. [21].

A more peculiar case is obtained for $p = 1$, which gives:

$$m_\nu \propto \epsilon^2 r_2 \begin{bmatrix} 1 & x \\ x & x^2 \end{bmatrix} \quad . \tag{16}$$

In this case the small parameter ϵ is completely factored out and for $x \sim 1$ the mixing is nearly maximal. The see-saw mechanism has created *large mixing from almost nothing* [22]: all relevant matrices entering the see-saw mechanism are nearly diagonal. Clearly, the crucial factorization of the small parameter ϵ^2 only arises for $p = 1$, that is the light Majorana eigenvalue is coupled to ν_μ and ν_τ with comparable strength. It is straightforward to extend the previous model to the 3 by 3 case [22]. In that case it is possible to reproduce both the SA and the LA MSW solutions. The required hierarchy among the matrix elements can be supported by a suitable Abelian flavour symmetry, which can be realized also at the level of an SU(5) grand unified theory. Moreover this hierarchy is not spoiled by the renormalization group evolution from the unification scale down to low energy.

3 An Explicit Model

We have seen that, in order to explain in a natural way widely split light neutrino masses together with large mixings, we need an automatic vanishing of the 23 subdeterminant. This in turn is most simply realized within mechanism *i*, by allowing some large left-handed mixing terms in the Dirac neutrino matrix. By left-handed mixing we mean non diagonal matrix elements that can only be eliminated by a large rotation of the left-handed fields. Thus the question is how to reconcile large left-handed mixings in the leptonic sector with the observed near diagonal form of V_{CKM}, the quark mixing matrix. Strictly speaking, since $V_{CKM} = U_u^\dagger U_d$, the individual matricesU_u and U_d need not be near diagonal, but V_{CKM} does, while the analogue for leptons apparently cannot be near diagonal. However nothing forbids for quarks that, in the basis where m_u is diagonal, the d quark matrix has large non diagonal terms that can be rotated away by a pure right-handed rotation. We suggest that this is so and that in some way right-handed mixings for quarks correspond to left-handed mixings for leptons.

In the context of (Susy) SU(5) [23] there is a very attractive hint of how the present mechanism can be realized [17,24]. In the $\bar{5}$ of SU(5) the d^c singlet appears together with the lepton doublet (ν, e). The (u, d) doublet and e^c belong to the 10 and ν^c to the 1 and similarly for the other families. As a consequence, in the simplest model with mass terms arising from only Higgs pentaplets, the Dirac matrix of down quarks is the transpose of the charged lepton matrix: $m_D^d = (m_D^l)^T$. Thus, indeed, a large mixing for right-handed down quarks corresponds to a large left-handed mixing for charged leptons. In the same simplest approximation with 5 or $\bar{5}$ Higgs, the up quark mass matrix is symmetric, so that left and right mixing matrices are equal in this case [b]. Then small mixings for up quarks and small left-handed mixings for down quarks are sufficient to guarantee small V_{CKM} mixing angles even for large d quark right-handed mixings. When the charged lepton matrix is diagonalized the large left-handed mixing of the charged leptons is transferred to the neutrinos. Note that in SU(5) we can diagonalize the u mass matrix by a rotation of the fields in the 10, the Majorana matrix M by a rotation of the 1 and the effective light neutrino matrix m_ν by a rotation of the $\bar{5}$. In this basis the d quark mass matrix fixes V_{CKM} and the charged lepton mass matrix fixes neutrino mixings. It is well known that a model where the down and the charged lepton mass matrices are exactly the transpose of one another cannot be exactly true because of the e/d and μ/s mass ratios [23]. It is also

[b] Up to a diagonal matrix of phases.

known that one remedy to this problem is to add some Higgs component in the 45 representation of SU(5) [25]. A different solution [26] will be described later. But the symmetry under transposition can still be a good guideline if we are only interested in the order of magnitude of the matrix entries and not in their exact values. Similarly, the Dirac neutrino mass matrix m_D is the same as the up quark mass matrix in the very crude model where the Higgs pentaplets come from a pure 10 representation of SO(10): $m_D = m_D^u$. For m_D the dominance of the third family eigenvalue as well as a near diagonal form could be an order of magnitude remnant of this broken symmetry. Thus, neglecting small terms, the neutrino Dirac matrix in the basis where charged leptons are diagonal could be directly obtained in the form of eq. (8).

We give here an explicit example of the mechanism under discussion in the framework of a unified Susy $SU(5)$ theory with an additional $U(1)_F$ flavour symmetry [27]. This model is to be taken as merely indicative, in that some important problems, like, for example, the cancellation of chiral anomalies are not tackled here. But we find it impressive that the general pattern of all what we know on fermion masses and mixings is correctly reproduced at the level of orders of magnitude. We regard the present model as a low-energy effective theory valid at energies close to $M_{GUT} \ll M_{Pl}$. We can think to obtain it by integrating out the heavy modes from an unknown underlying fundamental theory defined at an energy scale close to M_{Pl}. From this point of view the gauge anomalies generated by the light supermultiplets listed below can be compensated by another set of supermultiplets with masses above M_{GUT}, already eliminated from the low-energy theory. In particular, we assume that these additional supermultiplets are vector-like with respect to $SU(5)$ and chiral with respect to $U(1)_F$. Their masses are then naturally expected to be of the order of the $U(1)_F$ breaking scale, which, in the following discussion, turns out to be near M_{Pl}. We have explicitly checked the possibility of canceling the gauge anomalies in this way but, due to our ignorance about the fundamental theory, we do not find particularly instructive to illustrate the details here. In this model the known generations of quarks and leptons are contained in triplets Ψ_{10}^a and $\Psi_{\bar{5}}^a$, $(a = 1, 2, 3)$ transforming as 10 and $\bar{5}$ of $SU(5)$, respectively. Three more $SU(5)$ singlets Ψ_1^a describe the right-handed neutrinos. We assign to these fields the following F-charges:

$$\Psi_{10} \sim (3, 2, 0) \tag{17}$$

$$\Psi_{\bar{5}} \sim (3, 0, 0) \tag{18}$$

$$\Psi_1 \sim (1, -1, 0) \tag{19}$$

We start by discussing the Yukawa coupling allowed by $U(1)_F$-neutral Higgs multiplets φ_5 and $\varphi_{\bar{5}}$ in the 5 and $\bar{5}$ $SU(5)$ representations and by a pair θ

and $\bar{\theta}$ of $SU(5)$ singlets with $F = 1$ and $F = -1$, respectively.

In the quark sector we obtain [c]:

$$m_D^u = (m_D^u)^T = \begin{bmatrix} \lambda^6 & \lambda^5 & \lambda^3 \\ \lambda^5 & \lambda^4 & \lambda^2 \\ \lambda^3 & \lambda^2 & 1 \end{bmatrix} v_u \ , \qquad m_D^d = \begin{bmatrix} \lambda^6 & \lambda^5 & \lambda^3 \\ \lambda^3 & \lambda^2 & 1 \\ \lambda^3 & \lambda^2 & 1 \end{bmatrix} v_d \ , \quad (20)$$

from which we get the order-of-magnitude relations:

$$\begin{aligned} m_u : m_c : m_t &= \lambda^6 : \lambda^4 : 1 \\ m_d : m_s : m_b &= \lambda^6 : \lambda^2 : 1 \end{aligned} \quad (21)$$

and

$$V_{us} \sim \lambda \ , \qquad V_{ub} \sim \lambda^3 \ , \qquad V_{cb} \sim \lambda^2 \ . \quad (22)$$

Here $v_u \equiv \langle \varphi_5 \rangle$, $v_d \equiv \langle \varphi_{\bar{5}} \rangle$ and λ denotes the ratio between the vacuum expectation value of $\bar{\theta}$ and an ultraviolet cut-off identified with the Planck mass M_{Pl}: $\lambda \equiv \langle \bar{\theta} \rangle / M_{Pl}$. To correctly reproduce the observed quark mixing angles, we take λ of the order of the Cabibbo angle. For non-negative F-charges, the elements of the quark mixing matrix V_{CKM} depend only on the charge differences of the left-handed quark doublet [27]. Up to a constant shift, this defines the choice in eq. (17). Equal F-charges for $\Psi_{\bar{5}}^{2,3}$ (see eq. (18)) are then required to fit m_b and m_s. We will comment on the lightest quark masses later on.

At this level, the mass matrix for the charged leptons is the transpose of m_D^d:

$$m_D^l = (m_D^d)^T \quad (23)$$

and we find:

$$m_e : m_\mu : m_\tau = \lambda^6 : \lambda^2 : 1 \quad (24)$$

The O(1) off-diagonal entry of m_D^l gives rise to a large left-handed mixing in the 23 block which corresponds to a large right-handed mixing in the d mass matrix. In the neutrino sector, the Dirac and Majorana mass matrices are given by:

$$m_D = \begin{bmatrix} \lambda^4 & \lambda & \lambda \\ \lambda^2 & \lambda' & \lambda' \\ \lambda^3 & 1 & 1 \end{bmatrix} v_u \ , \qquad M = \begin{bmatrix} \lambda^2 & 1 & \lambda \\ 1 & \lambda'^2 & \lambda' \\ \lambda & \lambda' & 1 \end{bmatrix} \bar{M} \ , \quad (25)$$

[c] In eq. (20) the entries denoted by 1 in m_D^u and m_D^d are not necessarily equal. As usual, such a notation allows for O(1) deviations.

where $\lambda' \equiv \langle\theta\rangle/M_{Pl}$ and $\bar{M}$ denotes the large mass scale associated to the right-handed neutrinos: $\bar{M} \gg v_{u,d}$.

After diagonalization of the charged lepton sector and after integrating out the heavy right-handed neutrinos we obtain the following neutrino mass matrix in the low-energy effective theory:

$$m_\nu = \begin{bmatrix} \lambda^6 & \lambda^3 & \lambda^3 \\ \lambda^3 & 1 & 1 \\ \lambda^3 & 1 & 1 \end{bmatrix} \frac{v_u^2}{\bar{M}} \tag{26}$$

where we have taken $\lambda \sim \lambda'$. The O(1) elements in the 23 block are produced by combining the large left-handed mixing induced by the charged lepton sector and the large left-handed mixing in m_D. A crucial property of m_ν is that, as a result of the sea-saw mechanism and of the specific $U(1)_F$ charge assignment, the determinant of the 23 block is *automatically* of $O(\lambda^2)$ (for this the presence of negative charge values, leading to the presence of both λ and λ' is essential [17]).

It is easy to verify that the eigenvalues of m_ν satisfy the relations:

$$m_1 : m_2 : m_3 = \lambda^4 : \lambda^2 : 1 \quad . \tag{27}$$

The atmospheric neutrino oscillations require $m_3^2 \sim 10^{-3}$ eV^2. From eq. (26), taking $v_u \sim 250$ GeV, the mass scale $\bar{M}$ of the heavy Majorana neutrinos turns out to be close to the unification scale, $\bar{M} \sim 10^{15}$ GeV. The squared mass difference between the lightest states is of $O(\lambda^4)$ m_3^2, appropriate to the MSW solution to the solar neutrino problem. Finally, beyond the large mixing in the 23 sector, corresponding to $s_\gamma \sim c_\gamma$ in eq. (10), m_ν provides a mixing angle $s \sim (\lambda/2)$ in the 12 sector, close to the range preferred by the small angle MSW solution. In general U_{e3} is non-vanishing, of $O(\lambda^3)$.

In general, the charge assignment under $U(1)_F$ allows for non-canonical kinetic terms that represent an additional source of mixing. Such terms are allowed by the underlying flavour symmetry and it would be unnatural to tune them to the canonical form. We have checked that all the results quoted up to now remain unchanged after including the effects related to the most general kinetic terms, via appropriate rotations and rescaling in the flavour space (see also ref.[28]).

Obviously, the order of magnitude description offered by this model is not intended to account for all the details of fermion masses. Even neglecting the parameters associated with the CP violating observables, some of the relevant observables are somewhat marginally reproduced. For instance we obtain $m_u/m_t \sim \lambda^6$ which is perhaps too large. However we find it remarkable that in such a simple scheme most of the 12 independent fermion masses and the

6 mixing angles turn out to have the correct order of magnitude. Notice also that our model prefers large values of $\tan\beta \equiv v_u/v_d$. This is a consequence of the equality $F(\Psi^3_{10}) = F(\Psi^3_{\bar{5}})$ (see eqs. (17) and (18)). In this case the Yukawa couplings of top and bottom quarks are expected to be of the same order of magnitude, while the large m_t/m_b ratio is attributed to $v_u \gg v_d$ (there may be factors O(1) modifying these considerations, of course). We recall here that in supersymmetric grand unified models large values of $\tan\beta$ are one possible solution to the problem of reconciling the boundary condition $m_b = m_\tau$ at the GUT scale with the low-energy data [29]. Alternatively, to keep $\tan\beta$ small, one could suppress m_b/m_t by adopting different F-charges for the $\Psi^3_{\bar{5}}$ and Ψ^3_{10}.

Additional contributions to flavour changing processes and to CP violating observables are generally expected in a supersymmetric grand unified theory. However, a reliable estimate of the corresponding effects would require a much more detailed definition of the theory than attempted here. Crucial ingredients such as the mechanism of supersymmetry breaking and its transmission to the observable sector have been ignored in the present note. We are implicitly assuming that the omission of this aspect of the flavour problem does not substantially alter our discussion.

A common problem of all $SU(5)$ unified theories based on a minimal higgs structure is represented by the relation $m^l_D = (m^d_D)^T$ that, while leading to the successful $m_b = m_\tau$ boundary condition at the GUT scale, provides the wrong prediction $m_d/m_s = m_e/m_\mu$ (which, however, is an acceptable order of magnitude equality). We can easily overcome this problem and improve the picture [26] by introducing an additional supermultiplet $\bar{\theta}_{24}$ transforming in the adjoint representation of $SU(5)$ and possessing a negative $U(1)_F$ charge, $-n$ $(n > 0)$. Under these conditions, a positive F-charge f carried by the matrix elements $\Psi^a_{10}\Psi^b_{\bar{5}}$ can be compensated in several different ways by monomials of the kind $(\bar{\theta})^p(\bar{\theta}_{24})^q$, with $p + nq = f$. Each of these possibilities represents an independent contribution to the down quark and charged lepton mass matrices, occurring with an unknown coefficient of O(1). Moreover the product $(\bar{\theta}_{24})^q\varphi_{\bar{5}}$ contains both the $\bar{5}$ and the $\overline{45}$ $SU(5)$ representations, allowing for a differentiation between the down quarks and the charged leptons. The only, welcome, exceptions are given by the O(1) entries that do not require any compensation and, at the leading order, remain the same for charged leptons and down quarks. This preserves the good $m_b = m_\tau$ prediction. Since a perturbation of O(1) in the subleading matrix elements is sufficient to cure the bad $m_d/m_s = m_e/m_\mu$ relation, we can safely assume that $\langle\bar{\theta}_{24}\rangle/M_{Pl} \sim \lambda^n$, to preserve the correct order-of-magnitude predictions in the remaining sectors.

We have not dealt here with the problem of recovering the correct vacuum structure by minimizing the effective potential of the theory. It may be noticed that the presence of two multiplets θ and $\bar{\theta}$ with opposite F charges could hardly be reconciled, without adding extra structure to the model, with a large common VEV for these fields, due to possible analytic terms of the kind $(\theta\bar{\theta})^n$ in the superpotential. We find therefore instructive to explore the consequences of allowing only the negatively charged $\bar{\theta}$ field in the theory.

It can be immediately recognized that, while the quark mass matrices of eqs. (20) are unchanged, in the neutrino sector the Dirac and Majorana matrices get modified into:

$$m_D = \begin{bmatrix} \lambda^4 & \lambda & \lambda \\ \lambda^2 & 0 & 0 \\ \lambda^3 & 1 & 1 \end{bmatrix} v_u \ , \qquad M = \begin{bmatrix} \lambda^2 & 1 & \lambda \\ 1 & 0 & 0 \\ \lambda & 0 & 1 \end{bmatrix} \bar{M} \ . \tag{28}$$

The zeros are due to the analytic property of the superpotential that makes impossible to form the corresponding F invariant by using $\bar{\theta}$ alone. These zeros should not be taken literally, as they will be eventually filled by small terms coming, for instance, from the diagonalization of the charged lepton mass matrix and from the transformation that put the kinetic terms into canonical form. It is however interesting to work out, in first approximation, the case of exactly zero entries in m_D and M, when forbidden by F.

The neutrino mass matrix obtained via see-saw from m_D and M has the same pattern as the one displayed in eq. (26). A closer inspection reveals that the determinant of the 23 block is identically zero, independently from λ. This leads to the following pattern of masses:

$$m_1 : m_2 : m_3 = \lambda^3 : \lambda^3 : 1 \ , \qquad m_1^2 - m_2^2 = O(\lambda^9)\, m_3^2 \ . \tag{29}$$

Moreover the mixing in the 12 sector is almost maximal:

$$\frac{s}{c} = \frac{\pi}{4} + O(\lambda^3) \ . \tag{30}$$

For $\lambda \sim 0.2$, both the squared mass difference $(m_1^2 - m_2^2)/m_3^2$ and $\sin^2 2\theta_{sun}$ are remarkably close to the values required by the vacuum oscillation solution to the solar neutrino problem. We have also checked that this property is reasonably stable against the perturbations induced by small terms (of order λ^5) replacing the zeros, coming from the diagonalization of the charged lepton sector and by the transformations that render the kinetic terms canonical. We find quite interesting that also the just-so solution, requiring an intriguingly small mass difference and a bimaximal mixing, can be reproduced, at least at the level of order of magnitudes, in the context of a "minimal" model of flavour compatible with supersymmetric SU(5). In this case the role played by supersymmetry is essential, a non-supersymmetric model with $\bar{\theta}$ alone not being distinguishable from the version with both θ and $\bar{\theta}$, as far as low-energy flavour properties are concerned.

4 Conclusions

If we start from three light neutrinos and the see-saw mechanism then a natural interpretation of the present data on neutrino oscillations is in terms of hierarchical light neutrino masses and asymmetric mass matrices (at least for d quarks and charged leptons). This has the advantage that no conspiracy is required between the Dirac and the Majorana sectors. There is also the peculiar possibility that large neutrino mixing is only produced by the see-saw mechanism starting from all nearly diagonal matrices. Although this possibility is certainly rather special, models of this sort can be constructed without an unrealistic amount of fine tuning. Both scenarios are well compatible with Abelian flavour symmetries and with grand unification ideas and the related phenomenology for quark and lepton masses.

Acknowledgments

It is a pleasure to thank Susan Kovesi-Domokos and Gabor Domokos for the organization of a very stimulating meeting and the warm hospitality enjoyed in Baltimore. I would like to thank Guido Altarelli and Isabella Masina for the pleasant collaboration on which this talk is based.

References

1. Y. Fukuda et al., hep-ex/9805006, hep-ex/9805021 and hep-ex/9807003.
2. The data on solar neutrinos from Homestake, Gallex, Sage, Kamiokande and Superkamiokande are summarized and the list of references given in J. N. Bahcall, P. I. Krastev and A. Yu Smirnov, hep-ph/9807216.
3. C. Athanassopoulos et al., Phys. Rev. Lett. 77 (1996) 3082; nucl-ex/9706006, nucl-ex/9709006.
4. B. Armbruster et al., Phys. Rev. C57 (1998) 3414; G. Drexlin, talk at Wein'98.
5. R. Barbieri, L. J. Hall, D. Smith, A. Strumia and N. Weiner, hep/ph 9807235.
6. M. Apollonio et al., Phys. Lett. B 420 (1998) 397.
7. V. Barger, S. Pakvasa, T.J. Weiler and K. Whisnant, hep-ph/9806387; V. Barger, T.J. Weiler and K. Whisnant, hep-ph/9807319.

8. S.M. Bilenky and C. Giunti, hep-ph/9802201; A.J. Baltz, A.S. Goldhaber and M. Goldhaber, hep-ph/9806540.
9. L. Wolfenstein, Phys. Rev. D17 (1978) 2369; S.P. Mikheyev and A. Yu Smirnov, Sov. J. Nucl. Phys. 42 (1986) 913.
10. M. C. Gonzalez-Garcia, H. Nunokawa, O. L. G. Peres, J. W. F. Valle, hep-ph/9807305; G. L. Fogli, E. Lisi, A. Marrone and G. Scioscia, hep-ph/9808205; S. M. Bilenkii, C. Giunti, W. Grimus, hep-ph/9809368.
11. See for example L. K. Krauss, hep-ph/9807376.
12. L. Baudis et al., Phys. Lett. B407 (1997) 219.
13. F. Vissani, hep-ph/9708483; H. Georgi and S.L. Glashow, hep-ph/9808293.
14. N. Haba, N. Okamura and M. Sugiura, hep-ph/9810471; J. Ellis and S. Lola, hep-ph/9904279; J.A. Casas, J.R. Espinosa, A. Ibarra and I. Navarro, hep-ph/9904395, hep-ph/9905381, hep-ph/9906281; N. Haba and N. Okamura, hep-ph/9906481; E. Ma, hep-ph/9907400.
15. R. Barbieri, G.G. Ross and A. Strumia, hep-ph/9906470;
16. A. Ioannisian, J. W. F. Valle, Phys. Lett. B332 (1994) 93, hep-ph/9402333; M. Fukugita, M. Tanimoto, T. Yanagida, Phys. Rev. D57 (1998) 4429, hep-ph/9709388; M. Tanimoto, hep-ph/9807283 and hep-ph/9807517; H. Fritzsch, Z. Xing, hep-ph/9808272; R. N. Mohapatra, S. Nussinov, hep-ph/9808301 and hep-ph/9809415; M. Fukugita, M. Tanimoto, T. Yanagida, hep-ph/9809554; Yue-Liang Wu, hep-ph/9810491; J. I. Silva-Marcos, hep-ph/9811381; C. Wetterich, hep-ph/9812426.
17. G. Altarelli and F. Feruglio, Phys. Lett. B439 (1998) 112 [hep-ph/9807353]; JHEP 11 (1998) 021 [hep-ph/9809596]; Phys. Lett. B451 (1999) 388 [hep-ph/9812475]; hep-ph/9905536.
18. A. Y. Smirnov, Phys. Rev. D48 (1993) 3264.
19. Z. G. Berezhiani and A. Rossi, Phys. Lett. B367 (1996) 219.
20. S. F. King, Phys. Lett. B439 (1998) 350, hep-ph/9806440; S. F. King, hep-ph/9904210; S. Davidson and S. F. King, Phys. Lett. B445 (1998) 191, hep-ph/9808333; E. Ma and D. P. Roy, Phys. Rev. D59 (1999) 097702.
21. S. Lola and G. G. Ross, hep-ph/9902283; K. Babu, J. Pati and F. Wilczek, hep-ph/9912538.
22. G. Altarelli, F. Feruglio and I. Masina, hep-ph/9907532.
23. G.G.Ross, "Grand Unified Theories", Benjamin, 1985; R.N.Mohapatra, "Unification and Supersymmetry" Springer-Verlag, 1986. C. Kounnas, A.Masiero, D.V.Nanopoulos and K.A.Olive, "Grand Unification with and without Supersymmetry and Cosmolgical Implications", World Sci. 1984.
24. Z. Berezhiani and Z. Tavartkiladze, Phys. Lett. B409 (1997) 220; C.

H. Albright and S. M. Barr, Phys. Rev. D58 (1998) 013002, hep-ph/9712488; C. H. Albright, K. S. Babu and S. M. Barr, Phys. Rev. Lett. 81 (1998) 1167, hep-ph/9802314; Z. Berezhiani and A. Rossi, hep-ph/9811447; K.Hagiwara and N.Okamura, hep-ph/9811495.

25. H. Georgi and C. Jarlskog, Phys. Lett. B86(1979) 297.
26. J. Ellis and M. K. Gaillard, Phys. Lett. B88(1979) 315.
27. C. Froggatt and H. B. Nielsen, Nucl. Phys. B147 (1979) 277; J. K. Elwood, N. Irges and P. Ramond, hep-ph/9807228; P. Ramond, hep-ph/9808489, hep-ph/9809401 and these Proceedings; W. Buchmuller and T. Yanagida, hep-ph/9810308; P. Binetruy, S. Lavignac, S. Petcov and P. Ramond, Nucl. Phys. B496 (1997) 3, hep-ph/9610481; N. Irges, S. Lavignac, P. Ramond, Phys. Rev. D58 (1998) 5003, hep-ph/9802334; Y. Grossman, Y. Nir, Y. Shadmi, hep-ph/9808355; C. D. Froggart, M. Gibson and H. B. Nielsen, hep-ph/9811265; B. Stech, hep-ph/9905440 and these Proceedings.
28. M. Leurer, Y. Nir and N. Seiberg, Nucl. Phys. B420 (1994) 468, hep-ph/9310320.
29. M. Carena, S. Pokorski and C.E.M. Wagner, Nucl. Phys. B406 (1993) 59; P. Langacker and N. Polonsky, Phys. Rev. D49 (1194) 1451; N. Polonsky, Phys. Rev. D54 (1996) 4537.

ARE THE NEUTRINO MASSES AND MIXINGS CLOSELY RELATED TO THE MASSES AND MIXINGS OF QUARKS?

BERTHOLD STECH
Institut für Theoretische Physik
Universität Heidelberg
Philosophenweg 16, D-69120 Heidelberg
E-mail: B.Stech@thphys.uni-heidelberg.de

The mass matrices of quarks have a simple structure if expressed in powers of the small parameter $\sigma = (m_c/m_t)^{1/2}$. If there is a close relation between quarks and leptons, one would expect similar structures for the lepton matrices which involve the same parameter. To have a specific proposal, the see-saw mechanism is employed together with the stringent requirement that the singlet neutrinos carry non-zero generation charges. These charges then determine the powers of σ in the corresponding heavy neutrino mass matrix. As a result, the neutrino mixing matrix turns out to be of the bimaximal type. In addition, the mass splitting of the two lightest neutrinos is found to be tiny, and just of the correct magnitude necessary for the vacuum solution for solar neutrinos.

1 Introduction

Recent experiments on solar and atmospheric neutrinos indicate flavor mixing of neutrinos, and finite neutrino masses [1], [2]. Nevertheless, our knowledge on the mass spectrum and the mixing of neutrinos is still very limited. Several different scenarios can be envisaged [3]. In contrast to the neutrino properties, the masses of the quarks and the pattern of their mixings are well known. On a qualitative level only the validity and shape of the unitarity triangle is still an open problem. In this situation one may ask whether or not there exists a close relation between the masses and mixings of leptons and those of quarks. If an intimate relation could be discovered it would greatly reduce the number of free parameters and would shed light on the new physics behind the standard model of particle physics. Such a relation cannot be a trivial one since the large mixing angle observed in the study of atmospheric neutrinos [2] has no counterpart in the quark sector.

The mass matrices of quarks, if expressed in powers of a small parameter, have a simple structure, suggestive of a new generation symmetry [4]. A connection between quarks and leptons would be manifest, if the mass matrices of leptons are governed by powers of the same small parameter. In this work a very close relation of this type will be proposed. To this end two general suggestions of grand unified theories [5] are used: the existence of 3 two-component neutrino fields which are singlets with respect to the standard model gauge groups, and the similarity of the Dirac neutrino and charged lepton mass matrices with the up-quark and the down-quark mass matrix, respectively [6, 7]. (But a full gauge theory ($SO(10), E(6)$) with a necessarily complicated Higgs structure is not implied here, even though $SO(10)$ models gave the first hint for a large neutrino mixing angle [6, 8]). The remaining and decisive mass matrix is then the one for the 3 singlet neutrinos. Since these fields are not protected by a standard model symmetry, their mass values will be very large.

Because of the Majorana type selfcoupling of the singlet neutrinos, their generation charges play an important role in restricting the structure of the corresponding mass matrix. After exploring these restrictions, the straightforward use of the see-saw mechanism finally gives a rather definite form for the mass matrix of the light neutrinos [9]. It has interesting properties: i) The mixing matrix obtained from it is of the bimaximal type and also includes CP-violating phases. ii) By fixing the scale for the singlet neutrinos to obtain the correct mass scale for the atmospheric neutrinos, the mass splitting of the two lightest neutrinos turns out to be tiny ($\Delta m^2 \approx 10^{-11} - 10^{-10}(\mathrm{eV})^2$), thus favoring the vacuum oscillation solutions [10] for solar neutrinos.

2 Quark mass matrices

The 3x3 mass matrices of up and down quarks contain 36 parameters, but only 10 parameters are physical: 6 eigenvalues, 3 mixing angles and 1 CP-violating phase. Therefore, for different choices of independent parameters the mass matrices can look very different. From the point of view of assigning generation (familiy) charges to the quark fields it is preferable to use a basis in which each independent matrix element is represented by a different power of a small parameter. A natural choice for this parameter is $\sigma = (m_c/m_t)^{1/2} = 0.058 \pm 0.004$, because the top quark plays a dominant role among the quarks. Moreover, within experimental error limits, one finds for the up-quark mass ratios

$$m_u : m_c : m_t = \sigma^4 : \sigma^2 : 1 \tag{1}$$

and for the Kobayashi-Maskawa matrix elements [11]

$$|V_{us}| = 4\sigma, |V_{cb}| = \sigma/\sqrt{2} \quad , \quad |V_{ub}| = \sigma^2 \quad . \tag{2}$$

A suitable basis which involves just 5 parameters for the up-quark mass matrix m_U and 5 parameters for the down-quark mass matrix m_D, is to take m_U real and symmetric with $(m_U)_{11} = 0$ and m_D hermitian with $(m_D)_{13} =$

$(m_D)_{31} = (m_D)_{23} = (m_D)_{32} = 0$. The only complex matrix element is then $(m_D)_{12} = (m_D)^*_{21}$. The observed masses and mixing angles are well represented by taking

$$m_U = \begin{pmatrix} 0 & \sigma^3/\sqrt{2} & \sigma^2 \\ \sigma^3/\sqrt{2} & -\sigma^2/2 & \sigma/\sqrt{2} \\ \sigma^2 & \sigma/\sqrt{2} & 1 \end{pmatrix} m_t, \quad m_D = \begin{pmatrix} O(\sigma^4) & -i\sqrt{2}\sigma^2 & 0 \\ i\sqrt{2}\sigma^2 & -\sigma/3 & 0 \\ 0 & 0 & 1 \end{pmatrix} m_b \tag{3}$$

and $\sigma = 0.057$. However, for our purpose, it is convenient to transform to a basis in which m_U is diagonal and $m_D = V\ m_D^{diagonal} V^\dagger$, where V denotes the Cabibbo-Kobayashi-Maskawa matrix. To leading order in σ one finds

$$m_U \doteq \begin{pmatrix} \sigma^4 & 0 & 0 \\ 0 & \sigma^2 & 0 \\ 0 & 0 & 1 \end{pmatrix} \quad m_D = \begin{pmatrix} O(\sigma^3) & -i\sqrt{2}\sigma^2 & \sigma^2 \\ i\sqrt{2}\sigma^2 & -\sigma/3 & \sigma/\sqrt{2} \\ \sigma^2 & \sigma/\sqrt{2} & 1 \end{pmatrix} m_b \quad . \tag{4}$$

For definiteness "maximal CP-violation" has been used [7,12]. It is defined to maximize the area of the unitarity triangle with regard to phase changes of the off-diagonal elements [a]. It leads to a right-handed unitarity triangle with angles $\alpha \approx 70^o$, $\beta \approx 20^o$ and $\gamma \approx 90^o$ [11,12]. Irrespective of this choice the expressions (3) and (4) demonstrate that masses and mixings are governed by the same small parameter in a simple fashion.

3 The Dirac neutrino and charged lepton mass matrices

The 6x6 neutrino mass matrix (with zero entries for the light-light sector) has a block structure which contains a 3x3 Dirac neutrino mass matrix m_ν^{Dirac} and the 3x3 mass matrix M for the 3 singlet fields $\hat{\nu}_e, \hat{\nu}_\mu, \hat{\nu}_\tau$ [b]. Since there is no compelling theory which fixes these matrices and the charged lepton matrix m_E, one is forced to make assumptions. To have a very close connection between quark and leptons, and to be in line with the expectations from grand unified theories mentioned in section 1, I will assume that the mass matrices of quarks and leptons of the same isospin can be diagonalized simultaneously. For hermitian forms this implies

$$[m_\nu^{Dirac},\ m_U] = 0 \qquad [m_E,\ m_D] = 0 \quad . \tag{5}$$

More specifically, I will take

$$m_\nu^{Dirac} = m_U \tag{6}$$

[a] m_D in (4) can also be written in an equivalent form in which the off-diagonal matrix elements form an antisymmetric hermitian matrix [12].

[b] All fields are taken to be left-handed two-component fields.

and comment on deviations from (6) in section 5. These equations are assumed to hold at the scale of the heavy neutrinos.

Eq. (5) fixes the mass matrix for the charged leptons since m_D and the eigenvalues of m_E are known. To illustrate its form in terms of the parameter σ, one gets from $m_e : -m_\mu : m_\tau = \frac{3}{2}\sigma^3 : -\sigma : 1$ and $m_E = V\, m_E^{Diagonal} V^\dagger$ to leading order in σ

$$m_E = \begin{pmatrix} O(\sigma^3) & -i3\sqrt{2}\sigma^2 & \sigma^2 \\ i3\sqrt{2}\sigma^2 & -\sigma & \sigma/\sqrt{2} \\ i\sigma^2 & \sigma/\sqrt{2} & 1 \end{pmatrix} m_\tau \quad . \tag{7}$$

Notably, in the basis in which the up-quark mass matrix is diagonal, the matrix for charged leptons contains complex elements like m_D.

4 The mass matrix for the singlet neutrinos

It remains to discuss the mass matrix M for the singlet neutrino fields $\hat{\nu}_e, \hat{\nu}_\mu, \hat{\nu}_\tau$. It determines the mass matrix for the light neutrinos m_ν through the see-saw mechanism

$$m_\nu = -m_U \cdot M^{-1} \cdot m_U^T \quad . \tag{8}$$

This equation, in which (6) has already been used, is supposed to hold at the scale of the heavy neutrinos. Several suggestions for M can be found in the literature [13]. A close connection between charged and neutral fermions will exist if also M, in the basis in which m_U is diagonal, has a simple structure in terms of powers of the small parameter σ which occurs in (4), (6), and (7) [9]. These powers are restricted if the particle fields carry charges of a generation (family) symmetry [14].

Because of the self-coupling of the singlet fields, strong restrictions for the structure of M occur if the three $U(1)$ generation charges are all different from zero (and M does not vanish in the limit $\sigma \to 0$) [9]. By accepting this condition, two of the three fields must carry opposite $U(1)$ charges. They then form, for $\sigma \to 0$, a heavy singlet neutrino of the Dirac type. Giving opposite charges to $\hat{\nu}_e$ and $\hat{\nu}_\tau$ and allowing integer powers of σ^2 only, the matrix M has the structure[c]

$$M = \begin{pmatrix} \sim\sigma^6 & \sim\sigma^2 & 1 \\ \sim\sigma^2 & \sim\sigma^2 & \sim\sigma^4 \\ 1 & \sim\sigma^4 & \sim\sigma^6 \end{pmatrix} M_0 . \tag{9}$$

[c] As long as we dismiss matrices M which lead to a vanishing determinant when neglecting higher powers of σ than σ^2, the form (9) is unique apart from its mirror form corresponding to the charges $-3/2, -1/2, 3/2$.

The $U(1)$ charges for $\hat{\nu}_e, \hat{\nu}_\mu, \hat{\nu}_\tau$ are $-3/2, 1/2, 3/2$, respectively; they determine the powers of σ^2 in (9). As in the case of the quark mass matrices, the proportionality factors multiplying the powers of σ should be of order 1. In particular, if the correlation with m_U is strict, the factor of σ^2 in the first row and first column (p in eq. (10) should be equal or very close to one. Because of the smallness of σ^4 and σ^6, M can be approximated by the simpler form

$$M = \begin{pmatrix} 0 & p\sigma^2 & 1 \\ p\sigma^2 & r\sigma^2 & 0 \\ 1 & 0 & 0 \end{pmatrix} M_0 . \tag{10}$$

One can check that the approximation (10) is also applicable when calculating the mass matrix m_ν for the light neutrinos according to (8), even though the inverse of the matrix M enters here. Moreover, a simple consideration of the original 6×6 neutrino mass matrix (with zero entries in the light-light sector) shows that the coefficients p and r can be taken to be real.

5 The mass spectrum and the mixings of the light neutrinos

The consequences of the simple model considered here are now easily worked out. From Eqs. (4, 8, 10) one finds for the mass matrix of the light neutrinos

$$m_\nu = -\begin{pmatrix} 0 & 0 & r\sigma^2 \\ 0 & 1 & -p \\ r\sigma^2 & -p & p^2 \end{pmatrix} \frac{\sigma^2 m_t (M_0)^2}{r M_0} . \tag{11}$$

Apart from the mass scale M_0 we are left with only two important parameters (p, r). The remarkable point about this matrix is that it leads to a near degeneracy of the two lightest neutrinos, and for $p \approx 1$, to a neutrino mixing matrix which is of the bimaximal type as defined in [15]. To make contact with the atmospheric neutrino data [2] the heavy mass parameter M_0 can be adjusted to give the heaviest of the light neutrinos (ν_3) the mass $m_3 \approx 5.5 \times 10^{-2}$ eV. For $p = r = 1$ one then finds $M_0 = 10^{12}$ GeV, $m_3^2 - m_1^2 = 3 \times 10^{-3} (\text{eV})^2$ and $m_2^2 - m_1^2 = 1 \times 10^{-11}$ $(\text{eV})^2$. For $r = 2$ one gets $M_0 = 5 \times 10^{11}$ GeV and $m_2^2 - m_1^2 = 7 \times 10^{-11} (\text{eV})^2$. Thus, the neutrino mass matrix obtained by invoking an intimate relation between charged and neutral fermions favors large mixing angles for the atmospheric and for the solar neutrinos, and the vacuum oscillation solution [10] for the latter. But it seems not compatible with indications for $\bar{\nu}_\mu \to \bar{\nu}_e$ oscillations reported by the LSND collaboration [16].

For a more detailed treatment one has, of course, to take the charged lepton mass matrix m_E, which is nondiagonal in our basis, into account. In addition, the scale dependence of m_ν according to the renormalization group has to be considered.

Transforming therefore to a basis in which m_E is diagonal, the corresponding neutrino mass matrix at the scale M_0 is then

$$\tilde{m}_\nu(M_0) = V^T(M_0) m_\nu(M_0) V(M_0) \ . \tag{12}$$

For $m_\nu(M_0)$ we have to replace (11) by

$$m_\nu(M_0) = - \begin{pmatrix} 0 & 0 & r\frac{m_u(M_0)}{m_c(M_0)} \\ 0 & 1 & -p \\ r\frac{m_u(M_o)}{m_c(M_0)} & -p & p^2 \end{pmatrix} \frac{m_c(M_0) m_t(M_0)}{r M_0} \ . \tag{13}$$

Now, the renormalization group equation for $\tilde{m}_\nu(\mu)$ [17] can be used to calculate $\tilde{m}_\nu$ at the weak scale m_Z. After this calculation the neutrino mixing matrix U is obtained by diagonalizing the hermitian matrix $\tilde{m}_\nu \cdot \tilde{m}_\nu^*$:

$$\tilde{m}_\nu(m_Z) \cdot \tilde{m}_\nu^*(m_Z) = U D D^* U^\dagger \ . \tag{14}$$

The diagonal matrix D

$$D = U^\dagger \tilde{m}_\nu(m_Z) U^* \ . \tag{15}$$

gives the neutrino mass eigenvalues. These are complex because of the complex Cabibbo-Kobayashi-Maskawa matrix elements. Finally, U can be redefined by a diagonal phase matrix such that (15) contains now only positive definite eigenvalues. The new matrix U then expresses the light neutrino states ν_e, ν_μ, ν_τ in terms of the neutrino mass eigenstates ν_1, ν_2, ν_3 according to

$$\begin{pmatrix} \nu_e \\ \nu_\mu \\ \nu_\tau \end{pmatrix} = U \begin{pmatrix} \nu_1 \\ \nu_2 \\ \nu_3 \end{pmatrix} \ . \tag{16}$$

From the knowledge of U, the surviving and transition probabilities of the oscillating neutrino flavors can easily be calculated.

It turns out that the mass spectrum and mixings obtained from (14),(15) is not strikingly different from the ones found directly from the simple mass matrix, eq. (11). The small mixing angles occurring in V do not strongly influence $\tilde{m}_\nu$. The mixing matrix U has again – for $p \approx 1$ – a near bimaximal form, but contains now also CP-violating contributions [9]. Also the scale change from $M_0 \approx 10^{12}$ GeV down to the weak scale does not lead to a qualitatively different picture. In particular, the near degeneracy of the two lightest neutrino masses remains stable [9] in contrast to cases discussed in ref. [18]

To illustrate the results I give here the actual numbers taking $p = 1$ and $r = 2$. (They differ somewhat from the ones obtained in [9] since I use now in (12) the Cabibbo-Kobayashi-Maskawa Matrix V as a concequence of the condition (5)). With the input $m_3 = 0.055\ eV$ one gets $M_0 = 3 \times 10^{11}\ GeV$, $m_3^2 - m_1^2 = 3 \times 10^{-3}(eV)^2$ and $m_2^2 - m_1^2 = 9 \times 10^{-11}(eV)^2$. The survival and transition probabilities are (for notations see [9])

$$\begin{aligned} P(\nu_e \to \nu_e) &= 1 - 0.95\ S_{21} - 0.05\ S_{31} - 0.04\ S_{32} \\ P(\nu_\mu \to \nu_\mu) &= 1 - 0.32\ S_{21} - 0.49\ S_{31} - 0.49\ S_{32} \\ P(\nu_\tau \to \nu_\tau) &= 1 - 0.21\ S_{21} - 0.49\ S_{31} - 0.50\ S_{32} \\ P(\nu_e \to \nu_\mu) &= 0.53\ S_{21} + 0.02\ S_{31} + 0.02\ S_{32} \\ &+ 0.07\ T_{21} - 0.07\ T_{31} + 0.07\ T_{32} \\ P(\nu_e \to \nu_\tau) &= 0.42\ S_{21} + 0.02\ S_{31} + 0.03\ S_{32} \\ &- 0.07\ T_{21} + 0.07\ T_{31} - 0.07\ T_{32} \\ P(\nu_\mu \to \nu_\tau) &= -0.21\ S_{21} + 0.47\ S_{31} + 0.47\ S_{32} \\ &+ 0.07\ T_{21} - 0.07\ T_{31} + 0.07\ T_{32}\ . \end{aligned} \tag{17}$$

Our mass matrix for the light neutrinos is based on the eqs. (5), (6) and (13). If the eigenvalues of m_ν^{Dirac} differ from those of the up-quark mass matrix, but (5) still holds, the corresponding modification can be absorbed into the parameters r and M_0, which have anyhow to be adjusted to the data. A large effective value for r would give a significantly larger value for $m_2^2 - m_1^2$. In this case the model predicts an energy independent deficit of solar neutrinos, which is, however, barely compatible with the results of the Homestake collaboration [1].

The bimaximal mixing depends on the parameter p and gets spoiled if p is sizeably different from 1. The mixing angle for atmospheric neutrinos is sensitive to p. Deviations from $p = 1$ by more than 25 % would no more be compatible with the mixing angle extracted from present experiments.

6 Conclusion

A close relationship between leptons and quarks has been proposed. All mass matrices are expressed in terms of powers of the same small parameter $\sigma = (m_c/m_t)^{1/2}$. To obtain the neutrino mass matrix, the see-saw mechanism is employed. Rather detailed predictions on the spectrum and the mixing angles follow from an Ansatz for the mass matrix of the singlet neutrino fields, which is based on the assumption of non-zero generation (family) charges of these particles. The mixing matrix is of the bimaximal type, and the mass difference of the two lightest neutrinos has the right magnitude required for the vacuum oscillation solution for solar neutrinos. The spectrum and the mixing pattern is stable with respect to loop corrections.

Acknowledgement

The author likes to thank Susan and Gabor Domokos for the invitation to this workshop and for their kind hospitality. He also likes to thank his colleague Christof Wetterich for very helpful comments.

References

1. For the results of the collaborations Gallex, Homestake, Sage, Kamiokande and Superkamiokande see Bahcall et al., hep-ph 9807216.
2. Y. Fukuda et al., hep-ex/9807003.
3. A. Yu Smirnov, hep-ph/9901208;
   W. M. Alberico and S. M. Bilenky, hep-ph/9905254;
   P. Fisher, B. Kayser and K. S. McFarland, NSF-PT-99-1;
   G. Altarelli and F. Feruglio, hep-ph/9905536.
4. C. D. Froggatt and H. B. Nielsen, Nucl. Phys. **B147** (1979) 277;
   S. Dimopoulos, Phys. Lett. **B129** (1983) 417;
   J. Bijnens and C. Wetterich, Nucl. Phys. **B283** (1987) 237;
   Phys. Lett. **B199** (1987) 525;
   L. E. Ibanez and G. G. Ross, Phys. Lett. **B332** (1994) 100.
5. see G. G. Ross, "Grand Unified Theories", Benjamin 1985.
6. J. A. Harvey, P. Ramond, and D. B. Reiss, Nucl. Phys. **B199** (1982) 223.
7. B. Stech, in: Flavor Mixing in Weak Interactions, Ed. L. L. Chau, Plenum (1984) p. 735.
8. J. Bijnens and C. Wetterich, Nucl. Phys. **B292** (1987) 443.
9. B. Stech, hep-ph/9905440
10. J. H. Bahcall, P. I. Krastev, and A. Yu. Smirnov, Phys. Rev. **D58** (1998) 096016;
    V. Barger and K. Whisnant, hep-ph/9903262.
11. B. Stech, Phys. Lett. **B403** (1997) 114.
12. B. Stech, Phys. Lett. **B130** (1983) 189
13. e.g. M. Jezabek and Y. Sumino, Phys. Lett. **B440** (1998) 327, hep-ph/9807310, hep-ph/9904382;
    T. Blazek, S. Raby, and K. Tobe, hep-ph/9903340;
    C. H. Albright and S. M. Barr, Phys. Lett. **B452** (1999) 287;

R. Barbieri, P. Creminelli, and A. Romanino, hep-ph/9903460.
14. N. Irges, S. Lavignac, and P. Ramond, Phys. Rev. **D58**: (1998) 035003;
J. K. Ellwood, N. Irges, and P. Ramond, Phys. Rev. Lett. **81**: (1998) 5064 ;
G. Altarelli and F. Feruglio, Phys. Lett. **B451** (1999) 388;
S. Lola, hep-ph/9903203.
15. V. Barger, S. Pakvasa, T. J. Weiler, and K. Whisnant, Phys. Lett. **B437** (1998) 107.
16. C. Athanassopoulos et al., LSND Coll., Phys. Rev. **C58** (1998) 2511.
17. C. Wetterich, Nucl. Phys. **B187** (1981) 343;
K. Babu, C. N. Leung, and J. Pantalone, Phys. Lett. **B319** (1993) 191.
18. J. Ellis and S. Lola, hep-ph/9904279.

FOUR-NEUTRINO MIXING

CARLO GIUNTI
INFN, Sezione di Torino, and
Dipartimento di Fisica Teorica, Università di Torino,
Via P. Giuria 1, I–10125 Torino, Italy

It is shown that at least four massive neutrinos are needed in order to accommodate the evidences in favor of neutrino oscillations found in solar and atmospheric neutrino experiments and in the LSND experiment. Among all four-neutrino schemes, only two are compatible with the results of all neutrino oscillation experiments. These two schemes have a mass spectrum composed of two pairs of neutrinos with close masses separated by the "LSND gap" of the order of 1 eV.

1 Introduction

Neutrino oscillations have been proposed by B. Pontecorvo[1] more than forty years ago following an analogy with $K^0 \leftrightarrows \bar{K}^0$ oscillations. Today this beautiful quantum mechanical phenomenon[2,3] is subject to intensive experimental and theoretical research. Besides the intrinsic interest related to the investigation of the fundamental properties of neutrinos, it is considered to be one of the best ways to explore the physics beyond the Standard Model.

The best evidence in favor of the existence of neutrino oscillations has been recently provided by the measurement in the Super-Kamiokande experiment[4] of an up–down asymmetry of high-energy μ-like events generated by atmospheric neutrinos:

$$\mathcal{A}_\mu \equiv (D_\mu - U_\mu)/(D_\mu + U_\mu) = 0.311 \pm 0.043 \pm 0.01 \,. \tag{1}$$

Here D_μ and U_μ are, respectively, the number of downward-going and upward-going events, corresponding to the zenith angle intervals $0.2 < \cos\theta < 1$ and $-1 < \cos\theta < -0.2$. Since the fluxes of high-energy downward-going and upward-going atmospheric neutrinos are predicted to be equal with high accuracy on the basis of geometrical arguments,[5] the Super-Kamiokande evidence in favor of neutrino oscillations is model-independent and provides a confirmation of the indications in favor of oscillations of atmospheric neutrinos found in the Super-Kamiokande experiment itself[4,6] and in other experiments through the measurement of the ratio of μ-like and e-like events (Kamiokande, IMB, Soudan 2)[7] and through the measurement of upward-going muons produced by neutrino interactions in the rock below the detector (MACRO).[8] Large $\nu_\mu \leftrightarrows \nu_e$ oscillations of atmospheric neutrinos are excluded by the absence of a up–down asymmetry of high-energy e-like events generated by atmospheric neutrinos and detected in the Super-Kamiokande experiment ($\mathcal{A}_e = 0.036 \pm 0.067 \pm 0.02$)[4] and by the negative result of the CHOOZ long-baseline $\bar{\nu}_e$ disappearance experiment.[9] Therefore, the atmospheric neutrino anomaly consists in the disappearance of muon neutrinos and can be explained by $\nu_\mu \to \nu_\tau$ and/or $\nu_\mu \to \nu_s$ oscillations (here ν_s is a sterile neutrino that does not take part in weak interactions).

Other indications in favor of neutrino oscillations have been obtained in solar neutrino experiments (Homestake, Kamiokande, GALLEX, SAGE, Super-Kamiokande)[10] and in the LSND experiment.[11]

The flux of electron neutrinos measured in all five solar neutrino experiments is substantially smaller than the one predicted by the Standard Solar Model[12] and a comparison of the data of different experiments indicate an energy dependence of the solar ν_e suppression, which represents a rather convincing evidence in favor of neutrino oscillations.[3] The disappearance of solar electron neutrinos can be explained by $\nu_e \to \nu_\mu$ and/or $\nu_e \to \nu_\tau$ and/or $\nu_e \to \nu_s$ oscillations.[13]

The accelerator LSND experiment is the only one that claims the observation of neutrino oscillations in specific appearance channels: $\bar{\nu}_\mu \to \bar{\nu}_e$ and $\nu_\mu \to \nu_e$. Since the appearance of neutrinos with a different flavor represents the true essence of neutrino oscillations, the LSND evidence is extremely interesting and its confirmation (or disproof) by other experiments should receive high priority in future research. Four such experiments have been proposed and are under study: BooNE at Fermilab, I-216 at CERN, ORLaND at Oak Ridge and NESS at the European Spallation Source.[14] Among these proposals only BooNE is approved and will start in 2001.

Neutrino oscillations occur if neutrinos are massive and mixed particles,[2,3] *i.e.* if the left-handed components $\nu_{\alpha L}$ of the flavor neutrino fields are superpositions of the left-handed components ν_{kL} ($k = 1, \ldots, N$) of neutrino fields with definite mass m_k:

$$\nu_{\alpha L} = \sum_{k=1}^{N} U_{\alpha k} \, \nu_{kL} \,, \tag{2}$$

where U is a $N \times N$ unitary mixing matrix. From the measurement of the invisible decay width of the Z-boson it is known that the number of light active neutrino flavors is three,[15] corresponding to ν_e, ν_μ and ν_τ (active neutrinos are those taking part to standard weak interactions). This implies that the number N of massive neutrinos is bigger or equal to three. If $N > 3$, in the flavor basis there are $N_s = N - 3$ sterile neutrinos, $\nu_{s_1}, \ldots, \nu_{s_{N_s}}$, that do not take part to standard weak interactions. In this case the flavor index α in Eq. (2) takes the values $e, \mu, \tau, s_1, \ldots, s_{N_s}$.

2 The necessity of at least three independent Δm^2's

The three evidences in favor of neutrino oscillations found in solar and atmospheric neutrino experiments and in the accelerator LSND experiment imply the existence of at least three independent neutrino mass-squared differences. This can be seen by considering the general expression for the probability of $\nu_\alpha \to \nu_\beta$ transitions in vacuum, that can be written as[2,3]

$$P_{\nu_\alpha \to \nu_\beta} = \left| \sum_{k=1}^{N} U_{\alpha k}^{*} \, U_{\beta k} \, \exp\left(-i \, \frac{\Delta m^2_{kj} \, L}{2 \, E} \right) \right|^2 , \tag{3}$$

where $\Delta m^2_{kj} \equiv m^2_k - m^2_j$, j is any of the mass-eigenstate indices, L is the distance between the neutrino source and detector and E is the neutrino energy. The range of L/E characteristic of each type of experiment is different: $L/E \sim 10^{11} - 10^{12} \, \text{eV}^{-2}$ for solar neutrino experiments, $L/E \sim 10^2 - 10^3 \, \text{eV}^{-2}$ for atmospheric neutrino experiments and $L/E \sim 1 \, \text{eV}^{-2}$ for the LSND experiment. From Eq. (3) it is clear that neutrino oscillations are observable in an experiment only if there is at least one mass-squared difference Δm^2_{kj} such that

$$\frac{\Delta m^2_{kj} \, L}{2 \, E} \gtrsim 0.1 \tag{4}$$

(the precise lower bound depends on the sensitivity of the experiment) in a significant part of the energy and source-detector distance intervals of the experiment (if the condition (4) is not satisfied, $P_{\nu_\alpha \to \nu_\beta} \simeq \left| \sum_k U_{\alpha k}^{*} \, U_{\beta k} \right|^2 = \delta_{\alpha\beta}$). Since the range of L/E probed by the LSND experiment is the smaller one, a large mass-squared difference is needed for LSND oscillations:

$$\Delta m^2_{\text{LSND}} \gtrsim 10^{-1} \, \text{eV}^2 \, . \tag{5}$$

Specifically, the maximum likelihood analysis of the LSND data in terms of two-neutrino oscillations gives[11]

$$0.20 \, \text{eV}^2 \lesssim \Delta m^2_{\text{LSND}} \lesssim 2.0 \, \text{eV}^2 \, . \tag{6}$$

Furthermore, from Eq. (3) it is clear that a dependence of the oscillation probability from the neutrino energy E and the source-detector distance L is observable only if there is at least one mass-squared difference Δm^2_{kj} such that

$$\frac{\Delta m^2_{kj} \, L}{2 \, E} \sim 1 \, . \tag{7}$$

Indeed, all the phases $\Delta m^2_{kj} L/2E \gg 1$ are washed out by the average over the energy and source-detector ranges characteristic of the experiment. Since a variation of the oscillation probability as a function of neutrino energy has been observed both in solar and atmospheric neutrino experiments and the ranges of L/E characteristic of these two types of experiments are different from each other and different from the LSND range, two more mass-squared differences with different scales are needed:

$$\Delta m^2_{\text{sun}} \sim 10^{-12} - 10^{-11}\,\text{eV}^2 \qquad (\text{VO})\,, \tag{8}$$

$$\Delta m^2_{\text{atm}} \sim 10^{-3} - 10^{-2}\,\text{eV}^2\,. \tag{9}$$

The condition (8) for the solar mass-squared difference Δm^2_{sun} has been obtained under the assumption of vacuum oscillations (VO). If the disappearance of solar ν_e's is due to the MSW effect,[16] the condition

$$\Delta m^2_{\text{sun}} \lesssim 10^{-4}\,\text{eV}^2 \qquad (\text{MSW}) \tag{10}$$

must be fulfilled in order to have a resonance in the interior of the sun. Hence, in the MSW case Δm^2_{sun} must be at least one order of magnitude smaller than Δm^2_{atm}.

It is possible to ask if three different scales of neutrino mass-squared differences are needed even if the results of the Homestake solar neutrino experiment is neglected, allowing an energy-independent suppression of the solar ν_e flux. The answer is that still the data cannot be fitted with only two neutrino mass-squared differences because an energy-independent suppression of the solar ν_e flux requires large $\nu_e \to \nu_\mu$ or $\nu_e \to \nu_\tau$ transitions generated by Δm^2_{atm} or Δm^2_{LSND}. These transitions are forbidden by the results of the Bugey[17] and CHOOZ[9] reactor $\bar{\nu}_e$ disappearance experiments and by the non-observation of an up-down asymmetry of e-like events in the Super-Kamiokande atmospheric neutrino experiment.[4]

3 Four-neutrino schemes

The existence of three different scales of Δm^2 imply that at least four light massive neutrinos must exist in nature. Here we consider the schemes with four light and mixed neutrinos,[18–33] which constitute the minimal possibility that allows to explain all the existing data with neutrino oscillations. In this case, in the flavor basis the three active neutrinos ν_e, ν_μ, ν_τ are accompanied by a sterile neutrino ν_s that does not take part in standard weak interactions.

The six types of four-neutrino mass spectra with three different scales of Δm^2 that can accommodate the hierarchy $\Delta m^2_{\text{sun}} \ll \Delta m^2_{\text{atm}} \ll \Delta m^2_{\text{LSND}}$ are shown qualitatively in Fig. 1. In all these mass spectra there are two groups of close masses separated by the "LSND gap" of the order of 1 eV. In each scheme the smallest mass-squared difference corresponds to Δm^2_{sun} (Δm^2_{21} in

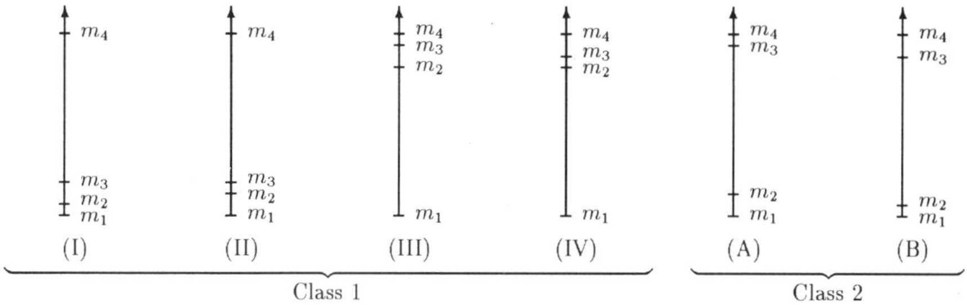

Figure 1

schemes I and B, Δm^2_{32} in schemes II and IV, Δm^2_{43} in schemes III and A), the intermediate one to $\Delta m^2_{\rm atm}$ (Δm^2_{31} in schemes I and II, Δm^2_{42} in schemes III and IV, Δm^2_{21} in scheme A, Δm^2_{43} in scheme B) and the largest mass squared difference $\Delta m^2_{41} = \Delta m^2_{\rm LSND}$ is relevant for the oscillations observed in the LSND experiment. The six schemes are divided into four schemes of class 1 (I–IV) in which there is a group of three masses separated from an isolated mass by the LSND gap, and two schemes of class 2 (A, B) in which there are two couples of close masses separated by the LSND gap.

4 The disfavored schemes of class 1

In the following we will show that the schemes of class 1 are disfavored by the data if also the negative results of short-baseline accelerator and reactor disappearance neutrino oscillation experiments are taken into account.[21,27,28] Let us remark that in principle one could check which schemes are allowed by doing a combined fit of all data in the framework of the most general four-neutrino mixing scheme, with three mass-squared differences, six mixing angles and three CP-violating phases as free parameters. However, at the moment it is not possible to perform such a fit because of the enormous complications due to the presence of too many parameters and to the difficulties involved in a combined fit of the data of different experiments, which are usually analyzed by the experimental collaborations using different methods. Hence, we think that it is quite remarkable that one can exclude the schemes of class 1 with the following relatively simple procedure.

Let us define the quantities d_α, with $\alpha = e, \mu, \tau, s$, in the schemes of class 1 as

$$d_\alpha^{(\rm I)} \equiv |U_{\alpha 4}|^2 \,, \qquad d_\alpha^{(\rm II)} \equiv |U_{\alpha 4}|^2 \,, \qquad d_\alpha^{(\rm III)} \equiv |U_{\alpha 1}|^2 \,, \qquad d_\alpha^{(\rm IV)} \equiv |U_{\alpha 1}|^2 \,. \tag{11}$$

Physically d_α quantifies the mixing of the flavor neutrino ν_α with the isolated neutrino, whose mass is separated from the other three by the LSND gap.

The probability of $\nu_\alpha \to \nu_\beta$ ($\beta \neq \alpha$) and $\nu_\alpha \to \nu_\alpha$ transitions (and the corresponding probabilities for antineutrinos) in short-baseline experiments are given by[3, 21]

$$P_{\nu_\alpha \to \nu_\beta} = A_{\alpha;\beta} \sin^2 \frac{\Delta m^2_{41} L}{4E} , \qquad P_{\nu_\alpha \to \nu_\alpha} = 1 - B_{\alpha;\alpha} \sin^2 \frac{\Delta m^2_{41} L}{4E} , \tag{12}$$

with the oscillation amplitudes

$$A_{\alpha;\beta} = 4\, d_\alpha\, d_\beta , \qquad B_{\alpha;\alpha} = 4\, d_\alpha \,(1 - d_\alpha) . \tag{13}$$

The probabilities (12) have the same form as the corresponding probabilities in the case of two-neutrino mixing, $P_{\nu_\alpha \to \nu_\beta} = \sin^2(2\vartheta) \sin^2(\Delta m^2 L/4E)$ and $P_{\nu_\alpha \to \nu_\alpha} = 1 - \sin^2(2\vartheta) \sin^2(\Delta m^2 L/4E)$, which have been used by all experimental collaborations for the analysis of the data in order to get information on the parameters $\sin^2(2\vartheta)$ and Δm^2 (ϑ and Δm^2 are, respectively, the mixing angle and the mass-squared difference in the case of two-neutrino mixing). Therefore, we can use the results of their analyses in order to get information on the corresponding parameters $A_{\alpha;\beta}$, $B_{\alpha;\alpha}$ and Δm^2_{41}.

The exclusion plots obtained in short-baseline $\bar{\nu}_e$ and ν_μ disappearance experiments imply that[21]

$$d_\alpha \leq a^0_\alpha \qquad \text{or} \qquad d_\alpha \geq 1 - a^0_\alpha \qquad (\alpha = e, \mu) , \tag{14}$$

with

$$a^0_\alpha = \frac{1}{2} \left(1 - \sqrt{1 - B^0_{\alpha;\alpha}} \right) \qquad (\alpha = e, \mu) , \tag{15}$$

where $B^0_{e;e}$ and $B^0_{\mu;\mu}$ are the upper bounds, that depend on Δm^2_{41}, of the oscillation amplitudes $B_{e;e}$ and $B_{\mu;\mu}$ given by the exclusion plots of $\bar{\nu}_e$ and ν_μ disappearance experiments. From the exclusion curves of the Bugey reactor $\bar{\nu}_e$ disappearance experiment[17] and of the CDHS and CCFR accelerator ν_μ disappearance experiments[34] it follows that $a^0_e \lesssim 3 \times 10^{-2}$ for $\Delta m^2_{41} = \Delta m^2_{\text{LSND}}$ in the LSND range (6) and $a^0_\mu \lesssim 0.2$ for $\Delta m^2_{41} \gtrsim 0.4\,\text{eV}^2$.[3]

Therefore, the negative results of short-baseline $\bar{\nu}_e$ and ν_μ disappearance experiments imply that d_e and d_μ are either small or large (close to one). Taking into account the unitarity limit $d_e + d_\mu \leq 1$, for each value of Δm^2_{41} above about $0.3\,\text{eV}^2$ there are three regions in the d_e–d_μ plane that are allowed by the results of disappearance experiments: region SS with small d_e and d_μ, region LS with large d_e and small d_μ and region SL with small d_e and large d_μ. These three regions are illustrated qualitatively by the three shadowed areas in Fig. 2. For $\Delta m^2_{41} \lesssim 0.3\,\text{eV}^2$ there is no constraint on the value of d_μ from

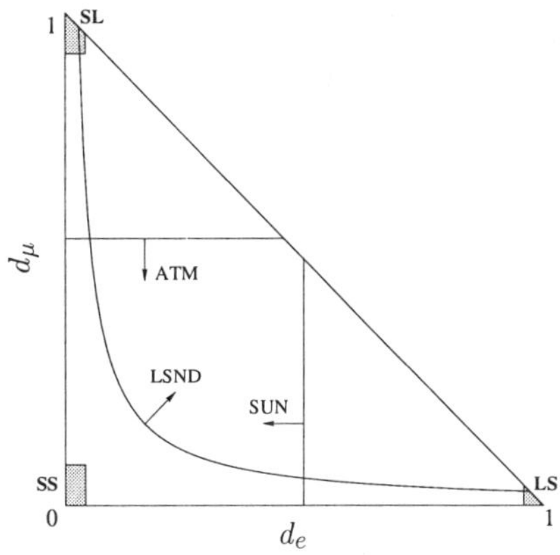

Figure 2

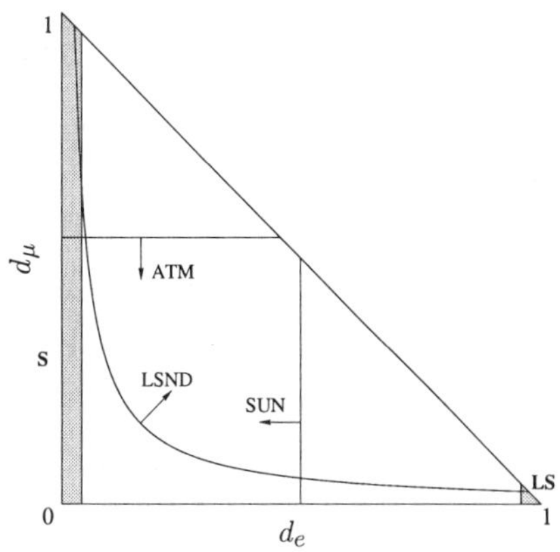

Figure 3

the results of short-baseline ν_μ disappearance experiments and there are two regions in the d_e–d_μ plane that are allowed by the results of $\bar{\nu}_e$ disappearance experiments: region S with small d_e and region LS with large d_e and small d_μ (the smallness of d_μ follows from the unitarity bound $d_e + d_\mu \leq 1$). These two regions are illustrated qualitatively by the two shadowed areas in Fig. 3.

Let us consider now the results of solar neutrino experiments, which imply a disappearance of electron neutrinos. The survival probability of solar ν_e's averaged over the fast unobservable oscillations due to Δm^2_{41} and Δm^2_{31} is bounded by[3, 21]

$$P^{\text{sun}}_{\nu_e \to \nu_e} \geq d_e^2 \,. \tag{16}$$

Therefore, only the possibility

$$d_e \leq a_e^0 \tag{17}$$

is acceptable in order to explain the observed deficit of solar ν_e's with neutrino oscillations. Indeed, the solar neutrino data imply an upper bound for d_e, that is shown qualitatively by the vertical lines in Figs. 2 and 3. It is clear that the regions LS in Figs. 2 and 3 are disfavored by the results of solar neutrino experiments.

In a similar way, since the survival probability of atmospheric ν_μ's and $\bar{\nu}_\mu$'s is bounded by[3, 21]

$$P^{\text{atm}}_{\nu_\mu \to \nu_\mu} \geq d_\mu^2 \,, \tag{18}$$

large values of d_μ are incompatible with the asymmetry (1) observed in the Super-Kamiokande experiment. The upper bound for d_μ that follows from

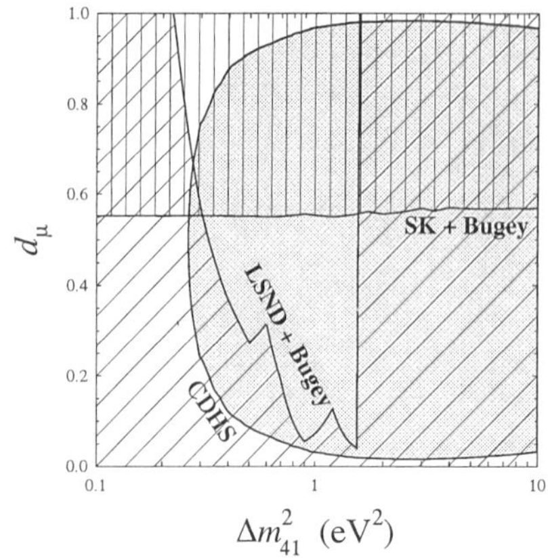

Figure 4

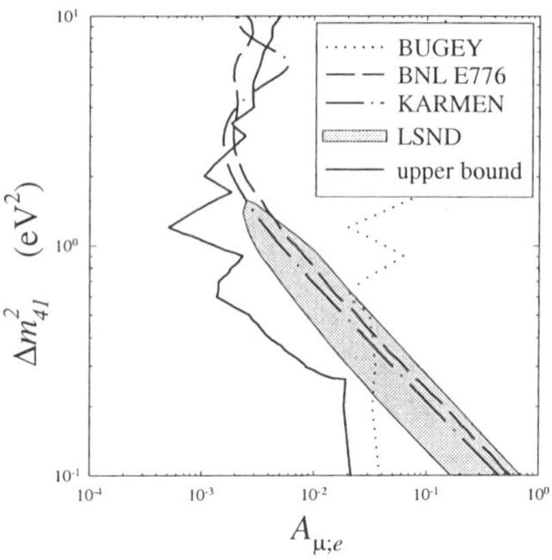

Figure 5

atmospheric neutrino data is shown qualitatively by the horizontal lines in Figs. 2 and 3. It is clear that the region SL in Fig. 2, that is allowed by the results of ν_μ short-baseline disappearance experiments for $\Delta m^2_{41} \gtrsim 0.3\,\mathrm{eV}^2$, and the large–$d_\mu$ part of the region S in Fig. 3 are disfavored by the results of atmospheric neutrino experiments. A precise calculation[28] shows that the Super-Kamiokande asymmetry (1) and the exclusion curve of the Bugey $\bar{\nu}_e$ disappearance experiment imply the upper bound

$$d_\mu \lesssim 0.55 \equiv a_\mu^{\mathrm{SK}} \,. \tag{19}$$

This upper bound is depicted by the horizontal line in Fig. 4 (the vertically hatched area above the line is excluded).

In Fig. 4 we have also shown the bound $d_\mu \leq a_\mu^0$ or $d_\mu \geq 1 - a_\mu^0$ obtained from the exclusion plot of the short-baseline CDHS ν_μ disappearance experiment, which exclude the shadowed region. It is clear that the results of short-baseline disappearance experiments and the Super-Kamiokande asymmetry (1) imply that $d_\mu \lesssim 0.55$ for $\Delta m^2_{41} \lesssim 0.3\,\mathrm{eV}^2$ and that d_μ is very small for $\Delta m^2_{41} \gtrsim 0.3\,\mathrm{eV}^2$. However, this range of d_μ is disfavored by the results of the LSND experiment, that imply a lower bound $A_{\mu;e}^{\mathrm{min}}$ for the amplitude $A_{\mu;e} = 4 d_e d_\mu$ of $\nu_\mu \to \nu_e$ oscillations. Indeed, we have

$$d_e\, d_\mu \geq A_{\mu;e}^{\mathrm{min}}/4 \,. \tag{20}$$

This bound, shown qualitatively by the LSND exclusion curves in Figs. 2 and 3, excludes region SS in Fig. 2 and the small-d_μ part of region S in Fig. 3. From Figs. 2 and 3 one can see in a qualitative way that in the schemes of class 1 the results of the solar, atmospheric and LSND experiments are incompatible with the negative results of short-baseline experiments.

A quantitative illustration of this incompatibility is given in Fig. 4. The curve in Fig. 4 labelled LSND + Bugey (the diagonally hatched area is excluded) represents the constraint

$$d_\mu \geq A_{\mu;e}^{\min}/4a_e^0\,, \tag{21}$$

derived from the inequality (20) using the bound (17). One can see that the results of the LSND experiment exclude the range of d_μ allowed by the results of short-baseline disappearance experiments and by the Super-Kamiokande asymmetry (1). Hence, in the framework of the schemes of class 1 there is no range of d_μ that is compatible with all the experimental data.

The incompatibility of the experimental results with the schemes of class 1 is shown also in Fig. 5, where we have plotted in the $A_{\mu;e}$–Δm^2_{41} plane the upper bound $A_{\mu;e} \leq 4\,a_e^0\,a_\mu^0$ for $\Delta m^2_{41} > 0.26\,\text{eV}^2$ and $A_{\mu;e} \leq 4\,a_e^0\,a_\mu^{\text{SK}}$ for $\Delta m^2_{41} < 0.26\,\text{eV}^2$ (solid line, the region on the right is excluded). One can see that this constraint is incompatible with the LSND-allowed region (shadowed area).

Summarizing, we have reached the conclusion that the four schemes of class 1 shown in Fig. 1 are disfavored by the data.

5 The favored schemes of class 2

The four-neutrino schemes of class 2 are compatible with the results of all neutrino oscillation experiments if the mixing of ν_e with the two mass eigenstates responsible for the oscillations of solar neutrinos (ν_3 and ν_4 in scheme A and ν_1 and ν_2 in scheme B) is large and the mixing of ν_μ with the two mass eigenstates responsible for the oscillations of atmospheric neutrinos (ν_1 and ν_2 in scheme A and ν_3 and ν_4 in scheme B) is large.[20,21,27,28] This is illustrated qualitatively in Figs. 6 and 7, as we are going to explain.

Let us define the quantities c_α, with $\alpha = e, \mu, \tau, s$, in the schemes A and B as

$$c_\alpha^{(\text{A})} \equiv \sum_{k=1,2} |U_{\alpha k}|^2\,, \qquad c_\alpha^{(\text{B})} \equiv \sum_{k=3,4} |U_{\alpha k}|^2\,. \tag{22}$$

Physically c_α quantify the mixing of the flavor neutrino ν_α with the two massive neutrinos whose Δm^2 is relevant for the oscillations of atmospheric neutrinos (ν_1, ν_2 in scheme A and ν_3, ν_4 in scheme B). The exclusion plots obtained in short-baseline $\bar{\nu}_e$ and ν_μ disappearance experiments imply that[21]

$$c_\alpha \leq a_\alpha^0 \qquad \text{or} \qquad c_\alpha \geq 1 - a_\alpha^0 \qquad (\alpha = e, \mu)\,, \tag{23}$$

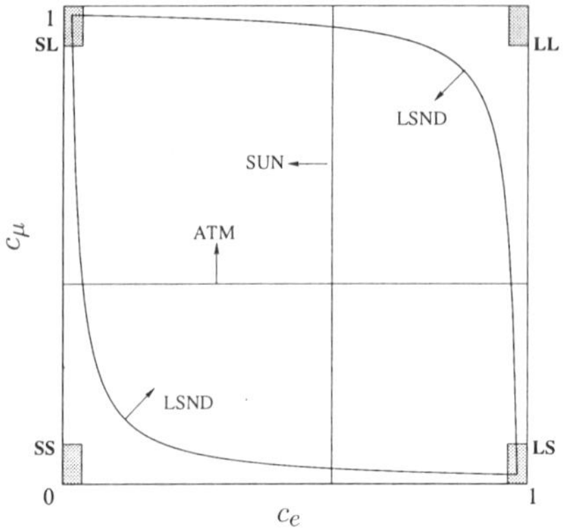

Figure 6

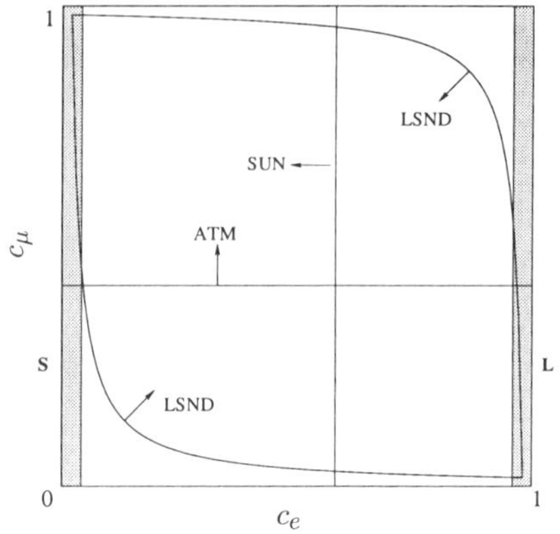

Figure 7

with a_α^0 given in Eq. (15).

The shadowed areas in Figs. 6 and 7 illustrate qualitatively the regions in the c_e–c_μ plane allowed by the negative results of short-baseline $\bar{\nu}_e$ and ν_μ disappearance experiments. Figure 6 is valid for $\Delta m^2_{41} \gtrsim 0.3\,\text{eV}^2$ and shows that there are four regions allowed by the results of short-baseline disappearance experiments: region SS with small c_e and c_μ, region LS with large c_e and small c_μ, region SL with small c_e and large c_μ and region LL with large c_e and c_μ. The quantities c_e and c_μ can be both large, because the unitarity of the mixing matrix imply that $c_\alpha + c_\beta \leq 2$ and $0 \leq c_\alpha \leq 1$ for $\alpha, \beta = e, \mu, \tau, s$. Figure 7 is valid for $\Delta m^2_{41} \lesssim 0.3\,\text{eV}^2$, where there is no constraint on the value of c_μ from the results of short-baseline ν_μ disappearance experiments. It shows that there are two regions allowed by the results of short-baseline $\bar{\nu}_e$ disappearance experiments: region S with small c_e and region L with large c_e.

Let us take now into account the results of solar neutrino experiments. Large values of c_e are incompatible with solar neutrino oscillations because in this case ν_e has large mixing with the two massive neutrinos responsible for atmospheric neutrino oscillations and, through the unitarity of the mixing matrix, small mixing with the two massive neutrinos responsible for solar neutrino oscillations. Indeed, in the schemes of class 2 the survival probability $P^{\text{sun}}_{\nu_e \to \nu_e}$ of solar ν_e's is bounded by[3, 21]

$$P^{\text{sun}}_{\nu_e \to \nu_e} \geq c_e^2/2\,, \tag{24}$$

and its possible variation $\Delta P^{\text{sun}}_{\nu_e \to \nu_e}(E)$ with neutrino energy E is limited by[3, 21]

$$\Delta P^{\text{sun}}_{\nu_e \to \nu_e}(E) \leq (1 - c_e)^2\,. \tag{25}$$

If c_e is large as in the LS or LL regions of Fig. 6 or in the L region of Fig. 7, we have

$$P^{\text{sun}}_{\nu_e\to\nu_e} \geq \frac{\left(1-a_e^0\right)^2}{2} \simeq \frac{1}{2}, \qquad \Delta P^{\text{sun}}_{\nu_e\to\nu_e}(E) \leq (a_e^0)^2 \lesssim 10^{-3}, \tag{26}$$

for $\Delta m^2_{41} = \Delta m^2_{\text{LSND}}$ in the LSND range (6). Therefore $P^{\text{sun}}_{\nu_e\to\nu_e}$ is bigger than 1/2 and practically does not depend on neutrino energy. Since this is incompatible with the results of solar neutrino experiments interpreted in terms of neutrino oscillations,[3,13] we conclude that the regions LS and LL in Fig. 6 and the region L in Fig. 7 are disfavored by solar neutrino data, as illustrated qualitatively by the vertical exclusion lines in Figs. 6 and 7.

Let us consider now the results of atmospheric neutrino experiments. Small values of c_μ are incompatible with atmospheric neutrino oscillations because in this case ν_μ has small mixing with the two massive neutrinos responsible for atmospheric neutrino oscillations. Indeed, the survival probability of atmospheric ν_μ's is bounded by[3,21]

$$P^{\text{atm}}_{\nu_\mu\to\nu_\mu} \geq \left(1-c_\mu\right)^2, \tag{27}$$

and it can be shown[28] that the Super-Kamiokande asymmetry (1) and the exclusion curve of the Bugey $\bar{\nu}_e$ disappearance experiment imply the upper bound

$$c_\mu \gtrsim 0.45 \equiv b_\mu^{\text{SK}}. \tag{28}$$

This limit is depicted qualitatively by the horizontal exclusion lines in Figs. 6 and 7. Therefore, we conclude that the regions SS and LS in Fig. 6 and the small-c_μ parts of the regions S and L in Fig. 7 are disfavored by atmospheric neutrino data.

Finally, let us consider the results of the LSND experiment. In the schemes of class 2 the amplitude of short-baseline $\nu_\mu \to \nu_e$ oscillations is given by

$$A_{\mu;e} = \left|\sum_{k=1,2} U_{ek}U^*_{\mu k}\right|^2 = \left|\sum_{k=3,4} U_{ek}U^*_{\mu k}\right|^2. \tag{29}$$

The second equality in Eq. (29) is due to the unitarity of the mixing matrix. Using the Cauchy–Schwarz inequality we obtain

$$c_e\,c_\mu \geq A^{\min}_{\mu;e}/4 \qquad \text{and} \qquad (1-c_e)\,(1-c_\mu) \geq A^{\min}_{\mu;e}/4, \tag{30}$$

where $A^{\min}_{\mu;e}$ is the minimum value of the oscillation amplitude $A_{\mu;e}$ observed in the LSND experiment. The bounds (30) are illustrated qualitatively in Figs.

6 and 7. One can see that the results of the LSND experiment confirm the exclusion of the regions SS and LL in Fig. 6 and the exclusion of the small-c_μ part of region S and of the large-c_μ part of region L in Fig. 7.

Summarizing, if $\Delta m^2_{41} \gtrsim 0.3\,\text{eV}^2$ only the region SL in Fig. 6, with

$$c_e \leq a_e^0 \qquad \text{and} \qquad c_\mu \geq 1 - a_\mu^0 \,, \tag{31}$$

is compatible with the results of all neutrino oscillation experiments. If $\Delta m^2_{41} \lesssim 0.3\,\text{eV}^2$ only the large-c_μ part of region S in Fig. 7, with

$$c_e \leq a_e^0 \qquad \text{and} \qquad c_\mu \geq b_\mu^{\text{SK}} \,, \tag{32}$$

is compatible with the results of all neutrino oscillation experiments. Therefore, in any case c_e is small and c_μ is large. However, it is important to notice that, as shown clearly in Figs. 6 and 7, the inequalities (30) following from the LSND observation of short-baseline $\nu_\mu \to \nu_e$ oscillations imply that c_e, albeit small, has a lower bound and c_μ, albeit large, has an upper bound:

$$c_e \gtrsim A_{\mu;e}^{\text{min}}/4 \qquad \text{and} \qquad c_\mu \lesssim 1 - A_{\mu;e}^{\text{min}}/4 \,. \tag{33}$$

6 Conclusions

We have seen that only the two four-neutrino schemes A and B of class 2 in Fig. 1 are compatible with the results of all neutrino oscillation experiments. Furthermore, we have shown that the quantities c_e and c_μ in these two schemes must be, respectively, small and large. Physically c_α, defined in Eq. (22), quantify the mixing of the flavor neutrino ν_α with the two massive neutrinos whose Δm^2 is relevant for the oscillations of atmospheric neutrinos (ν_1, ν_2 in scheme A and ν_3, ν_4 in scheme B).

The smallness of c_e implies that electron neutrinos do not oscillate in atmospheric and long-baseline neutrino oscillation experiments. Indeed, one can obtain rather stringent upper bounds for the probability of ν_e transitions into any other state[22] and for the size of CP or T violation that could be measured in long-baseline experiments in the $\nu_\mu \leftrightarrows \nu_e$ and $\bar\nu_\mu \leftrightarrows \bar\nu_e$ channels.[23]

Let us consider now the effective Majorana mass in neutrinoless double-β decay,

$$|\langle m \rangle| = \left| \sum_{k=1}^{4} U_{ek}^2 \, m_k \right| . \tag{34}$$

In scheme A, since c_e is small, the effective Majorana mass is approximately given by

$$|\langle m \rangle| \simeq \left| U_{e3}^2 + U_{e4}^2 \right| m_4 \simeq \left| U_{e3}^2 + U_{e4}^2 \right| \sqrt{\Delta m^2_{\text{LSND}}} \,. \tag{35}$$

Therefore, in scheme A the effective Majorana mass can be as large as $\sqrt{\Delta m^2_{\mathrm{LSND}}}$.[29–32] On the other hand, in scheme B the contribution of the "heavy" neutrino masses m_3 and m_4 to the effective Majorana mass is strongly suppressed:[29–32]

$$|\langle m \rangle|_{34} \equiv \left| U_{e3}^2 \, m_3 + U_{e4}^2 \, m_4 \right| \lesssim c_e \, m_4 \leq a_e^0 \sqrt{\Delta m^2_{\mathrm{LSND}}} \,. \tag{36}$$

Finally, if the upper bound $N_\nu^{\mathrm{BBN}} < 4$ for the effective number of neutrinos in Big-Bang Nucleosynthesis is correct,[35] the mixing of ν_s with the two mass eigenstates responsible for the oscillations of atmospheric neutrinos must be very small.[24–26] In this case atmospheric neutrinos oscillate only in the $\nu_\mu \to \nu_\tau$ channel and solar neutrinos oscillate only in the $\nu_e \to \nu_s$ channel. This is very important because it implies that the two-generation analyses of solar and atmospheric neutrino data give correct information on neutrino mixing in the two four-neutrino schemes A and B. Otherwise, it will be necessary to reanalyze the solar and atmospheric neutrino data using a general formalism that takes into account the possibility of simultaneous transitions into active and sterile neutrinos in solar and atmospheric neutrino experiments.[33]

Acknowledgments

I would like to thank S.M. Bilenky, W. Grimus and T. Schwetz for enjoyable and stimulating collaboration on the topics presented in this report.

References

1. B. Pontecorvo, J. Exptl. Theoret. Phys. **33**, 549 (1957) [Sov. Phys. JETP **6**, 429 (1958)]; B. Pontecorvo, J. Exptl. Theoret. Phys. **34**, 247 (1958) [Sov. Phys. JETP **7**, 172 (1958)].
2. S.M. Bilenky and B. Pontecorvo, Phys. Rep. **41**, 225 (1978); S.M. Bilenky and S.T. Petcov, Rev. Mod. Phys. **59**, 671 (1987); C.W. Kim and A. Pevsner, *Neutrinos in Physics and Astrophysics*, Contemporary Concepts in Physics, Vol. 8, Harwood Academic Press, Chur, Switzerland, 1993.
3. S.M. Bilenky, C. Giunti and W. Grimus, hep-ph/9812360 [Prog. Part. Nucl. Phys. **43**, in press].
4. Y. Fukuda *et al.* (Super-Kamiokande Coll.), Phys. Rev. Lett. **81**, 1562 (1998); K. Scholberg (Super-Kamiokande Coll.), hep-ex/9905016.
5. P. Lipari, hep-ph/9904443.
6. Y. Fukuda *et al.* (Super-Kamiokande Coll.), Phys. Rev. Lett. **82**, 2644 (1999); A. Habig (Super-Kamiokande Coll.), hep-ex/9903047.

7. Y. Fukuda *et al.* (Kamiokande Coll.), Phys. Lett. **B335**, 237 (1994); R. Becker-Szendy *et al.* (IMB Coll.), Nucl. Phys. B (Proc. Suppl.) **38**, 331 (1995); W.W.M. Allison *et al.* (Soudan Coll.), Phys. Lett. **B449**, 137 (1999).
8. M. Ambrosio *et al.* (MACRO Coll.), Phys. Lett. **B434**, 451 (1998); P. Bernardini (MACRO Coll.), hep-ex/9906019.
9. M. Apollonio *et al.* (CHOOZ Coll.), Phys. Lett. **B420**, 397 (1998); hep-ex/9907037.
10. B.T. Cleveland *et al.*, Astrophys. J. **496**, 505 (1998); K.S. Hirata *et al.* (Kamiokande Coll.), Phys. Rev. Lett. **77**, 1683 (1996); W. Hampel *et al.* (GALLEX Coll.), Phys. Lett. **B447**, 127 (1999); J.N. Abdurashitov *et al.* (SAGE Coll.), astro-ph/9907113; Y. Fukuda *et al.* (Super-Kamiokande Coll.), Phys. Rev. Lett. **81**, 1158 (1998); Phys. Rev. Lett. **82**, 2430 (1999); M.B. Smy (Super-Kamiokande Coll.), hep-ex/9903034.
11. C. Athanassopoulos *et al.* (LSND Coll.), Phys. Rev. Lett. **75**, 2650 (1995); Phys. Rev. Lett. **77**, 3082 (1996); Phys. Rev. Lett. **81**, 1774 (1998); G. Mills (LSND Coll.), Talk presented at the XXXIV$^{\rm th}$ Rencontres de Moriond *Electroweak Interactions and Unified Theories*, Les Arcs, March 1999 (http://moriond.in2p3.fr/EW/transparencies).
12. J.N. Bahcall *et al.*, Phys. Lett. **B433**, 1 (1998).
13. J.N. Bahcall, P.I. Krastev and A.Yu. Smirnov, Phys. Rev. **D58**, 096016 (1998); Y. Fukuda *et al.*, Phys. Rev. Lett. **82**, 1810 (1999); V. Barger and K. Whisnant, Phys. Lett. **B456**, 54 (1999); M.C. Gonzalez-Garcia *et al.*, hep-ph/9906469.
14. Booster Neutrino Experiment (BooNE), http://nu1.lampf.lanl.gov/-BooNE; I-216 $\nu_\mu \to \nu_e$ proposal at CERN, http://chorus01.-cern.ch/~pzucchel/loi/; Oak Ridge Large Neutrino Detector, http://-www.phys.subr.edu/orland/; NESS: Neutrinos at the European Spallation Source, http://www.isis.rl.ac.uk/ess/neut%5Fess.htm.
15. C. Caso *et al.* (Particle Data Group), Eur. Phys. J. **C3**, 1 (1998).
16. S.P. Mikheyev and A.Yu. Smirnov, Yad. Fiz. **42**, 1441 (1985) [Sov. J. Nucl. Phys. **42**, 913 (1985)]; Il Nuovo Cimento **C9**, 17 (1986); L. Wolfenstein, Phys. Rev. **D17**, 2369 (1978); Phys. Rev. **D20**, 2634 (1979).
17. B. Achkar *et al.* (Bugey Coll.), Nucl. Phys. **B434**, 503 (1995).
18. J.T. Peltoniemi *et al.*, Phys. Lett. **B298**, 383 (1993); E.J. Chun *et al.*, Phys. Lett. **B357**, 608 (1995); S.C. Gibbons *et al.*, Phys. Lett. **B430**, 296 (1998); B. Brahmachari and R.N. Mohapatra, Phys. Lett. **B437**, 100 (1998); S. Mohanty *et al.*, Phys. Lett. **B445**, 185 (1998); J.T. Peltoniemi and J.W.F. Valle, Nucl. Phys. **B406**, 409 (1993); Q.Y.

Liu and A.Yu. Smirnov, Nucl. Phys. **B524**, 505 (1998); D.O. Caldwell and R.N. Mohapatra, Phys. Rev. **D48**, 3259 (1993); E. Ma and P. Roy, Phys. Rev. **D52**, R4780 (1995); A.Yu. Smirnov and M. Tanimoto, Phys. Rev. **D55**, 1665 (1997); N. Gaur *et al.*, Phys. Rev. **D58**, 071301 (1998); E.J. Chun *et al.*, Phys. Rev. **D58**, 093003 (1998); K. Benakli and A.Yu. Smirnov, Phys. Rev. Lett. **79**, 4314 (1997); Y. Chikira, N. Haba and Y. Mimura, hep-ph/9808254; C. Liu and J. Song, Phys. Rev. **D60**, 036002 (1999); W. Grimus, R. Pfeiffer and T. Schwetz, hep-ph/9905320.

19. J.J. Gomez-Cadenas and M.C. Gonzalez-Garcia, Z. Phys. **C71**, 443 (1996); S. Goswami, Phys. Rev. **D55**, 2931 (1997); V. Barger, T.J. Weiler and K. Whisnant, Phys. Lett. **B427**, 97 (1998); V. Barger *et al.*, Phys. Rev. **D59**, 113010 (1999); C. Giunti, hep-ph/9906456.
20. S.M. Bilenky *et al.*, Phys. Rev. **D54**, 4432 (1996).
21. S.M. Bilenky, C. Giunti and W. Grimus, Eur. Phys. J. **C1**, 247 (1998), hep-ph/9607372; Proc. of Neutrino '96, Helsinki, June 1996, p. 174, hep-ph/9609343.
22. S.M. Bilenky, C. Giunti and W. Grimus, Phys. Rev. **D57**, 1920 (1998).
23. S.M. Bilenky, C. Giunti and W. Grimus, Phys. Rev. **D58**, 033001 (1998).
24. N. Okada and O. Yasuda, Int. J. Mod. Phys. A **12**, 3669 (1997).
25. S.M. Bilenky *et al.*, Astropart. Phys. **11**, 413 (1999).
26. S.M. Bilenky *et al.*, in *New Trends in Neutrino Physics*, Proc. of the Ringberg Euroconference (Tergernsee, May 1998), p. 117, hep-ph/9807569; in *New Era in Neutrino Physics*, Proc. of a Satellite Symposium after Neutrino '98 (Tokyo, June 1998), p. 179, hep-ph/9809466.
27. V. Barger *et al.*, Phys. Rev. **D58**, 093016 (1998).
28. S.M. Bilenky *et al.*, Phys. Rev. **D60**, 073007 (1999).
29. S.M. Bilenky *et al.*, Phys. Rev. **D57**, 6981 (1998).
30. S.M. Bilenky, C. Giunti and W. Grimus, Nucl. Phys. B (Proc. Suppl.) **77**, 151 (1999), hep-ph/9809368; S.M. Bilenky and C. Giunti, hep-ph/9904328.
31. C. Giunti, hep-ph/9906275.
32. S.M. Bilenky *et al.*, hep-ph/9907234 [Phys. Lett. B, in press].
33. D. Dooling, C. Giunti, K. Kang and C.W. Kim, hep-ph/9908513.
34. F. Dydak *et al.* (CDHS Coll.), Phys. Lett. **B134**, 281 (1984); I.E. Stockdale *et al.* (CCFR Coll.), Phys. Rev. Lett. **52**, 1384 (1984).
35. S. Burles *et al.*, Phys. Rev. Lett. **82**, 4176 (1999); E. Lisi, S. Sarkar and F.L. Villante, Phys. Rev. **D59**, 123520 (1999); S. Sarkar, astro-ph/9903183; K.A. Olive, G. Steigman and T.P. Walker, astro-ph/9905320.

QCD AND ELECTROWEAK STUDIES AT NEUTRINO FACTORIES

ALEXEY A. PETROV
Department of Physics and Astronomy
Johns Hopkins University
Baltimore, MD 21218

I give a brief overview of QCD and electroweak studies at the neutrino factories, i.e. facilities providing high energy and high intensity neutrino beams. These facilities would offer a unique opportunity to perform new precision studies of QCD and electroweak interactions. New types of measurements, like charmonium production and extractions of V_{cb} and V_{ub} CKM matrix elements will become possible. Interesting New Physics scenarios can also be explored.

1 Motivation

The purpose of this talk is to overview physics goals for the experiments utilizing high energy and high intensity neutrino beams. These include the Standard Model electroweak physics, novel tests of QCD, and sensitivity to physics beyond the Standard Model (SM).

The Standard Model electroweak parameters that are conventionally measured in neutrino experiments include $\sin^2\theta_W$ and Cabibbo-Kobayashi-Maskawa (CKM) quark mixing matrix elements V_{cd} and V_{cs}. On the QCD side, neutrino-nucleon interactions are potentially the best probes of various valence parton distribution functions and/or intrinsic strangeness or charm of the nucleon, as well as the strong coupling constant α_s. One can also study various non-perturbative parameters, such as fragmentation functions. Finally, there are interesting New Physics scenarios that can be tested with νN interactions¿ These include supersymmetric extensions of the Standard Model with broken $\mathcal{R}$ parity (or any models with leptoquarks), and new heavy neutral leptons or gauge bosons.

There are several currently running or proposed high energy neutrino experiments. All of the currently running experiments, such as NuTeV at FNAL and NOMAD and CHORUS at CERN, produce neutrino beams with the energies of hundreds GeV in the decays of K or π mesons. This makes determination of the initial neutrino energy quite a formidable task. Several proposed experiments, such as COSMOS (MINOS) and NGS, would utilize beams of neutrinos produced by similar methods, while a new proposal for a Muon Collider should provide highly collimated, high-intensity ν beam from muons decaying in the accelerator tunnel.

2 QCD and Electroweak studies

2.1 Measurements of $\sin^2\theta_W$

One of the most important parameters of the Standard Model is a weak mixing angle θ_W which represents the angle of rotation from the "gauge" basis to the "physical" basis where the mass matrix of gauge Z boson and a photon is diagonal. One of the many possible definitions is

$$\sin^2\theta_W^{os} \equiv 1 - \frac{M_W^2}{M_Z^2}, \tag{1}$$

which corresponds to the on-shell definition of $\sin^2\theta_W$. In neutrino-nucleon interactions $\sin^2\theta_W$ can be extracted using the Llewellyn Smith [1] or Pascos-Wolfenstein relations [2]. These methods involve measurement of three total cross sections and forming three ratios,

$$R_\nu = \frac{\sigma(\nu N \to \nu X)}{\sigma(\nu N \to \mu^- X)}, \qquad R_{\bar\nu} = \frac{\sigma(\bar\nu N \to \bar\nu X)}{\sigma(\bar\nu N \to \mu^+ X)}, \tag{2}$$

and

$$r = \frac{\sigma(\nu N \to \mu^+ X)}{\sigma(\nu N \to \mu^- X)}. \tag{3}$$

In the approach of Llewellyn Smith, these can be combined to obtain $\sin^2\theta_W$

$$\begin{aligned} R_\nu &= \frac{1}{2} - \sin^2\theta_W + \frac{5}{9}(1+r)\sin^4\theta_W + C_\nu \\ R_{\bar\nu} &= \frac{1}{2} - \sin^2\theta_W + \frac{5}{9}(1+r^{-1})\sin^4\theta_W + C_{\bar\nu}, \end{aligned} \tag{4}$$

where C_ν and $C_{\bar\nu}$ represent known QCD and electroweak corrections. Alternatively, a Pascos-Wolfenstein construction can be used to extract $\sin^2\theta_W$

$$\begin{aligned} R^\pm &= \frac{R_\nu \pm r R_{\bar\nu}}{1 \pm r}, \\ R^- &= \frac{1}{2} - \sin^2\theta_W + \frac{C_\nu - rC_{\bar\nu}}{1 \pm r} \\ R^+ &= \frac{1}{2} - \sin^2\theta_W + \frac{10}{9}\sin^2\theta_W + \frac{C_\nu + rC_{\bar\nu}}{1 \pm r}. \end{aligned} \tag{5}$$

This method is actually cleaner since the QCD and electroweak corrections partially cancel out in Eq. (5). These relations are now used by CCFR/NuTeV collaboration. Future determinations of $\sin^2\theta_W$ from neutrino factory at the Muon Collider should be comparable or better than the projected result of SLAC E158 Moller scattering experiment.

2.2 CKM and quark densities

Extraction of the matrix elements of the Cabibbo-Kobayashi-Maskawa quark mixing matrix is one of the outstanding challenges in particle phenomenology of the Standard Model. It is most likely that Nature has chosen only three generations of quarks, so

$$V_{CKM} = S_L^{u\dagger} S_L^d = \begin{pmatrix} V_{ud} & V_{us} & V_{ub} \\ V_{cd} & V_{cs} & V_{cb} \\ V_{td} & V_{ts} & V_{tb} \end{pmatrix}. \tag{6}$$

One of the most useful parameterizations of the CKM matrix is due to L. Wolfenstein. It highlights a hierarchal structure of this matrix by representing its matrix elements as expansions in terms of a small parameter $\lambda = V_{us} \simeq 0.22$

$$V_{CKM} = \begin{pmatrix} 1-\lambda^2/2 & \lambda & A\lambda^3(\rho - i\eta) \\ -\lambda & 1-\lambda^2/2 & A\lambda^2 \\ A\lambda^3(1-\rho-i\eta) & -A\lambda^2 & 1 \end{pmatrix}. \tag{7}$$

Currently, *charm* quark sector of CKM is best studied in νN charged current interactions, where neutrino interacts with valence and sea quarks of the nucleon. For example, CCFR collaboration has provided a direct measurement of $|V_{cd}| = 0.232^{+0.017}_{-0.019}$. Independent knowledge of the strange see quark density should also provide an independent measurement of $|V_{cs}|$ CKM matrix element as well. In the framework of parton model

$$\begin{aligned}\frac{d^2\sigma(\nu N \to \mu^+\mu^- X)}{d\xi dy} &= \frac{G_F^2 M E_\nu}{\pi}\{[\xi u(\xi) + \xi d(\xi)]\,|V_{cd}|^2 \\ &+ 2\xi s(\xi)\,|V_{cs}|^2\}\left[1 - \frac{m_c^2}{2ME_\nu\xi}\right] D(z) B_c,\end{aligned} \tag{8}$$

where $\xi = x(1 + m_c^2/Q^2)$ in a 'slow rescaling' model of Georgi and Politzer [3] and $D(z)$ is a charm fragmentation function. Thus, measuring $d^2\sigma$ at different values of ξ provides an independent measurement of $|V_{cd}|$ and $|V_{cs}|$, if quark densities are known well enough [4]. Otherwise, a multiparameter fit can be performed to determine both $q(x)$ and CKM matrix elements.

In the framework of QCD factorization theorems, the essential problem of having heavy quark in the final state is the fact that its mass brings about an additional scale to the problem at hand. In its simplest form, the factorization theorem of QCD separates high scale set by the value of transferred momentum $\mu_h^2 \sim Q^2$ and low scale of hadronic physics $\mu_l^2 \sim \Lambda_{QCD}^2$. This formalism is based on the Operator Product Expansion (OPE) techniques and allows to sum up large logarithms of the form $\log\left(\mu_h^2/\mu_l^2\right)$ that multiply the expansion parameter (α_s) and might render perturbation theory expansion useless.

The presence of additional scale introduced by the heavy quark mass brings back large logarithms of the form $\alpha_s \log\left(\mu_{ph}^2/m_Q^2\right)$ which have to be resummed without loosing control over the region where $\mu_{ph}^2 \sim m_Q^2$. Here μ_{ph} is a "physical" scale of the process set by, for example, the value of transferred momentum Q^2. A practical recipe for such resummation is provided by the prescription of Aivazis *et. al.* (ACOT) [5]. Yet, a model-dependent assumption regarding the functional form of the heavy quark fragmentation function still remains.

Future measurement utilizing high intensity neutrino beams would provide accurate determinations of various $q(x)$. It is interesting to note that future neutrino factories would be able to study production of the b-flavored mesons, which would allow for an accurate determination of the intrinsic charm content of the nucleon. If x-dependent high statistics measurement of b-quark production becomes available, an independent determination of $|V_{ub}|$ and $|V_{cb}|$ CKM matrix elements will be possible as well [6].

2.3 Charmonium production and gluon density

High intensity neutrino beams would also allow studies of *charmonium* production, a sensitive probe of gluon distribution function in the nucleon. Contrary to the open-flavor meson production, the production of charmonium states can be described in a model-independent fashion using the formalism of Non-Relativistic QCD (NRQCD). Factorization theorem of Non-Relativistic QCD states

$$\sigma(A + B \to H_{c\bar{c}} + X) = \sum_n \frac{F_n}{m_c^{d_n - 4}} \langle 0|\mathcal{O}_n^H|0\rangle, \tag{9}$$

which separates short-distance physics, represented by the coefficients F_n (which might be sensitive to various parton distribution functions) from the long-distance physics, represented by the NRQCD matrix elements

$$\langle 0|\mathcal{O}_n^H|0\rangle = \sum_X \sum_{m_J} \langle 0|\mathcal{K}_n|H_{m_J} + X\rangle \langle H_{m_J} + X|\mathcal{K}'_n|0\rangle \tag{10}$$

For instance, matrix elements relevant for the production of the J/ψ include

$$\langle 0|\mathcal{O}_1^{J/\psi}(^3S_1)|0\rangle = \langle 0|\chi^\dagger\sigma\psi|J/\psi+X\rangle\langle J/\psi+X|\psi^\dagger\sigma\chi|0\rangle,$$
$$\langle 0|\mathcal{O}_8^{J/\psi}(^3S_1)|0\rangle = \langle 0|\chi^\dagger\sigma T^a\psi|J/\psi+X\rangle\langle J/\psi+X|\psi^\dagger\sigma T^a\chi|0\rangle, \quad (11)$$

and determine the probabilities of charm quarks produced in the $^3S_1$ color singlet and color octet states to evolve into a physical J/ψ state. Model-independent computation of these matrix elements calls for genuine non-perturbative methods (such as lattice calculations). However, these matrix elements are *universal* (i.e. process-independent), so they can be extracted from other experiments.

Due to its clean experimental signature, one of the most important charmonium production processes is J/ψ production. Future neutrino facilities (such as the one at Muon Collider) would allow the studies of the total cross section and the differential Q^2 distribution of J/ψ's. These studies can help to determine several important nonperturbative parameters (matrix elements of (NRQCD) operators) that describe hadronization of $c\bar{c}$ quark pair into charmonium state. These matrix elements are usually determined from hadron collider experiments, but the processes involved are sensitive only to various linear combinations of NRQCD matrix elements. The problem is aggravated by the usually very large theoretical uncertainties involved in these calculations (on the order of 50 – 100%), due to higher twist effects, uncalculated and/or incalculable higher order perturbative and nonperturbative contributions. In this situation any independent determination of these quantities should be welcome. Clearly, J/ψ produced in sufficiently high numbers can be used to study gluon distribution function in the wide range of x [7].

A major advantage of using the neutrino beam is that, at leading order in α_s, the spin structure of the νZ coupling selects a certain combination of octet operators. The largest contribution is from the one with the quantum numbers $^3S_1^{(8)}$. Of course, order of magnitude measurements of the size of the matrix elements of this operator have already been performed for the J/ψ and ψ', as well as for the χ_{cJ} states. The estimates of these matrix elements mostly come from Tevatron fits to hadroproduction cross sections for the J/ψ and χ_{cJ} and yield, with large theoretical errors [8],

$$\langle 0|\mathcal{O}_8^{J/\psi}(^3S_1)|0\rangle \sim 0.01 \text{ GeV}^3, \quad \text{and}$$
$$\langle 0|\mathcal{O}_8^{\chi_{c0}}(^3S_1)|0\rangle \sim 0.01 \text{ GeV}^3. \quad (12)$$

These values are consistent, within a $\sim 50\%$ accuracy level, with the value found from Z decay at LEP [9] (the latter does not separate cascade and direct production, so the value of $0.019\,GeV^3$ is understandably larger than the one in Eq. (12)).

The differential cross section was calculated in [7], and we quote the result in the form of the following structure functions $h_n\left(Q^2\right)$:

$$\frac{d\sigma\left(s,Q^2\right)}{dQ^2} = \frac{\pi^2\alpha^2\alpha_s}{3\sin^4 2\theta_W}\,\frac{1}{\left(Q^2+{m_Z}^2\right)^2} \times \sum_n \frac{\langle 0|\mathcal{O}_n|0\rangle}{m_c^3}\int_{\frac{Q^2+4m_c^2}{s}}^{1} dx\, f_{g/N}\left(x,Q^2\right) h_n\left(y,Q^2\right), \qquad (13)$$

where s is the total invariant mass of the νN system, x is the momentum fraction of the incoming gluon, $-Q^2$ is the momentum-squared transferred from the leptonic system, $y = \frac{Q^2+4m^2}{sx}$, and $f_{g/N}(x,Q^2)$ is a gluon distribution function in the nucleon. The charmonium structure functions are given by

$$h_{{}^1S_0^{(8)}}\left(y,Q^2\right) = \left(g_V^c\right)^2 \times 6\,\frac{Q^2\,m_c^2}{\left(Q^2+4m_c^2\right)^2}\left(y^2-2y+2\right)$$

$$h_{{}^3S_1^{(8)}}\left(y,Q^2\right) = \left(g_A^c\right)^2 \times 2m_c^2\,\frac{Q^2(y^2-2y+2)+16(1-y)m_c^2}{\left(Q^2+4m_c^2\right)^2}$$

$$h_{{}^3P_0^{(8)}}\left(y,Q^2\right) = \left(g_V^c\right)^2 \times 2\,Q^2\,\frac{\left(Q^2+12m_c^2\right)^2}{\left(Q^2+4m_c^2\right)^4}\left(y^2-2y+2\right) \qquad (14)$$

$$h_{{}^3P_1^{(8)}}\left(y,Q^2\right) = \left(g_V^c\right)^2 \times 4\,Q^4\,\frac{Q^2(y^2-2y+2)+16(1-y)m_c^2}{\left(Q^2+4m_c^2\right)^4}$$

$$h_{{}^3P_2^{(8)}}\left(y,Q^2\right) = \left(g_V^c\right)^2 \times \frac{4}{5}\,Q^2\left[\frac{(y^2-2y+2)Q^4}{\left(Q^2+4m_c^2\right)^2} + \frac{48(1-y)Q^2m_c^2+96(y^2-2y+2){m_c}^4}{\left(Q^2+4m_c^2\right)^2}\right],$$

where $g_A^c = \frac{1}{2}$ and $g_V^c = \frac{1}{2}\left(1-\frac{8}{3}\sin^2\Theta_W\right)$ are the vector and axial couplings of the c-quark. Clearly, the coupling constants favor the ${}^3S_1^{(8)}$ contribution, which is due to the large axial coupling (a similar contribution is, of course, absent in the case of J/ψ lepto and photoproduction). Indeed a numerical estimate [7] shows that this matrix element dominates the total cross section, and also the differential cross section unless $Q^2 \gg m_c^2$. At large Q^2, the relative Q^4 enhancement of the P-wave structure functions makes them dominant. These structure functions should be incorporated in the specific Monte Carlo generators built for the particular detector design.

Using approximate values for the NRQCD matrix elements satisfying the restrictions imposed by heavy quark symmetry and compatible with the existing extractions [8,10,11] the behavior of differential cross section can be studied. The energy dependence of the total cross section, taken at various representative incident neutrino energies, is shown in Table 1. The total cross section is clearly very sensitive to the neutrino beam energy E_ν. The origin of this strong dependence can be traced to the fact that the main contribution to the integral in Eq. (13) comes from the smallest available $x \approx \frac{Q^2+4m_c^2}{s}$, so that larger s probes a smaller region of x where the gluon PDF's are enhanced. It becomes obvious that in order to have a reasonably large event number, the neutrino beams with the highest possible energy should be employed. As seen

$E_\nu[GeV]$	7.5	25	120	450
$\sigma[nb]$	7.8×10^{-13}	6.9×10^{-10}	1.3×10^{-8}	5.5×10^{-8}

Table 1. Total cross sections for the J/ψ production in $\nu N \to J/\psi X$ for various representative incident neutrino energies.

from the Eqs. (13) and (14), the differential cross section quickly decreases with Q^2. However, the decrease is not as fast as in the leptoproduction case. Empirically,

$$\left(\frac{d\sigma}{dQ^2}\right)_{\nu N} \propto \left(\frac{d\sigma}{dQ^2}\right)_{eN} \times \frac{Q^2(Q^2+\beta m_c^2)}{m_Z^4} \tag{15}$$

where $\beta \approx 6$. The upshot is that the absence of a photon propagator results in a much wider tail of the Q^2 distribution than in leptoproduction, allowing for better discrimination between S- and P-wave contributions.

Fig. 1 shows the ratio of the contributions from these matrix elements as a function of Q^2. At low Q^2, where the contribution into the cross section is large, the ${}^3S_1^{(8)}$ dominates but a determination of the differential cross section in the range $\sim 2\,GeV < Q < 4\,GeV$, where the event number is still sizable, would allow a reliable extraction of both ${}^3S_1^{(8)}$ and ${}^3P_J^{(8)}$ matrix elements. Higher Q^2's will only measure a linear combination of these, as the ratio in Fig. 1 becomes Q^2-independent.

An important question to address is the expected event rate of J/ψ production. A simple calculation shows that currently running neutrino experiments NOMAD and NuTeV could collect a few J/ψ events (due to either low energy of the neutrino beam or particular detector configuration) and 'confirm' color octet mechanism. On the contrary, a neutrino experiment at

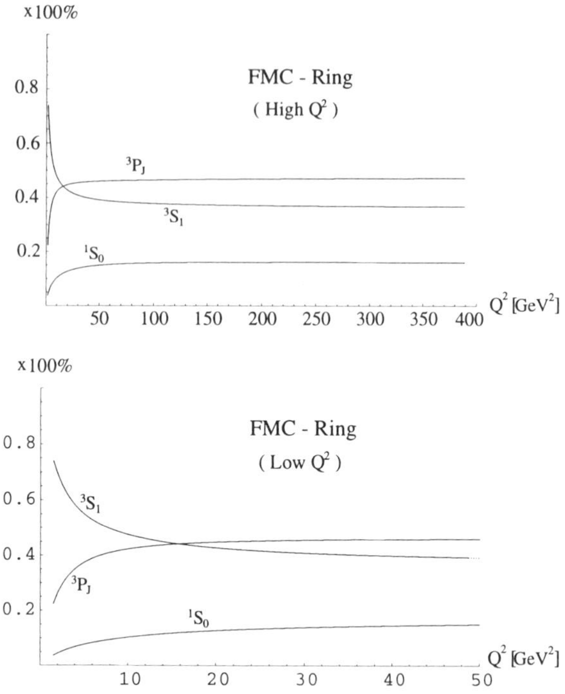

Figure 1. The contributions of the various matrix elements as a function of Q^2 for FMC-Ring, with baseline-sized matrix elements. Observe that the contributing linear combination of matrix elements drastically changes with Q^2 within a region where the rate is still sizable, so one can realistically extract both the ${}^3S_1^{(8)}$ and the ${}^3P_J^{(8)}$ matrix elements even without requiring exceedingly high statistics.

the future Muon Collider would collect about 3×10^3 events/year and provide precise measurement of various NRQCD matrix elements and/or gluon distribution function.

3 New Physics sensitivity

Neutrino-nucleon processes at low momentum transfer are sensitive to generic four-fermion contact terms produced by high energy neutral current interactions (other, flavor changing couplings, are well constrained by the processes like $\pi \to e\nu$). For instance, low energy remnant of a generic high energy electron-quark neutral current interaction can be represented by

$$\mathcal{L}_{NC} = \sum_q \Big[\eta^{eq}_{LL}\left(\overline{e_L}\gamma_\mu e_L\right)\left(\overline{q_L}\gamma^\mu q_L\right) + \eta^{eq}_{RR}\left(\overline{e_R}\gamma_\mu e_R\right)\left(\overline{q_R}\gamma^\mu q_R\right)$$
$$+ \eta^{eq}_{LR}\left(\overline{e_L}\gamma_\mu e_L\right)\left(\overline{q_R}\gamma^\mu q_R\right) + \eta^{eq}_{RL}\left(\overline{e_R}\gamma_\mu e_R\right)\left(\overline{q_L}\gamma^\mu q_L\right)\Big] . \quad (16)$$

One can use $SU(2)$ symmetry to relate ν and e couplings

$$\eta^{\nu u}_{LL} = \eta_1 + \frac{1}{4}\eta_2 = \eta^{ed}_{LL} ,$$
$$\eta^{\nu d}_{LL} = \eta_1 - \frac{1}{4}\eta_2 = \eta^{eu}_{LL} ,$$
$$\eta^{\nu u}_{LR} = \eta_3 = \eta^{eu}_{LR} ,$$
$$\eta^{\nu d}_{LR} = \eta_4 = \eta^{ed}_{LR} ,$$

so νN interactions can be used to constrain η's of the Lagrangian of Eq. (16). A particular example of these interactions is provided by $\not{R}$ SUSY

$$\mathcal{L}_{\not{R}} = \lambda'_{ijk}\Big(\tilde{e}^i_L \overline{d^k_R} u^j_L + \tilde{u}^j_L \overline{d^k_R} e^i_L + \tilde{d}^{k*}_R \overline{e^i_L}^c u^j_L$$
$$- \tilde{\nu}^i_L \overline{d^k_R} d^j_L - \tilde{d}^j_L \overline{d^k_R} \nu^i_L - \tilde{d}^{k*}_R \overline{\nu^i_L}^c d^j_L\Big) + \text{h.c.} . \quad (17)$$

At low energies (transferred momenta) Lagrangian can be rewritten in terms of local four-fermion interactions

$$\mathcal{L}_{ed} = \frac{(\lambda'_{1j1})^2}{m^2_{\tilde{u}^j_L}} \overline{e_L} d_R \overline{d_R} e_L + \frac{(\lambda'_{1j1})^2}{m^2_{\tilde{d}^j_L}} \overline{\nu_L} d_R \overline{d_R} \nu_L$$
$$= \left(-\frac{(\lambda'_{1j1})^2}{2m^2_{\tilde{u}^j_L}} \overline{e_L}\gamma^\mu e_L - \frac{(\lambda'_{1j1})^2}{2m^2_{\tilde{d}^j_L}} \overline{\nu_L}\gamma^\mu \nu_L \right) \overline{d_R}\gamma_\mu d_R . \quad (18)$$

Assuming that the squarks of first two generations are degenerate and imposing $SU(2)$ symmetry constraints,

$$\eta^{ed}_{LR} = -\frac{(\lambda'_{1j1})^2}{2m^2_{\tilde{u}^j_L}} = -\frac{(\lambda'_{1j1})^2}{2m^2_{\tilde{d}^j_L}} = \eta^{\nu d}_{LR} \quad (19)$$

Indeed, the best constraint on this coupling, $\eta^{ed}_{LR} < 0.07^{+0.24}_{-0.24}$ comes from the analysis of neutrino nucleon scattering experiments [12]. Other new physics scenarios involve new heavy neutral leptons (models with $H^0_L - \nu_\mu$ mixing) [13] or new neutral gauge bosons like Z' which appears in many superstring-motivated models.

4 Conclusions

New experiments utilizing high energy and intensity neutrino beams would offer a unique opportunity to perform new precision studies of QCD and electroweak interactions. New types of measurements, like charmonium production and extractions of V_{cb} and V_{ub} CKM matrix elements will become possible. Interesting New Physics scenarios can also be explored. As a result, high intensity neutrino facility could prove to be a very useful addition to the Muon Collider physics program.

References

1. C.H. Llewellyn Smith, Nucl. Phys. **B228**, 205 (1983).
2. E.A. Pascos and L. Wolfenstein, Phys. Rev. **D7**, 91 (1973).
3. H. Georgi and H.D. Politzer, Phys. Rev. **D14**, 1829 (1976).
4. J.M. Conrad, M.H. Shaevitz and T. Bolton, Rev. Mod. Phys. **70**, 1341 (1998) hep-ex/9707015.
5. M.A. Aivazis, J.C. Collins, F.I. Olness and W. Tung, Phys. Rev. **D50**, 3102 (1994).
6. I.I. Bigi *et. al.* "An overview of the potential for neutrino physics at future muon collider complexes". Work in progress.
7. A.A. Petrov and T. Torma, Phys. Rev. **D60**, 093009 (1999) hep-ph/9906254.
8. P. Cho and A.K. Leibovich, Phys. Rev. **D53**, 150 (1996), hep-ph/9505329; **ibid D53**, 6203 (1996), hep-ph/9511315; M. Beneke and M. Kramer, Phys. Rev. **D55**, 5269 (1997), hep-ph/9611218.
9. C.G. Boyd, A.K. Leibovich and I.Z. Rothstein, Phys. Rev. **D59**, 054016 (1999), hep-ph/9810364.
10. M. Beneke, F. Maltoni and I.Z. Rothstein, Phys. Rev. **D59**, 054003 (1999), hep-ph/9808360.
11. J. Amundson, S. Fleming and I. Maksymyk, Phys. Rev. **D56**, 5844 (1997), hep-ph/9601298.
12. D. Zeppenfeld and K. Cheung, hep-ph/9810277.
13. M. Gronau, C.N. Leung and J.L. Rosner, Phys. Rev. **D29**, 2539 (1984).

THE HOT DARK MATTER

DAVID O. CALDWELL
Institute for Nuclear and Particle Astrophysics and Cosmology
and
Physics Department, University of California, Santa Barbara, CA 93106, USA
E-mail: caldwell@slac.stanford.edu

There is a puzzling contradiction: direct observations favor a low-mass-density universe ($0.2 \leq \Omega_m \leq 0.6$), but the only model which fits universe structure over more than three orders of magnitude in distance scale has a mix of hot (neutrino) and cold dark matter providing a critical density universe. Models of an open universe (low Ω_m) or one adding a cosmological constant (Λ) to provide a critical energy density ($\Omega_m + \Omega_\Lambda = 1$) have probabilities of $< 10^{-3}$. Two-neutrino dark matter works better than having the needed $\sim$ 5 eV of neutrino mass in one species of neutrino, and this is consistent with the only model which fits all present indications for neutrino mass: $\nu_\mu \to \nu_\tau$ accounting for the atmospheric anomaly (with ν_μ and ν_τ being the hot dark matter), $\bar{\nu}_\mu \to \bar{\nu}_e$ being observed by LSND, and $\nu_e \to \nu_s$ explaining the solar ν_e deficit. The LSND/KARMEN results are consistent with the needed mass of hot dark matter. Further support for this mass pattern is provided by the need for the sterile neutrino, ν_s, to make possible heavy-element nucleosynthesis in supernovae. It is a fascinating question as to whether the hot dark matter paradox will be resolved by better measurements or by the introduction of new physics.

1 One-, Two, or Three-Neutrino Dark Matter?

Since there are about $100/\text{cm}^3$ of neutrinos of each type left over from the Big Bang, if they have mass they surely are part of the dark matter of the universe. While they cannot be the major component of this missing mass, the neutrinos can have profound effects on universe structure if they have sufficient mass. So far there is only evidence for differences in mass between neutrino types, and with one exception, those differences are so small that if they are representative of mass values, the neutrinos would have little effect. We shall see, however, that there are several types of evidence that neutrino hot dark matter is quite significant.

The common view, especially among astronomers, has been that if there is hot dark matter it is due mainly to one neutrino, presumably the ν_τ. This would be ruled out if, as fits the Super-Kamiokande data[1] best, the atmospheric anomalous ν_μ/ν_e ratio is due to $\nu_\mu \to \nu_\tau$, since the mass-squared difference required is $\Delta m^2_{\mu\tau} \sim 10^{-3}\text{eV}^2$, whereas the needed neutrino mass is $94\ \Omega_\nu h^2 \sim$ eV (with h the Hubble constant in units of 100 $\text{km}{\cdot}\text{s}^{-1} \cdot \text{Mpc}^{-1}$).

Other processes to explain the atmospheric results are very unlikely: $\nu_\mu \to \nu_e$ does not fit the Super-Kamiokande angular distributions, and the CHOOZ ν_e disappearance experiment[2] rules out almost all the parameter space; $\nu_\mu \to \nu_s$ (a sterile neutrino) likely has a problem with the nucleosynthesis limit,[3] since near maximal mixing is required, and the Super-Kamiokande results rule it out at the 95% C.L.

If hot dark matter being one active neutrino is essentially ruled out, what about using all three active neutrinos? In this case $\nu_\mu \to \nu_\tau$ explains the atmospheric anomaly, $\nu_e \to \nu_\mu$ (with $\Delta m^2_{e\mu} \lesssim 10^{-5}$ eV^2) provides the solar ν_e deficit, and the three nearly mass degenerate neutrinos could provide the dark matter. When this scheme was first suggested,[4] there was a possible problem with neutrinoless double beta decay. While limits on that process have improved, theoretical ways have been found to ameliorate the problem. If results from the LSND experiment[5] are correct, however, three-neutrino dark matter is also ruled out, since this requires $\Delta m^2_{e\mu} > 0.3$ eV^2, making three quite distinct mass differences, necessitating more than three neutrinos.

That leaves two-neutrino dark matter. This scheme[4,6] requires four neutrinos, with the solar deficit explained by $\nu_e \to \nu_s$, both neutrinos being quite light, the atmospheric effect due to $\nu_\mu \to \nu_\tau$, which share the dark matter role, and the LSND $\nu_\mu \to \nu_e$ demonstrating the mass difference between these two nearly mass-degenerate doublets. Note that the solar $\nu_e \to \nu_s$ is for the small mixing angle (or "just-so" vacuum oscillation) solution, so ν_s does not affect nucleosynthesis. The original motivation for this mass pattern preceded LSND and was simply to provide some hot dark matter, given the solar and atmospheric phenomena. If LSND is correct, it becomes the unique pattern.

This neutrino scheme was the basis for simulations[7] which showed that two-neutrino dark matter fits observations better than the one-neutrino variety. The latter produces several problems at a distance scale of the order of $10h^{-1}$ Mpc, particularly overproducing clusters of galaxies. Whether the ~ 5 eV of neutrino mass is in the form of one neutrino species or two makes no difference at very large or very small scales, but at $\sim 10h^{-1}$ Mpc the larger free streaming length of $\sim 5/2$ eV neutrinos washes out density fluctuations and hence lowers the abundance of galactic clusters. In every aspect of simulations done subsequently, the two-neutrino dark matter has given the best results. For example, a single neutrino species, as well as low universe density models, overproduce void regions between galaxies, whereas the two-neutrino model agrees well with observations.[8]

2 Evidence from Universe Structure

Cosmic microwave background radiation observations of the first Doppler peak favor the total energy density of the universe having the critical value ($\Omega = 1$) of a flat universe. Such a flat universe has the only time-stable value of density and is expected in all but rather contrived models of an early era of exponential expansion, or "inflation". Until recently it has usually been assumed that $\Omega = \Omega_m = 1$; i.e., the energy density is the matter density, and the universe will expand forever at an ever decreasing rate. Now evidence points to $0.3 \leq \Omega_m \leq 0.6$, however, based on a number of observations: high-redshift Type Ia supernovae, evolution of galactic clusters, high baryon content of clusters, lensing arcs in clusters, and dynamical estimates from infrared galaxy surveys. On this basis it has become popular to assume $\Omega_m \approx 0.3$, but $\Omega = 1$ through the addition of a vacuum energy density, often designated as a cosmological constant, Λ. The model with $\Omega_m = 0.3$, $\Omega_\Lambda = 0.7$ is in trouble with some determinations of the age of the universe, and lensing measurements require $\Omega_\Lambda < 0.74$ at the two standard deviation level.

The evidence for low Ω_m does not include a global look at universe structure. Gawiser and Silk[9], however, used all the published data from the cosmic microwave background and galaxy surveys which covered three orders of magnitude in distance scale and came to a different conclusion. They compared the data with ten models of universe structure, but of concern here are only three of these, the other seven giving extremely poor fits. In two low-density models the parameters were varied to get best fits, resulting in $\Omega_m = 0.5$, of which the baryons contribute $\Omega_b = 0.05$, and the rest is cold dark matter. One of these is an open universe model (OCDM) having $\Omega = \Omega_m = 0.5$, and the other (ΛCDM) has $\Omega = 1$ with $\Omega_\Lambda = 0.5$. The third model (CHDM) has $\Omega_m = 1$, of which $\Omega_\nu = 0.2$ is in neutrinos, and $\Omega_b = 0.1$ in baryons, with the main component being cold dark matter, that which was nonrelativistic at the time it dropped out of equilibrium in the early universe. The probabilities of the fits were CHDM = 0.09, OCDM = 2.9×10^{-5}, and ΛCDM = 1.1×10^{-5}. If one dubious set of data is removed, the APM cluster survey (which disagrees with galaxy power spectra), these probabilities become CHDM = 0.34, OCDM = 6.7×10^{-4}, and ΛCDM = 4.3×10^{-4}.

Had it been possible to extend the fit to even smaller scales, the discrepancy between CHDM and the others would have been even greater, but this is the non-linear regime requiring simulations. The CHDM model with two neutrinos gives an excellent fit[10] to the data at this extended scale, whereas the others deviate even more strongly than in the linear region.

Clearly there is a serious conflict between the observations indicating a low value of Ω_m and the degree of structure in the universe as a function of distance scale. Two recent developments raise some doubts about the present popular interpretation of the data. The most compelling evidence for Λ comes from the observation of distant supernovae Ia,[11] but recent measurements on nearby SNIa shows[12] they take over two days longer to reach peak brightness than do distant SNIa. This indicates that these may not be the "standard candles" required for the conclusions reached, but rather that evolutionary effects mimic the need for Λ. The second straw in the wind is a geometric measurement of the distance to galaxy NGC4258 which disagrees with the standard Cepheid ladder of distances by $\sim 15\%$.[13] This lowering of the distance scale would reduce the age of the universe, more in line with larger values of Ω_m. It is, however, a single determination. Better measurements will be available soon, and these may resolve the conundrum, but if the difference sharpens, this could lead to important new physics.

3 Evidence from the LSND Experiment

While new galaxy surveys and cosmic microwave background experiments will be so precise that they can provide a good measure of neutrino mass and the number of neutrino types contributing to dark matter, the ultimate answer must come from laboratory experiments. Neutrino oscillation experiments can provide only mass differences, but if the difference is sufficient, the case for cosmologically significant neutrino dark matter will be settled. Only one experiment[5] has given evidence for such dark matter, but so far the range of possible mass difference between ν_e and ν_μ is quite large: $0.3\ \mathrm{eV}^2 \leq \Delta m^2_{e\mu} \leq 10\mathrm{eV}^2$.

In its 1996 publication,[5] LSND claimed a signal in $\bar{\nu}_\mu \to \bar{\nu}_e$ on the basis that 22 events of the type $\bar{\nu}_e p \to e^+ n$ were seen, using a stringent criterion to reduce accidental coincidences between e^- or e^+ and γ rays mimicking the 2.2-MeV γ from $np \to d\gamma$, whereas only 4.6 ± 0.06 events were expected. The probability of this being a fluctuation is 4×10^{-8}. Note especially that these data were restricted to the energy range 36 to 60 MeV to stay below the $\bar{\nu}_\mu$ endpoint and to stay above the region where backgrounds are high due to the $\nu_e {}^{12}\mathrm{C} \to e^- X$ reaction. In plotting Δm^2 vs. $\sin^2 2\theta$, however, events down to 20 MeV were used to increase the range of E/L, the ratio of the neutrino's energy to its distance from the target to detection. This was done because the plot employed was intended to show the favored regions of Δm^2, and all information about each event was used. The likelihood analysis applied did not have a Gaussian likelihood distribution, since its integral is infinite, but the likelihood contour labeled "90%" was obtained by going down a factor of 10 from the maximum, as in the Gaussian case. The contours in the LSND plot have been widely misinterpreted as confidence levels—which they certainly are not—because they were plotted along with confidence-level limits from other experiments.

Recently the difficult, computer-intensive analysis in terms of real confidence levels has been done.[14] The likelihood for a grid in $(\sin^2 2\theta,\ \Delta m^2)$ space, including backgrounds, has been computed and compared with numerous Monte Carlo experiments to obtain a 90% confidence region. While the equivalency varies from point to point in the $\Delta m^2 - \sin^2 2\theta$ plane, a typical value for the 90% confidence level is down a factor of 20 from the likelihood maximum. Thus the LSND allowed regions are considerably broader in $\sin^2 2\theta$ than in the plots published so far, and other experiments are less constraining of allowed Δm^2 regions.

The confusion of comparing likelihood levels for LSND with confidence levels from other experiments may be exacerbated by using the 20–36 MeV region for the LSND data. While this higher background energy range makes some difference for the 1993–5 data, it could have had an appreciable effect for the parasitic 1996–7 runs, which were at a low event rate. This decreased the ratio of signal/background events, since the main background is from cosmic rays. This could raise the low end of the supposed signal energy spectrum, making the higher Δm^2 values desirable for dark matter appear less likely.

Nevertheless, when a joint analysis is made of the LSND and KARMEN[14] experiments even using the 20–36 MeV range for LSND, the region around 6 eV^2 is as probable as the banana-shaped region at lower Δm^2, as shown in Fig. 1. Frequently theorists consider only the latter, whereas the $\nu_\mu \to \nu_e$ LSND data favors the higher mass region. Of course the $\nu_\mu \to \nu_e$ data,[5] which uses ν_μ from π^+ decay-in-flight and detects ν_e by $\nu_e{}^{12}C \to e^- X$ has higher backgrounds and hence much poorer statistics than the $\bar{\nu}_\mu \to \bar{\nu}_e$ with $\bar{\nu}_\mu$ from μ^+ at rest. In addition to the Δm^2 issue, the important point of Fig. 1 is that although the KARMEN data are consistent with background, the joint analysis of the $\bar{\nu}_\mu \to \bar{\nu}_e$ data from the two experiments shows an appreciable region for a signal. KARMEN is continuing to take data, and LSND should have an improved analysis available soon.

4 Evidence from Supernova Nucleosynthesis

Support for the double doublet of neutrinos with sufficient mass in two of the neutrinos to provide significant hot dark matter comes from an unusual source: the creation of heavy nuclei by supernovae. Initially the reverse appeared to be the case, since this r process of rapid neutron capture, which occurs in the

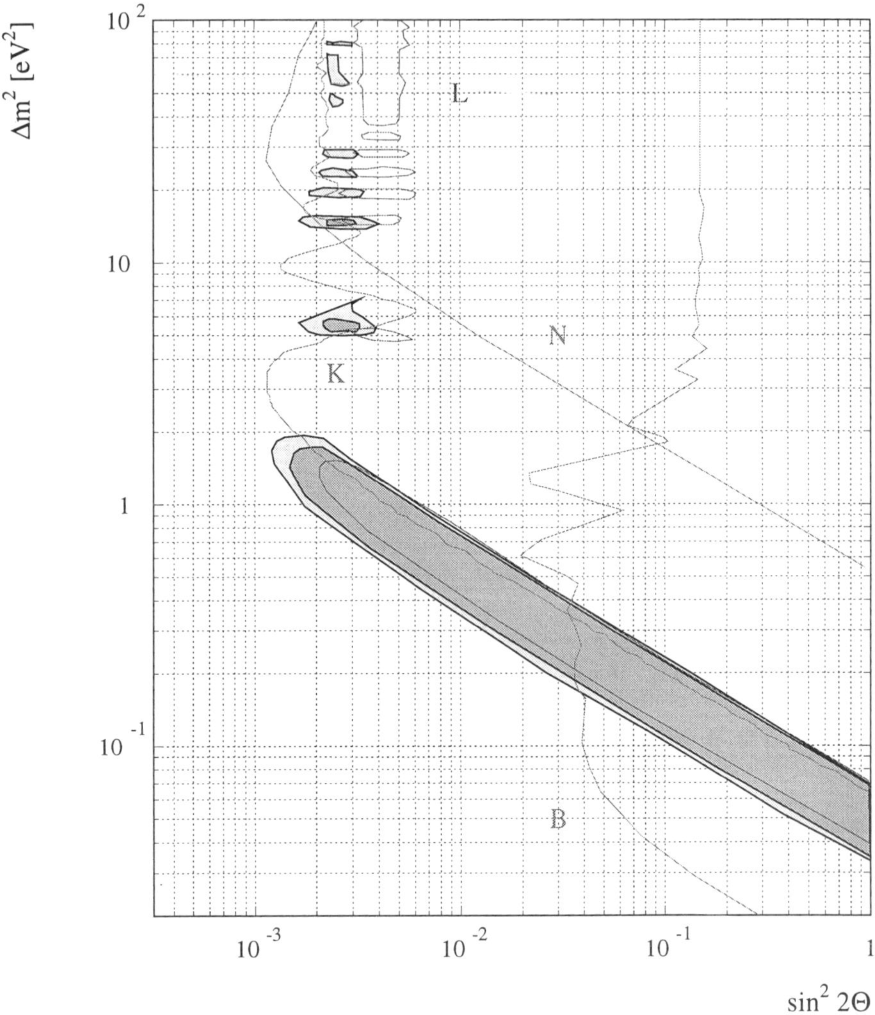

Figure 1. Filled in areas are 90% and 95% confidence regions based on the product of the KARMEN and LSND Feldman-Cousins likelihood ratios. Also shown is the Feldman-Cousins 90% confidence region for LSND alone ("L"). Left of the "K", "N", and "B" curves are exclusion regions of KARMEN, NOMAD, and Bugey.

outer neutrino-heated ejecta of Type II supernovae, seemed to place a limit on the mixing of ν_μ and ν_e. Energetic ν_μ ($\langle E\rangle \approx 25$ MeV) coming from deep in the supernova core could convert via an MSW transition to ν_e inside the region of the r-process, producing ν_e of much higher energy than the thermal ν_e ($\langle E\rangle \approx 11$ MeV). The latter, because of their charge-current interactions, emerge from farther out in the supernova where it is cooler. Since the cross section for $\nu_e n \to e^- p$ rises as the square of the energy, these converted energetic ν_e would deplete neutrons, stopping the r-process. Calculations[15] of this effect limit $\sin^2 2\theta$ for $\nu_\mu \to \nu_e$ to $\lesssim 10^{-4}$ for $\Delta m^2_{e\mu} \gtrsim 2$ eV^2, in conflict with compatibility between the LSND result and a neutrino component of dark matter.

Since the work of reference 15, serious problems have been found with the r process itself. First, recent simulations have revealed the r-process region to be insufficiently neutron-rich, since about 10^2 neutrons is required for each seed nucleus, such as iron. This was bad enough, but the recent realization of the full effect of α-particle formation has created a disaster for the r process.[16] At a radial region inside where the r process should occur, all available protons swallow up neutrons to form the very stable α particles, following which $\nu_e n \to e^- p$ reactions reduce the neutrons further and create more protons which make more α particles, and so on. The depletion of neutrons by making α particles and by $\nu_e n \to e^- p$ rapidly shuts off the r process, and essentially no nuclei above $A = 95$ are produced.

What is required to solve this problem is to remove the ν_e flux at the r process site, but there still has to be a very large ν_e flux at a smaller radius for material heating and ejection. This apparent miracle can be accomplished[17] if there is (1) a sterile neutrino, (2) approximately maximal $\nu_\mu \to \nu_\tau$ mixing, (3) small $\nu_\mu \to \nu_e$ mixing, and (4) an appreciable ($\gtrsim 2$ eV^2) mass-squared difference between ν_s and the ν_μ–ν_τ. This is precisely the neutrino mass pattern required to explain the solar and atmospheric anomalies and the LSND result, plus providing some hot dark matter!

Such a mass-mixing pattern creates two level crossings. The inner one, which is outside the neutrinosphere (beyond which neutrinos can readily escape) is near where the $\nu_{\mu,\tau}$ potential $\propto (n_{\nu_e} - n_n/2)$ goes to zero. Here n_{ν_e} and n_n are the numbers of ν_e and neutrons, respectively. The $\nu_{\mu,\tau} \to \nu_s$ transition which occurs depletes the dangerous high-energy $\nu_{\mu,\tau}$ population. Outside of this level crossing, another occurs where the density is appropriate for a matter-enhanced MSW transition corresponding to whatever $\Delta m^2_{e\mu}$ LSND is observing. Because of the $\nu_{\mu,\tau}$ reduction at the first level crossing, the dominant process in the MSW region reverses from the deleterious $\nu_{\mu,\tau} \to \nu_e$, becoming $\nu_e \to \nu_{\mu,\tau}$ and dropping the ν_e flux going into the r-

process region. For an appropriate value of $\Delta m^2_{e\mu}$, the two level crossings are separate but sufficiently close so that the transitions are coherent. Then in the limits of adiabatic transitions and near maximal ν_μ–ν_τ mixing, the neutrino flux emerging from the second level crossing is 1/4 ν_μ, 1/4 ν_τ, and 1/2 ν_s, with no ν_e at all. Calculations show the transitions to be adiabatic, and the atmospheric observations require near maximal mixing, so the ν_e flux is certainly sufficiently depleted to allow a successful r process, especially as the $\bar{\nu}_e$ flux is unaffected, so that $\bar{\nu}_e p \rightarrow e^+ n$ enhances the neutron number. It should be emphasized that this mechanism is quite robust, not depending on details of the supernova dynamics, especially as it occurs quite late in the explosive expansion.

It is essential that the two level crossings be in the correct order, and this provides a requirement on $\Delta m^2_{e\mu}$, since the MSW transition depends on density and hence on radial distance from the protoneutron star. Detailed calculations have been made for $\Delta m^2_{e\mu} \sim 6$ eV^2, which works very well. Possibly $\Delta m^2_{e\mu}$ as low as 2 eV^2 or maybe even 1 eV^2 would work, but that is speculative. At any rate, the mass difference needed in this scheme, which is the only one surely consistent with all manifestations of neutrino mass and which rescues the r process,[18] implies appreciable hot dark matter.

5 Conclusions

A neutrino component of dark matter appears very probable, both from the astrophysics and particle physics standpoints. Despite abundant evidence for $\Omega_m < 1$, the one model which fits universe structure has $\Omega_m = 1$, with 20% neutrinos and most of the rest as cold dark matter. Open universe and low-density models with a cosmological constant give extremely bad fits. This conflict should be the source of future progress, but since there are $10^2/\text{cm}^3$ of neutrinos of each active species left over from the early universe, the ultimate answer on neutrino dark matter will come from determinations of neutrino mass. While the solar and atmospheric evidences for neutrino mass are important, the crucial issue is the much larger mass-squared difference observed by the LSND experiment. In the mass region needed for dark matter, no other experiment excludes the LSND result, and a joint analysis of the LSND and KARMEN experiments shows this region has good probability.

The resulting mass pattern, $\nu_e \rightarrow \nu_s$ for solar, $\nu_\mu \rightarrow \nu_\tau$ for atmospheric, and $\nu_\mu \rightarrow \nu_e$ for LSND, requires a sterile neutrino and provides two-neutrino (ν_μ and ν_τ) dark matter. This form of dark matter fits observational data better than the one-neutrino variety. Furthermore, the four-neutrino pattern, and especially the sterile neutrino, provides a robust way to make possible the production of heavy elements by supernovae.

Acknowledgments

This work was supported in part by the U.S. Department of Energy. Portions of this paper were done in enjoyable past collaborations with R.N. Mohapatra, J.R. Primack, G.M. Fuller, and Y.-Z. Qian. Assistance from S.J. Yellin is much appreciated.

References

1. Y. Fukuda *et al*, *Phys. Rev. Lett.* **81**, 1562 (1998); *Phys. Lett.* B **436**, 33 (1998); M. Nakahata, Sixth International Workshop on Topics in Astroparticle and Underground Physics (Paris, France, 1999).
2. M. Apollonio *et al*, *Phys. Lett.* B **420**, 397 (1998).
3. At the time of writing there is considerable controversy about the nucleosynthesis limit, not only regarding concordance of light-element abundances, but also about the possible effects of neutrino particle-antiparticle asymmetries.
4. D.O. Caldwell, *Perspectives in Neutrinos, Atomic Physics and Gravitation* (Editions Frontières, Gif-sur-Yvette, France, 1993), 187; D.O. Caldwell and R.N. Mohapatra, *Phys. Rev.* D **48**, 3259 (1993); *ibid.* **D50**, 3477 (1994).
5. C. Athanassopoulos *et al*, *Phys. Rev. Lett.* **75**, 2650 (1995); *Phys. Rev.* C **54**, 2685 (1996); *Phys. Rev. Lett.* **77**, 3082 (1996); *Phys. Rev. Lett.* **81**, 1774 (1998).
6. J.T. Peltoniemi and J.W.F. Valle, *Nucl. Phys.* B **406**, 409 (1993).
7. J.R. Primack, J. Holtzman, A. Klypin, and D.O. Caldwell, *Phys. Rev. Lett.* **74**, 2160 (1995).
8. S. Ghigna *et al*, *Astrophys. J.* **479**, 580 (1997); *ibid.* **437**, L71 (1994); J.R. Primack, astro-ph/9707285.
9. E. Gawiser and J. Silk, *Science* **280**, 1405 (1998).
10. J.R. Primack and A. Klypin, *Nucl. Phys. (Proc. Suppl.)* **51B**, 30 (1996).
11. S. Perlmutter *et al*, *Nature* **391**, 51 (1998); astro-ph/9812133 (to be published in *Astrophys. J.*); A.G. Riess *et al*, *Astron. J.* **509**, 74 (1998); S. Perlmutter and A. Riess, Proceedings of COSMO-97, ed. D.O. Caldwell (Am. Inst. of Phys., New York, 1999), 129.
12. A.G. Riess *et al*, astro-ph/9907037.
13. J.R. Herrnstein *et al*, *Nature* **400**, 539 (1999).
14. K. Eitel, hep-ex/9909036.

15. Y.-Z. Qian *et al*, *Phys. Rev. Lett.* **71**, 1965 (1993); Y.-Z. Qian and G.M. Fuller, *Phys. Rev.* D **51**, 1479 (1995); G. Sigl, *Phys. Rev.* D **51**, 4035 (1995).
16. G.M. Fuller and B.S. Myer, *Astrophys. J.* **453**, 202 (1995); B.S. Myer, G.C. McLaughlin and G.M. Fuller, *Phys. Rev.* C **58**, 3696 (1998).
17. D.O. Caldwell, G.M. Fuller and Y.-Z. Qian, astro-ph/9910175.
18. G.C. McLaughlin, J.M. Fetter, A.B. Balantekin, and G.M. Fuller, *Phys. Rev.* C **59**, 2873 (1999) employs a neutrino scheme with a heavy sterile neutrino and also rescues the *r* process, but it is likely to be in trouble with supernova nucleosynthesis, as well as with the Super-Kamiokande atmospheric up-down asymmetry [see S.M. Bilenky, C. Giunti, W. Grimus, and T. Schwatz, *Phys. Rev.* D **60**, 073007 (1999)].

IMPLICATIONS OF NEUTRINO FLAVOR CONVERSION FOR COSMOLOGY AND SUPERNOVAE

GEORGE M. FULLER

Department of Physics, University of California, San Diego, La Jolla, CA 92093-0319, USA

E-mail: gfuller@ucsd.edu

Present data represents strong evidence for nonzero neutrino masses and (sometimes) large vacuum mixings between flavors. This data, taken as a whole, suggests the existence of at least one light sterile neutrino species which mixes with active flavors. The impact of these results on astrophysics could be considerable, since neutrino flavor mixing can be matter-enhanced in many environments in which neutrinos and their interactions determine dynamics and nucleosynthesis. Considerations of matter-enhanced neutrino flavor conversion either could force us to alter our picture for the synthesis of the light elements in the early universe, or allow the observed light element abundances to become a precision tool for constraining neutrino mass and mixing schemes. Likewise, the synthesis of the heaviest elements in the post-supernova-explosion environment could be sensitive to neutrino flavor conversion. Matter-enhanced production of relatively light sterile neutrinos in the early universe could allow these particles to behave as Cold Dark Matter, a possibility that demolishes a long-held tenet of cosmological thought.

1 Introduction

In this paper we examine the surprising and serrendipitous relationship between the neutrino mass and mixing schemes suggested by current data, and the dynamics and nucleosynthesis associated with the early universe and supernovae. The conclusions at this point are not clear cut. In fact, the results are both disturbing and tantalizing: it seems very difficult to reconcile the nucleosynthesis record with existing neutrino mass and mixing models, unless the latter are finely tuned. Furthermore, we may have to surrender long held notions of the way that neutrinos can behave as Dark Matter particles.

What is the driving force behind these new conclusions? Ultimately, it is the fact that the vacumm mixing of neutrino flavors can be modified, even amplified by matter effects (*e.g.*, the MSW effect of Wolfenstein[1], and Mikheyev and Smirnov[2]). In the adiabatic limit the MSW effect represents a way whereby a neutrino of one flavor can be converted with high probability at resonance to another flavor. Though most physicists are familiar with MSW in the context of solar neutrinos, this process can also happen in supernovae, the early universe, or almost anywhere where neutrinos propagate through a time/position varying weak potential. Moreover, the flavor conversion process is relatively robust, since the MSW mechanism itself is a mass level crossing, requiring coherence of neutrino states only in the small resonance width region. Proposed neutrino mass and mixing schemes could allow for such neutrino flavor conversion in important astrophysical sites.

As a result, there is a revolution brewing in the way we look at the role of neutrinos in astrophysical environments where these particles dominate the dynamics and determine the thermodynamic conditions. Such astrophysical venues include the early universe, the core collapse supernova phenomenon, and possibly the gamma-ray burst environment. In the first two examples we are fortunate to have a "fossil record" of what the neutrinos may have done and how they may have behaved: the nucleosynthesis products.

More specifically, we know that the light elements ^{2}H, ^{3}He, ^{4}He, and ^{7}Li are cooked (principally) in Big Bang Nucleosynthesis (BBN). There is growing observational evidence that the heaviest elements, the rapid neutron capture, or r-process elements (*e.g.*, uranium, platinum) were synthesized in the neutrino-heated ejecta at late times following a supernova explosion. Ironically, each of these nucleosynthsis processes is essentially similar in that they are freeze-outs from statistical equilibrium among nuclear reactions. The key quantities determining the nucleosynthesis products of these freeze outs are all set by the flavor states and the interactions of the neutrinos.

We summarize the BBN and neutrino-heated supernova ejecta models for nucleosynthesis in Section **2**. We give an overview of the present status of the neutrino data and its implications for neutrino masses and mixngs in Section **3**. In Section **4** we examine how, through the MSW mechanism, these proposed neutrino mass and mixing schemes could spell trouble for our models of nucleosynthesis and even our notions on neutrino Dark Matter.

2 Freeze-Out Nucleosynthesis in the Early Universe and in Neutrino-Heated Ejecta from Supernovae

At high enough temperature and density all reactions that build up and break down down nuclei from constituent neutrons and protons will come into equilibrium. This is Nuclear Statitical Equilibrium, or NSE. Both the very early universe (for temperatures $T > 3\,\mathrm{MeV}$) and the region very near the surface of the hot proto-neutron star remnant of a supernova explosion are examples of environments where complete or partial NSE obtains. In both cases, the material is undergoing fairly rapid (near) adiabatic expansion. In the first case, this is just the expansion of the universe, driven by high total mass-energy density and overall spherical symmetry. In the second case, the intense fluxes of neutrinos of all kinds from the proto-neutron star deposit enough energy to drive a "wind" which flows radially outward from the neutron star surface.

These otherwise disparate astrophysical sites are similar in other ways as well. Each environment is extremely disordered, that is, characterised by high entropy per baryon, s. In the early universe, $s \approx 2.53 \times 10^8 {\Omega_b}^{-1} h^{-2}$, corresponding to a baryon-to-photon ratio $\eta \approx 2.79 \times 10^{-8} \Omega_b h^2$. In these expressions the baryon rest mass contribution to the closure density is Ω_b and the Hubble expansion rate at the present epoch in units of $100\,\mathrm{km\,s^{-1}\,Mpc^{-1}}$ is h. We know from the Burles and Tytler deuterium abundance derived from QSO absorption systems that $\Omega_b \approx 0.02$, implying that $s_{BBN} \sim 10^{10}$ in units of Boltzmann's constant.[3] We can also estimate the entropy per baryon in the neutrino-driven wind from a hot neutron star and we find $s_{SNN} \sim 100$. Although the entropies in these environments look different, they are actually similar on a *nuclear physics* scale. In broad brush, nucleosynthesis in each environment is largely similar, the environments for BBN and neutrino-heated ejecta are nearly isospin mirrors of one another. The salient difference in these environments is the ratio of neutrons to protons, and it is this quantity which is especially affected by neutrino flavor mixing.

In each environment, there will be a parallel succession of freeze out events as expansion proceeds. First, the weak interactions will will cease to be able to mediate rapid energy exchange between neutrinos and the charged particles in the plasma. This is what is termed "Weak Decoupling" in the BBN case. Note, however, that this thermal equilibrium condition can be different from chemical equilibrium among the weakly interacting constituent particles. For example, the neutron-to-proton ratio n/p (sometimes parameterized by the electron fraction $Y_e = 1/(1+n/p)$) is initially in steady state equilibrium with the fluxes of ν_e and $\bar{\nu}_e$, and the ambient distributions of electron/positron pairs, through the reactions:

$$\nu_e + n \rightleftharpoons p + e^-, \tag{1}$$

$$\bar{\nu}_e + p \rightleftharpoons n + e^+, \tag{2}$$

$$n \rightleftharpoons p + e^- + \bar{\nu}_e. \tag{3}$$

Eventually, in each of our examples, the expansion will proceed to the point where the neutrino energies (early universe) or the fluxes (neutrino-heated ejecta) fall so low that the rates corresponding to these weak reactions fall below the local material expansion rate. This is what is usually referred to in BBN as "Weak Freeze-Out."

Subsequently, n/p in the BBN case drops slowly (owing to free neutron decay) until the epoch where nearly all of the available free neutrons are incorporated into alpha particles. This last event occurs abruptly in BBN when the temperature has dropped to $\sim 100\,\mathrm{keV}$. In turn, this event causes the electromagnetic reaction $n + p \rightleftharpoons {}^2H + \gamma$ to drop out (freeze out) of equilibrium, as it deprives the forward rate of this reaction of its neutron supply. This is nearly the whole story in BBN, as all nuclear reactions freeze out rapidly at this point, so that only of order one in 10^5 nucleons ends up in a deuterium nucleus and only one in 10^{10} in ^{7}Li.

Examination of the Saha equation for the conditions relevant for BBN at the Weak Freeze-Out point, $T_{\mathrm{WFO}} \approx 0.7\,\mathrm{MeV}$, shows that the n/p ratio at this point is $n/p \approx exp(-\delta m_{\mathrm{np}}/T_{\mathrm{WFO}}) \approx 1/6$, where the neutron-proton mass difference is $\delta m_{\mathrm{np}} \approx 1.293\,\mathrm{MeV}$. The process in Eq. (3) causes this ratio to fall to 1/7 by the time of alpha particle formation. It is clear from this simple description that the faster the expansion rate of the universe, the earlier the Weak Freeze-Out event and so the larger will be T_{WFO}, the larger n/p and, therefore, the larger will be the ^{4}He yield in BBN.

It is an unfortunate and confusing custom to express the Hubble expansion rate $H \approx \left(8\pi^3/90\right)^{1/2} g^{1/2} T^2/m_{\mathrm{pl}}$ in the radiation-dominated conditions of the early universe in terms of an effective number of "Neutrino Families," N_ν. Here N_ν reresents *all* the relativistic degrees of freedom beyond those stemming from photons and electron/positron pairs. In the expression for H, m_{pl} is the Planck mass and the statistical weight in relativistic particles of all kinds is $g \approx 11/2 + 7/8\,(2N_\nu)$. If we have only the three active neutrino species present in the plasma, and no other kinds of relativistic particles besides photons and $e^{\pm}$-pairs, then $N_\nu = 3$ and $g \approx 10.75$. Taking $N_\nu = 3$, the best value of η from the Burles and Tytler deuterium abundance,[3] and computing the light element abundances results in a ^{4}He yield of some 24.5% by mass, a value barely compatable with observational constraints.[4] (Put another way, observational constraints imply that $N_\nu < 3.3$.) There is apparently no "room" for extra neutrino degrees of freedom, a faster expansion rate, more neutrons, and hence, more ^{4}He.

It is instructive to examine more closely how the n/p ratio is set in BBN near Weak Freeze-Out. The neutrino distribution functions in the early universe can be taken to be thermal, Fermi-Dirac spectra. The normalized Fermi-Dirac energy distribution function for each neutrino species will be of the form:

$$f(E_\nu) = \frac{1}{T_\nu^3 F_2(\eta_\nu)} \cdot \frac{E_\nu^2}{\exp\left(E_\nu/T_\nu - \eta_\nu\right) + 1}, \tag{4}$$

where T_ν is the appropriate neutrino distribution function temperature (roughly the same as the matter temperature in the BBN case for epochs near T_{WFO}), and where the neutrino degeneracy parameters are $\eta_\nu = \mu_\nu/T_\nu$

with μ_ν the appropriate neutrino species chemical potential. Here $F_k(\eta_\nu)$ is the Fermi integral of order k, defined to be,

$$F_k(\eta_\nu) \equiv \int_0^\infty \frac{x^k dx}{\exp(x - \eta_\nu) + 1}. \tag{5}$$

In BBN the $\alpha = e, \mu, \tau$ net lepton number carried by each neutrino flavor ν_α is defined to be

$$L_{\nu_\alpha} \equiv \frac{n_{\nu_\alpha} - n_{\bar{\nu}_\alpha}}{n_\gamma}, \tag{6}$$

where $n_{\bar{\nu}_\alpha}$, $n_{\bar{\nu}_\alpha}$, and n_γ, are the proper number densities of ν_α, $\bar{\nu}_\alpha$, and photons, respectively.

There are two **assumptions** typically made in BBN studies. First, it is usually assumed that the entropy is uniformly and homogeneously distributed spatially. Second, it is usually assumed that all net lepton numbers (neutrino flavor asymmetries) are small, that is $L_{\nu_\alpha} \sim \eta$, or smaller. This latter assumption is functionally equivalent to taking all the neutrino chemical potentials and degeneracy parameters to be very near zero, so that all active neutrino species have nearly identical energy distribution functions.

Were this not the case and there were to be an appreciable asymmetry in, for example, ν_e over $\bar{\nu}_e$ (*i.e.*, positive L_{ν_e}), then it is clear that at Weak Freeze-Out the inherent competition between the processes in Eq. (1) and Eq. (2) would be unbalanced in favor of the neutron destroying reaction in Eq. (1). The result would be a smaller ratio n/p than in the standard case and, hence, a smaller ^{4}He yield. Therein resides the great sensitivity of BBN to neutrino transformation-produced lepton number. Note, however, that the problem is complicated by the fact that a positive L_{ν_e} can compensate for a larger N_ν! It is both the expansion rate (N_ν) and the lepton numbers/asymmetries which determine the ^{4}He yield.

The story for neutrino-heated ejecta nucleosynthesis is similar to the BBN case. However, the neutrino energy spectra in this case are essentially set at the edge of the neutron star and they remain fixed above tne neutron star (more or less), even while the neutrino fluxes and number densities are "diluted" by inverse square law effects as one moves out further from the neutron star. Though they will be roughly Fermi-Dirac in form, as in Eq. (4), the neutrino energy spectra are expected to be different for different neutrino species on general grounds. The average (that is average over the distribution function f_{ν_α}) neutrino energy for the various species will satisfy: $\langle E_{\nu_\tau}\rangle \approx \langle E_{\bar{\nu}_\tau}\rangle \approx \langle E_{\nu_\mu}\rangle \approx \langle E_{\bar{\nu}_\mu}\rangle > \langle E_{\bar{\nu}_e}\rangle > \langle E_{\nu_e}\rangle$. As in the BBN case, we expect all neutrino chemical potentials and degeneracy parameters to be low, near zero.

As material flows away from the neutron star the local n/p ratio is again set by a competition between the rates of the processes in Eq. (1) and Eq. (2), but this time it is really only the forward processes which contribute, as the electron/positron (wind) temperature eventually falls well below those characterizing the ν_e and $\bar{\nu}_e$ distribution functions. Since we expect $\langle E_{\bar{\nu}_e} \rangle > \langle E_{\nu_e} \rangle$, the steady state equilibrium with the neutrino fluxes in the wind will be neutron-rich, $n/p > 1$.

In fact, simple hydrodynamic models of the wind suggest that the flow is self similar and homologous, with the outflow speed proportional to radius. We expect this wind to develop some 10 s after the bounce of the core and the generation of the supernova explosion. In turn, the wind-like solution may persist for of order a few seconds, during which time each outgoing fluid element will experience roughly the same time, density, temperature, radius, and Y_e history. If this approximation is accurate, then radius in the wind is analagous to the age of the universe in the BBN problem. Furthermore, if the flow is adiabatic and radiation dominated, then the radius and temperature will be related inversely.

As a fluid element recedes from the neutron star, its n/p ratio will freeze out at the local equilibrium value when the forward rates for the process in Eq (1) and Eq. (2) fall below the material expansion rate. This is the Weak Freeze-Out radius and occurs where the temperature of the plasma has reached about $T \approx 0.8\,\mathrm{MeV}$. The charged particle nuclear reactions begin to freeze out of NSE at this point as well.

Just as in BBN, roughly every nucleon that can be incorporated into an alpha particle **is** so incorporated. Unlike the BBN case, the alpha particle formation process in neutron-rich conditions will leave a sea of free neutrons around the soup of alphas and a very few "seed" nuclei with masses between 50 and 100 nucleons. Of course, the neutrons experience no Coulomb barriers, so that as the material cools further the free neutrons can capture on the seeds to produce the r-process nuclides.[5]

If we want to produce, for example, the actinide nuclei, we will require more than 100 neutrons per seed nucleus. Therein lies the rub. If the supernova modelers can manage to produce neutrino energy spectra conducive to a large n/p ratio at Weak Freeze-Out (it is not at all clear that this can be accomplished[6]), then when the alpha particles form many neutrons will be isolated outside nuclei and these will capture ν_e's through the process in Eq. (1). The protons which result will immediately catalyze the formation of more alpha particles, removing still more free neutrons. The net result of this "alpha effect" will be to greatly reduce the number of available free neutrons, therefore wrecking any prospects of obtaining the heavier r-process nuclei.[7]

3 Neutrino Masses and Mixings

At the present writing the LSND data, the SuperK atmospheric neutrino data, and the results of the solar neutrino experiments are impossible to reconcile with just three active neutrino flavors. Attempts to fit this data with three neutrino flavors have failed.[8] This is because of (a) the observed zenith angle distribution of atmospheric muon neutrinos in SuperKamiokande (SuperK) [9], and (b) the apparent energy dependence of the solar neutrino deficit.[10,11,12]

The choice is stark and clear. Either some of these data must have a explanation which doesn't involve neutrino mixing, or we must introduce a fourth kind of neutrino which, because of the limit on the Z^0 width, must be "sterile."

The SuperK atmospheric neutrino data is consistent with ν_μ disappearance resulting from vacuum neutrino oscillations in $\nu_\mu \rightleftharpoons \nu_\tau$ channel, with a mass-squared difference $\delta m^2 \sim 10^{-2}\,\mathrm{eV}^2$ and with an almost maximal vacuum mixing angle, $\sin^2 2\theta > 0.82$.[9] This experiment really does seem solid in its conclusions, thereby establishing once and for all that neutrinos have mass and the flavors mix. What is a surprise for particle physics is the large mixing between ν_μ and ν_τ.

The solar neutrino deficit seems to be unexplainable in terms of simple alterations in the standard solar model.[10,11,12]. And, major alterations in the density/composition structure and run of sound speed in the sun seem incompatable with recent helioseismological data. At present active-active ($\nu_e \rightleftharpoons \nu_{\mu,\tau}$) or active-sterile ($\nu_e \rightleftharpoons \nu_s$) matter-enhanced MSW (Mikheyev, Smirnov, Wolfenstein [?]) neutrino flavor conversion can solve the problem. These solutions would envision vacuum mass-squared differences in the range $\delta m^2 \sim 10^{-5}\,\mathrm{eV}^2$. Alternatively, a "just so" vacuum oscillation solution in the active-active (maybe also the active-sterile) channel with $\delta m^2 \sim 10^{-10}\,\mathrm{eV}^2$ remains a possibilty for reconciling the solar neutrino data as well.

Last, but of pivotal importance, the Los Alamos Liquid Scintillator Neutrino Detector (LSND) persists in repoting an excess of $\bar{\nu}_e$ events which are not explainable with known backgrounds.[13] These excess events could be interpreted as evidence for vacuum oscillations in the channel $\bar{\nu}_\mu \rightleftharpoons \bar{\nu}_e$ with mass-squared difference $\delta m^2 \approx 0.2\,\mathrm{eV}^2$ to $10\,\mathrm{eV}^2$.[13] Though this experiment is the one that many would like to dismiss, the $\delta m^2 \sim 1\,\mathrm{eV}^2$ mass splitting is one which this author feels has potentially beneficial effects in curing the alpha effect problem for r-process nucleosynthesis outlined in the last section.

There are two widely discussed four-neutrino mass and mixing schemes which could, possibly, fit all of this data.[14] One of these schemes we shall term the MFBF scheme, after the authors who first examined its implications for the alpha effect and the r-process.[15] In this scheme the active neutrinos are bunched together at an arbitrary (but presumably low) mass, with the ν_τ and ν_μ nearly maximally mixed and with mass eigenvalues split by $\delta m^2 \sim 10^{-2}\,\text{eV}^2$, with a lighter ν_e corresponding to a mass eigenvalue split from the higher mass doublet by $\delta m^2 < 10^{-4}\,\text{eV}^2$. A fourth mass state, corresponding closely to the sterile neutrino, resides a higher mass than the nearly degenerate active species. In this scheme the LSND result must arise from an indirect neutrino oscillation $\bar{\nu}_\mu \to \bar{\nu}_s \to \bar{\nu}_e$, and therein lies a potentially fatal problem. The *effective* two-neutrino vacuum mixing angle for this channel in terms of the 4×4 CKM mixing matrix elements would be $\sin^2 2\theta_{\text{eff}} \approx 4|U_{es}|^2|U_{\mu s}|^2$. From disappearance experiments this is limited to be $< 10^{-3}$. However, to explain LSND we would require $\sin^2 2\theta_{\text{eff}} > 10^{-3}$. Here I have been generous in the upper limits, so this channel is likely ruled out, or, at best, only very tenuously still allowed.[16]

The competing mass scheme would envision a near maximally mixed doublet of ν_τ and ν_μ again, but this time split from a lower mass doublet of ν_e and ν_s.[17] The inter-doublet mass-squared splitting and mixing could be adjusted to be conducive to a direct vacuum $\bar{\nu}_\mu \rightleftharpoons \bar{\nu}_e$ solution for LSND ($\delta m^2_{\text{doublets}} \approx 0.2\,\text{eV}^2$ to $8\,\text{eV}^2$). We will term this mass and mixing arrangement, the CFQ scheme, again after the authors who first examined its role in defeating the alpha effect in models of the r-process sited in neutrino-heated supernova ejecta.[18] In this scheme the solar neutrino deficit is explained by either invoking matter-enhanced $\nu_e \rightleftharpoons \nu_s$ with $\delta m^2 \sim 10^{-5}\,\text{eV}^2$ and effective two-neutrino vacuum mixing $\sin^2 2\theta_{\text{eff}} > 10^{-2}$; or else, by invoking an active-sterile "just so" solution with near maximal mixing of ν_e and ν_s and an appropriate mass-squared splitting $\delta m^2 \sim 10^{-10}\,\text{eV}^2$.

4 Effects of Neutrino Flavor Conversion on Nucleosynthesis

Let us now examine what would happen to the nucleosynthesis models outlined in Section 2 were we to adopt the neutrino mass/mixing schemes of the last section. Fotunately, or unfortunately, depending on your outlook, these mass schemes tuned to explain LSND, solar and atmospheric neutrinos will give matter-enhanced level crossings between neutrino species at significant epochs in both supernovae and BBN. (The problem of MSW in the supernova environment is more complicated than that associated with the sun, but nucleosynthesis and explosion physics effects have been considered.[19,20,21,15,18])

First, let's tackle the supernova neutrino-heated ejecta issue. The essential problem in producing an acceptable r-process in current models is obtaining a high enough neutron to seed nucleus ratio. In turn, the alpha effect is the biggest impediment to this. But the alpha effect is generic: if we invoke a large enough neutrino (ν_e and $\bar{\nu}_e$) flux to effect lifting baryons out of the gravitational potential well of a neutron star, then necessarily there will be a ferocious alpha effect.

The MFBF and CFQ schemes can be tuned to circumvent this seeming paradox. Recent work has shown that the neutrino mass splittings and mixing angles in these schemes can be adjusted to leave the material heating unaffected near the neutron star surface, yet cause the disappearance of the ν_e flux near or just outside the Weak Freeze Out radius.[15,18] The disappearance of the ν_e's in just the right place to defeat the alpha effect comes about from MSW active-sterile conversion plus the interplay of this process with the electron fraction in a rapidly outflowing medium.

Is this the only way in which the alpha effect can be circumvented? We do not know the answer at this time. There may be purely "astrophysical" fixes for this problem involving extremely rapid or inhomogeneous outflow, or high entropies, or ejection mechanisms which do not rely on neutrino heating. However, the weight of current observational evidence favors models where the bulk of the r-process material in the Galaxy is produced by neutrino-heated ejecta from supernovae. Likewise, it has been argued that observed solar system r-process abundance patterns also support neutrino-heated supernova ejecta as the site of origin for at least some of this material.[22,23] Nature has found a solution, but we don't know enough *yet* to say whether or not active-sterile neutrino transformation is at its root.

The BBN situation is also ambiguous, but not because we don't understand the nucleosynthesis process. Here the problem is following neutrino transformation numerically in models of the expanding universe. Though it remains controversial, it has been known for some time that lepton number in the various neutrino flavors can be produced in the early universe by neutrino mixing via the oscillation and matter-enhanced channels.[24,25,26,27] In a beautiful series of papers Foot and Volkas have shown how matter-enhanced active-sterile neutrino transformation in the $\nu_{\mu,\tau}$, ν_s sector in the early universe can build up a significant asymmetry $|L_{\nu_\mu,\nu_\tau}|$.[28,29] They showed how this can be transferred to the $\nu_e/\bar{\nu}_e$ sector by a subsequent matter-enhanced active-sterile transformation $\nu_{\mu,\tau} \rightleftharpoons \nu_e$ or $\bar{\nu}_{\mu,\tau} \rightleftharpoons \bar{\nu}_e$.

From the discussion in Section 2, it should be clear that there are two ways in which neutrino mixing-driven (oscillation, or MSW) lepton number generation can affect BBN. First, of course, any oscillation or MSW process goes in the direction of increasing the energy density by (partially) populating a sterile neutrino sea. This represents an increase in N_ν, and hence, an increase in the expansion rate. In turn, this would give an increase in the neutron number and helium yield. Second, MSW can create and alter an asymmetry in the $\nu_e/\bar{\nu}_e$ sectors, thereby affecting the neutron to proton ratio directly, and so influencing the helium yield in a different manner.

At this point we should be able to choose parameters for either the CFQ or MFBF neutrino mass/mixing schemes and then solve the Boltzmann transport equations with the MSW evolution equations in an expanding universe, all coupled with the BBN nuclear reaction network. Unfortunately there is as yet no quantitative agreement on how to carry out this procedure, and many approximations and shortcuts are made. Some groups claim that it is possible to compensate for a larger N_ν with a large L_{ν_e}, therefore leaving the neutron abundance about the same, and so perhaps easing the ^{4}He constraint.[28,29,30]. Others find that the increased expansion rate nearly always wins out over an increased L_{ν_e}, with the end result that it seems to be easy to *increase* the ^{4}He BBN yield, but not to decrease it.[31]

However, there is a crucial point which has been overlooked until recently. Namely, the lepton number generation process attendant to MSW conversion in the early universe can be, for some mixing parameters, essentially "chaotic" in the sign of the generated asymmetry.[27,32] If this is the case, then the sign of the lepton number could be inhomogeneously distributed on any scale larger than the particle horizon at the epoch of the lepton number generation process.[33] By the epoch of BNN, this will lead to a domain structure for lepton number sign. (Depending on the "percolation" processes involved, these domains of opposite lepton number sign could have complicated "spaghetti-like" geometries, but we expect typical domain boundaries to be separated from one another by a scale smaller than or of order the horizon scale at the generation event, comoving to the BBN epoch.) The problem is that once neutrinos can freely stream to the horizon (*i.e.*, after Weak Decoupling), they can cross these domain boundaries. As they do so, they will encounter a lepton number gradient which can cause further MSW production of sterile neutrinos.[33] This could result in a significantly larger sterile neutrino sea population, concomitant increase in the expansion rate and, hence, a bigger ^{4}He yield. In short, a kind of "helium catastrophe." If this effect holds up, it would greatly facilitate the ability of BBN to constrain neutrino mixing parameters.

Over the range of mixing parameters that are relevant for the CFQ and MFBF schemes we expect unpredictability of lepton number sign in the early universe. The result is startling. The MFBF scheme parameter space tuned to fit *all* of the neutrino experimental data would be in trouble with overproduction of ^{4}He in BBN.[34] Likewise, the CFQ scheme is similarly challenged. This scheme also overproduces helium, save for the case of a narrowly constrained scheme where the mixing between the upper ν_μ/ν_τ doublet and the ν_e in the lower doublet is some 50 to 100 times larger than the mixing between the upper doublet and the sterile partner in the lower doublet. This large ratio in mixings would have to hold over the entire "LSND" range for the splitting of the doublets. Furthermore, to dodge the helium overproduction bound in this way, and just as in the r-process case [18], the ν_e/ν_s mixing in the lower doublet would have to be very close to maximal.[34]

What does all this mean? It could imply one or more of three alternatives. First, one of the experiments/problems in the neutrino data outlined in the last section may not involve neutrino mixing afterall. Second, it is entirely possible that we do not yet know how to calculate the effects of these neutrino mass/mixing schemes in the early universe. Finally, one has to wonder what the nature of the sterile neutrino state is.

The first alternative speaks for itself. The second alternative may turn out to be the key. In fact, recent work has suggested to one group that all of the previous estimates of neutrino asymmetry production resulting from neutrino mixing in the early universe are wildly wrong, being grossly overestimated.[35] Nevertheless, it appears that there could be considerable sterile neutrino production in this work, resulting, again, in a challenge to BBN limits.

The last alternative should be handled with care. Different "models" for sterile neutrinos can have widely varying consequences in astrophysical settings. For example, Kaluza-Klein neutrinos propagating in a compactified extra dimension in the "bulk" behave differently from the sterile neutrinos considered here when they propagate through a matter-enhanced level crossing.[36] Other schemes for producing singlet sterile neutrinos could allow them to have some leptonic couplings and this would change the results presented here.[37]

With these disturbing notions of the nature of sterile neutrinos in mind, one may ask if there are additional sterile neutrino states beyond the fourth one seemingly required to fit the experimental data. Perhaps these could have higher rest masses. If so, then the possibility of MSW production of these states could completely alter our notions of neutrino Dark Matter.

Let's begin by assuming that the sterile neutrinos are "sterile enough" (have small enough couplings) that they have thermally decoupled prior to the cosmic QCD transition, $T \sim 100\,\mathrm{MeV}$. This would guarantee that the sterile neutrino sea could be regarded as unpopulated from the standpoint of primordial nucleosynthesis. at least at the QCD epoch.[38]

However, matter-enhanced neutrino transformation subsequently could partially populate a sterile neutrino sea. Furthermore, it could do so in a way that left the resulting sterile neutrino energy distribution function populated only at quite low energies.[39] Thereby, sterile neutrinos with rest masses between $\sim 100\,\text{eV}$ and $\sim 10\,\text{keV}$ could behave like Cold Dark Matter (CDM). This is a new kind of CDM. The trick in its production is to exploit the fact that resonant neutrino transformation is only efficient when the flavor amplitude evolution through the resonance is adiabatic. In turn, for the large values of δm^2 contemplated here, only lower energy neutrinos will propagate through MSW resonances adiabatically. This is partly because lower energy neutrinos have smaller effective oscillation lengths at resonance, and the adiabaticity is measured by the ratio of of resonance width to oscillation length.

This mass range for sterile neutrino dark matter has been considered previously, but where the production mechanism was ordinary vacuum oscillations.[40] As a result, the sterile neutrinos so produced had more or less ordinary thermal spectra, and these particles comprised what was termed "Warm Dark Matter." By contrast our MSW production scenario renders the energy spectrum of the sterile neutrinos with similar masses grossly nonthermal and quite "cold."

Remember that the unattractive feature of neutrino Dark Matter has always been that light neutrinos freely stream to the horizon so long as they are relativistic. Therefore, density fluctuations are damped until the epoch where the neutrinos go nonrelativistic. Observed structures in the universe at high redshift then can be employed to put constraints on the light neutrino contribution to the closure density.[41] This last step is, of course, dependent on an assumption that the neutrinos have thermal energy spectra. Violating this assumption is how the Shi & Fuller scenario can avoid constraint.

Nevertheless, the Shi & Fuller scenario does have some requirements that may prove its undoing. First, we require a net lepton number in some neutrino flavor α of order $L_{\nu_\alpha} \sim 10^{-2}$ to 10^{-3}, for sterile neutrino masses between several hundred keV and 10 keV. This lepton number by itself will not run afoul of BBN constraints, but it is nevertheless large. Where could it come from? It could possibly result from a "Foot & Volkas" style lepton number generation process (see above). It can't be a byproduct of ordinary electroweak baryogenesis, unless there are lepton numbers for several neutrino species generated with comparable magnitudes but opposite signs so that the *net* lepton number is $\sim \eta$. It is possible, however, that the baryon number and associated lepton numbers arise in scenarios where there is no requirement that $B - L = 0$, *e.g.*, Affleck/Dine Baryogenesis.[42]

Acknowledgments

I would like to acknowledge discussions with K. Abazajian, A. B. Balantekin, J. Fetter, G. C. McLaughlin, A. Mezzacappa, M. Patel, Y.-Z. Qian, X. Shi, and J. R. Wilson. This work was partially supported by the National Science Foundation under NSF Grant PHY98-00980 at UCSD.

References

1. Wolfenstein, L., *Phys. Rev. D* **17**, 2369 (1978); **20**, 2634 (1979).
2. Mikeyev, S. P., and Smirnov, A. Yu., *Sov. Phys. JETP*, **69**, 4 (1986).
3. S. Burles and D. Tytler, in *Proceedings of the Second Oak Ridge Symposium on Atomic and Nuclear Astrophysics*, edited by A. Mezzacappa (Institute of Physics, Bristol, 1998); and references therein.
4. K. A. Olive, E. Skillman, and G. Steigman, Astrophys. J., **483**, 788 (1997); Y. I. Izotov and T. X. Thun, Astrophys, J., **500**, 188 (1998).
5. B. S. Meyer, G. J. Mathews, W. M. Howard, S. E. Woosley, and R. D. Hoffman, Astrophys. J., **399**, 656 (1992); S. E. Woosley, G. J. Mathews, J. R. Wilson, R. D. Hoffman and B. S. Meyer, Astrophys. J., **433**, 229 (1994); K. Takahashi, J. Witti and H.-T. Janka, Astron. Astrophys. **286**, 857 (1994).
6. B. S. Meyer and J. S. Brown, Astrophys. J. Suppl. **112**, 199 (1997); R. D. Hoffman, S. E. Woosley, and Y.-Z. Qian, Astrophys. J. **482**, 951 (1996).
7. G.M. Fuller and B.S. Meyer, Astrophys. J., **453**, 792 (1995); B. S. Meyer, G. C. McLaughlin, and G. M. Fuller, Phys. Rev. C, **58**, 3696 (1998).
8. C. Y. Cardall and G. M. Fuller, Phys. Rev. D**53**, 4421 (1996); G. L. Fogli, E. Lisi, and G. Scioscia, Phys. Rev. D**52**, 5334 (1995); A. Acker and S. Pakvasa, Phys. Lett. **B397**, 209 (1997); P. Harrison, D. Perkins, and W.Scott, Phys. Rev. **B349**, 137 (1995); T. Teshima and T. Sakai hep-ph/9805386 and hep-ph/9801276.
9. Y. Fukuda, *et al.*, Phys. Rev. Lett. **81**, 1562 (1998); Phys. Lett. B **436**, 33 (1998).
10. J. N. Bahcall, these proceedings.
11. B.T. Cleveland, T. Daily, R. Davis, J.R. Distel, K. Lande, C.K. Lee, P.S. Wildenhain, and J. Ullman, Astrophys. J. **496**, 505 (1998); W. Hampel, *et al.*, Phys. Lett. B **388**, 384 (1996), **420**, 114 (1998), *ibid.* **436**, 158 (1998); J.N. Abdurashitov, *et al.*, Nucl. Phys. Proc. Suppl. **48**, 299 (1996), hep-ph/9803418.
12. J.N. Bahcall, P.I. Krastev, and A.Yu. Smirnov, Phys. Rev. D **58**, 096016 (1998); N. Hata and P. Langacker, Phys. Rev. D **56**, 6107 (1997).
13. C. Athanassopoulos, *et al.*, Phys. Rev. Lett. **75**, 2650 (1996), **77**, 3082 (1996), Phys. Rev. C **54**, 2685 (1996), Phys. Rev. C **58**, 2489 (1998).
14. See for example V. Barger, S. Pakvasa, T. J. Weiler, and K. Whisnant,

Phys. Rev. D **58**, 093016 (1998).
15. G. C. McLaughlin, J. Fetter, A. B. Balantekin, and G. M. Fuller, Phys. Rev. C **59**, 2873 (1999).
16. C. Giunti, these proceedings. See also S.M. Bilenky, G. Giunti, and W. Grimus, Eur. Phys. J., **C1**, 247 (1998).
17. D. O. Caldwell and R. N. Mohapatra, Phys. Rev. D**48**, 3259 (1993); D. O. Caldwell and R. N. Mohapatra, Phys. Lett. B **354**, 371 (1995); J. T. Peltoniemi and J. W. F. Valle, Nucl. Phys. B **406**, 409 (1993); J. M. Gelb and S. P. Rosen, hep-ph/9909293; G. M. Fuller, J. R. Primack, and Y.-Z. Qian, Phys. Rev. D**52**, 1288 (1995).
18. D. O. Caldwell, G. M. Fuller, and Y.-Z. Qian, astro-ph/9910175, Phys. Rev. D, *submitted* (1999).
19. G. M. Fuller, R. Mayle, J. R. Wilson, and D. N. Schramm, Astrophys. J., **322**, 795 (1987); G. M. Fuller, R. Mayle, B. S. Meyer, and J. R. Wilson, Astrophys. J., **389**, 517 (1992); Y.-Z. Qian, *et el.*, Phys. Rev. Lett. **71**, 1965 (1993); Y.-Z. Qian and G.M. Fuller, Phys. Rev. D **51**, 1479 (1995), **52**, 656 (1995).
20. J. T. Peltoniemi, Astron. Astrophys., **254**, 121 (1992); J. T. Peltoniemi, preprint hep-ph/9511323, unpublished (1995); H. Nunokawa, J. T. Peltoniemi, A. Rossi, and J. W. Valle, Phys. Rev. D **56**, 1704 (1997).
21. A. Mezzacappa and S. W. Bruenn, in *Proceedings of Future Directions of Supernova Research: Progenitors to Remnants*, Memoirs of the Italian Astronomical Society, in press (1999); Proceedings of the *Second International Workshop on the Identification of Dark Matter* (World Scientific, Singapore), in press (1999).
22. Y.-Z. Qian, W. C. Haxton, K. Langanke, and P. Vogel, Phys. Rev. C **55**, 1532 (1997).
23. Y.-Z. Qian, P. Vogel, and G. J. Wasserburg, Astrophys. J., **494**, 285 (1998).
24. A. Dolgov, Sov. J. Nucl. Phys. **33**, 700 (1981); D. Fargion and M. Shepkin, Phys. Lett **B 146**, 46 (1984); P. G. Langacker, Univ. Penn. Preprint UPR0401T (unpublished).
25. R. Barbieri and A. Dolgov, Phys. Lett., **B 349**, 743 (1991); K. Enqvist, K. Kainulainen, and M. Thomson, Nucl. Phys. **B 373**, 498 (1992); K. Enqvist, K. Kainulainen, and J. Maalampi, Nucl. Phys. **B 349**, 754 (1991); A. D. Dolgov, J. H. Hansen, and D. Y. Semikoz, Nucl. Phys. **B503**, 426 (1997); Nucl. Phys. **B543**, 269 (1999); D.P. Kirilova and M.V. Chizhov, Phys. Lett **B393**, 375 (1997); Phys. Rev. **D 58**, 073004 (1998); Nucl. Phys. **B534**, 447 (1998).
26. X. Shi, D. N. Schramm, and B. D. Fields, Phys. Rev. D **48**, 2563 (1993).

27. X. Shi, Phys. Rev. D **54**, 2753 (1996).
28. R. Foot and R. R. Volkas, Phys. Rev. Lett., **75**, 4350 (1995); R. Foot, M. J. Thomson, and R. R. Volkas, Phys. Rev. D **53**, 5349 (1996); R. Foot and R. R. Volkas, Phys. Rev. D **56**, 6653 (1997).
29. R. Foot and R. R. Volkas, Phys. Rev. D **55**, 5147 (1997).
30. P. Di Bari, P. Lipari, and M. Lusignol, hep-ph/9907548.
31. X. Shi, G. M. Fuller, and K. Abazajian, Phys. Rev. D **60**, 063002 (1999).
32. K. Enqvist, K. Kainulainen, and A. Sorri, hep-ph/9906452.
33. X. Shi and G. M. Fuller, astro-ph/9904041, Phys. Rev. Lett., in press (1999).
34. X. Shi, G. M. Fuller, and K. Abazajian, submitted to Phys. Rev. D (1999), astro-ph/9908081.
35. A.D. Dolgov, S.H. Hansen, S. Pastor, and D.V. Semikoz, preprint, TAC-1999-018.
36. S. Dvali and A. Yu. Smirnov, preprint (1999); M. Patel and G.M. Fuller, preprint (1999).
37. G. C. McLaughlin and J. N. Ng, Physics Letters B, in press (1999), hep-ph/9907449.
38. G. M. Fuller and R. A. Malaney, Phys. Rev. D, **43**, 3136 (1991).
39. X. Shi and G. M. Fuller, Phys. Rev. Lett. , **82**, 2832 (1999).
40. S. Dodelson and L. M. Widrow, Phys. Rev. Lett. **72**, 17 (1994).
41. See for example, A. Klypin, J. Holtzman, J. Primack, and E. Regos, Astrophys, J., **416**, 1 (1993).
42. L. Affleck and M. Dine, Nucl. Phys. **B249**, 361 (1985).

NEUTRINOS FROM THE NEXT GALACTIC SUPERNOVA

J. F. BEACOM
California Institute of Technology, Physics 161-33
1200 E. California Blvd., Pasadena, CA 91125 USA
E-mail: beacom@citnp.caltech.edu

The next core-collapse supernova in our Galaxy will be a spectacular event, with some 10^4 neutrino detections in total expected among several detectors. This data will allow unprecedented tests of neutrino properties and new opportunities in astrophysics. In this paper, I focus on two main topics: (1) Measurement of the ν_μ and ν_τ masses by time-of-flight, with an emphasis on introducing as little supernova model dependence as possible, and (2) Methods for locating a supernova by its neutrinos in advance of the light, which may allow improved astronomical observations. In the latter, I also discuss the recent result that the positrons from $\bar{\nu}_e + p \rightarrow e^+ + n$ are *not* isotropically emitted, as commonly thought.

1 Supernova Neutrino Mass Measurements

1.1 Current Knowledge About Neutrino Masses

What is known about the neutrino masses, in terms of direct experimental evidence? After several decades of experiments on the tritium beta spectrum, the published limit on the mass of the $\bar{\nu}_e$ (and hence also ν_e) is now about 3 eV [1]. These results are still attended by some controversy, due to apparent systematic errors near the endpoint (discussed in detail at a recent workshop [2] at the Institute for Nuclear Theory in Seattle). Much less is known about the ν_μ mass (< 170 keV from π decay) [3], and the ν_τ mass (< 18 MeV from τ decay) [4]. It would be a great advance if the latter two limits could be improved to the eV or tens of eV level, and it seems unlikely that any terrestrial experiment could do that.

There are two indirect arguments which *may* restrict the ν_μ and ν_τ masses. The first says that the predicted background of relic neutrinos ($\approx 100/\text{cm}^3$) should not overclose the universe. This leads to a limit on the sum of the neutrino masses of about 100 eV in the most conservative case, and about 10 eV if reasonable values for the non-baryonic dark matter density and the Hubble constant are used [5]. The second is based on the recent evidence for neutrino oscillations (sensitive only to *differences* of neutrino masses). For three neutrino flavors, there are two independent mass differences, which can be chosen to explain the solar and atmospheric neutrino data. The overall scale can be fixed by using the limit from tritium beta decay, and since the observed mass differences are small, then all masses are below a few eV [6].

These indirect arguments sound compelling, but they are not hard to evade. The cosmological bound doesn't apply if the neutrinos are allowed to decay. There are many such models for heavy ν_τ masses, motivated by particle physics [7] or astrophysics [8]. The oscillation bound doesn't apply if there are more free mass differences than there are positive signals of neutrino oscillations. For example, suppose the atmospheric neutrinos are undergoing maximal $\nu_\mu \leftrightarrow \nu_\tau$ mixing, LSND is ruled out, and the solar neutrinos are undergoing $\nu_e \leftrightarrow \nu_{sterile}$ mixing. Then nothing prevents the ν_2 and ν_3 masses from having a tiny splitting near 50 or even 500 eV, and leaving the ν_1 mass below a few eV.

It may be possible in the future to make new astrophysical tests of the neutrino masses, using data on large-scale structure[9,10], the cosmic microwave background [11], weak lensing [12], and the Lyman-alpha forest [13]. While these techniques claim sensitivities of order 1 - 10 eV for the sum of the neutrino masses, none are based on direct detection of neutrinos, and hence may be vulnerable to uncertainties in their assumptions.

Thus while indirect evidence will be valuable, there is still a need for a *direct* measurement of the ν_μ and ν_τ masses (or, in the presence of mixing, the masses of the heavy mass eigenstates).

1.2 Supernova Neutrino Emission

When the core of a large star ($M \geq 8M_\odot$) runs out of nuclear fuel, it collapses, with a change in the gravitational binding energy of about 3×10^{53} ergs. This huge amount of energy must be removed from the proto-neutron star, but the high density prevents the escape of any radiation. Inside of the hot proto-neutron star, neutrino-antineutrino pairs of all flavors are produced. Despite their weak interactions, even the neutrinos are trapped and diffuse out over several seconds, in the end carrying away about 99% of the supernova energy. (I have neglected the flux of ν_e neutrinos necessary to change the core from $N \simeq Z$ nuclei into a neutron star, because the total energy release of this flux is of order 1% of the pair emission phase).

When the neutrinos are about one mean free path from the edge, they escape freely, with a thermal spectrum (approximately Fermi-Dirac) characteristic of the surface of last scattering. Because different flavors have different interactions with the matter, and because the neutron star temperature is decreasing with increasing radius, the neutrino decoupling temperatures are different. The ν_μ and ν_τ neutrinos and their antiparticles have a temperature of about 8 MeV, the $\bar{\nu}_e$ neutrinos about 5 MeV, and the ν_e neutrinos about 3.5 MeV. Equivalently, the average energies are about $\langle E \rangle \simeq 25$ MeV, $\simeq 16$ MeV, and $\simeq 11$ MeV, respectively. The luminosities of the different neutrino flavors are approximately equal at all times. The neutrino luminosity rises quickly over 0.1 s or less, and then falls off over several seconds. The SN1987A data can be reasonably fit to a decaying exponential with time constant $\tau = 3$ s. The detailed form of the neutrino luminosity used below is less important than the general shape features and their characteristic durations.

The estimated core-collapse supernova rate in the Galaxy is about 3 times per century [14]. The present neutrino detectors can easily observe a supernova anywhere in the Galaxy or its immediate companions (e.g., the Magellanic Clouds). Unfortunately, the present detectors do not have large enough volumes to observe a supernova in even the nearest galaxy (Andromeda, about 700 kpc away).

1.3 Time-of-Flight Concept

Even a tiny neutrino mass will make the velocity slightly less than for a massless neutrino, and over the large distance to a supernova will cause a measurable delay in the arrival time. A neutrino with a mass m (in eV) and energy E (in MeV) will experience an energy-dependent delay (in s) relative to a massless neutrino in traveling over a distance D (in 10 kpc, approximately the distance to the Galactic center) of

$$\Delta t(E) = 0.515 \left(\frac{m}{E}\right)^2 D\,, \tag{1}$$

where only the lowest order in the small mass has been kept. A distance of 10 kpc corresponds to a travel time of about 10^{12} s, and the delays of interest are less than 1 s.

If the neutrino mass is nonzero, lower-energy neutrinos will arrive later, leading to a correlation between neutrino energy and arrival time. This is exploitable in some of the charged-current detection reactions for ν_e and $\bar{\nu}_e$ since the neutrino energy can be reasonably measured. Using this idea, the next supernova will allow sensitivity to a $\bar{\nu}_e$ mass down to about 3 eV [15]. However, the terrestrial experiments now claim [1] to rule out a mass that large, so I ignore the ν_e mass in describing the supernova neutrino signal.

For measuring the masses of the ν_μ and ν_τ neutrinos, one must realize that only neutral-current detection reactions are possible (the charged-current thresholds are too high). Since the neutrino energy is *not* measured in neutral-current interactions, an energy-time correlation technique cannot be used (the incoming neutrino energy is not determined since a complete kinematic reconstruction of the reaction products is not possible).

Instead, the strategy for measuring the ν_τ mass is to look at the difference in time-of-flight between the neutral-current events (mostly ν_μ,ν_τ,$\bar{\nu}_\mu$, and $\bar{\nu}_\tau$, because of the higher temperature of those flavors) and the charged-current events (just ν_e and $\bar{\nu}_e$). We assume that the ν_μ is massless and will ask what limit can be placed on the ν_τ mass (other cases will be discussed below). There are three major complications to a simple application of Eq. (1): (i) The neutrino energies are not fixed, but are characterized by spectra; (ii) The neutrino pulse has a long intrinsic duration of about 10 s, as observed for SN1987A; and (iii) The statistics are finite. In some early work [16], the duration of the pulse was thought to be of order 100 milliseconds, nearly a delta-function. Here (considering the case of the smallest detectable mass), the delay is much less than the width of the pulse, which is what makes the problem much more difficult (the large-mass case is covered elsewhere [17,18]). Given that the energy-time correlation method cannot be used, we must turn to considerations of the shape of the event rate as a function of time.

1.4 Neutrino Scattering Rate

For a time-independent spectrum $f(E)$, and luminosity $L(t_i)$, the double-differential number distribution at the source is

$$\frac{d^2 N_\nu}{dE_\nu dt_i} = f(E_\nu)\frac{L(t_i)}{\langle E_\nu \rangle}\,. \tag{2}$$

The neutrino spectrum and average energy $\langle E_\nu \rangle$ are different for each flavor, but the luminosities $L(t_i)$ of the different flavors are roughly equal [19], so that each flavor carries away about 1/6 of the total binding energy E_B. For a massive neutrino, the double-differential number distribution at the detector is

$$\frac{d^2 N_\nu}{dE_\nu dt} = f(E_\nu)\frac{L(t \quad \Delta t(E_\nu))}{\langle E_\nu \rangle} \tag{3}$$

and the scattering rate for a given reaction is

$$\frac{dN_{sc}}{dt} = C\int dE_\nu f(E_\nu)\left[\frac{\sigma(E_\nu)}{10 \;\; ^{42}\mathrm{cm}^2}\right]\left[\frac{L(t \quad \Delta t(E_\nu))}{E_B/6}\right]\,. \tag{4}$$

The overall constant for an H_2O target is

$$C = 9.21\left[\frac{E_B}{10^{53}\ \mathrm{ergs}}\right]\left[\frac{1\ \mathrm{MeV}}{T}\right]\left[\frac{10\ \mathrm{kpc}}{D}\right]^2\left[\frac{\mathrm{det.\ mass}}{1\ \mathrm{kton}}\right] n\,, \tag{5}$$

where n is the number of targets per molecule for the given reaction. For a D_2O target, the leading factor becomes 8.28 instead of 9.21.

Table 1: Calculated numbers of events expected in SK (32 kton H_2O) with a 5 MeV threshold and a supernova at 10 kpc. The other parameters (e.g., neutrino spectrum temperatures) are given in the text. In rows with two reactions listed, the number of events is the total for both. The second row is a subset of the first row that is an irreducible background to the reactions in the third and fourth rows [20]. The detected final-state particles are e^+, e^-, γ.

reaction	number of events
$\bar{\nu}_e + p \rightarrow e^+ + n$	8300
$\bar{\nu}_e + p \rightarrow e^+ + n \quad (E_{e^+} \leq 10$ MeV$)$	530
$\nu_\mu + {}^{16}\mathrm{O} \rightarrow \nu_\mu + \gamma + X$ $\bar{\nu}_\mu + {}^{16}\mathrm{O} \rightarrow \bar{\nu}_\mu + \gamma + X$	355
$\nu_\tau + {}^{16}\mathrm{O} \rightarrow \nu_\tau + \gamma + X$ $\bar{\nu}_\tau + {}^{16}\mathrm{O} \rightarrow \bar{\nu}_\tau + \gamma + X$	355
$\nu_e + e^- \rightarrow \nu_e + e^-$ $\bar{\nu}_e + e^- \rightarrow \bar{\nu}_e + e^-$	200
$\nu_\mu + e^- \rightarrow \nu_\mu + e^-$ $\bar{\nu}_\mu + e^- \rightarrow \bar{\nu}_\mu + e^-$	60
$\nu_\tau + e^- \rightarrow \nu_\tau + e^-$ $\bar{\nu}_\tau + e^- \rightarrow \bar{\nu}_\tau + e^-$	60

For a massless neutrino, $\Delta t(E_\nu) = 0$, so the luminosity comes outside of the integral as $L(t)$ and completely specifies the time dependence. For a massive neutrino, the time dependence from the luminosity is modified in an energy-dependent way by the mass effects, as above.

1.5 Separation of the Massive Signal

The scattering rates of the massless neutrinos (ν_e or $\bar{\nu}_e$) can then be used to measure the shape of the luminosity as a function of time. The scattering rate for the massive neutrinos (ν_τ) can be tested for additional time dependence due to a mass. We define two rates: a Reference $R(t)$ containing only massless events, and a Signal $S(t)$ containing some fraction of massive events.

The Reference $R(t)$ can be formed in various ways, for example from the charged-current reaction $\bar{\nu}_e + p \rightarrow e^+ + n$ in the light water of SK or SNO. (The numbers of events expected for all reactions are given in Table I for SK and Table II for SNO).

The Signal $S(t)$ can be based on various neutral-current reactions [17,18,21], though here I will focus on the results for SNO [18]. Thus the primary component of the Signal $S(t)$ are the 485 neutral-current events on deuterons. With the hierarchy of temperatures assumed here, these events are 18% ($\nu_e + \bar{\nu}_e$), 41%

Table 2: Calculated numbers of events expected in SNO for a supernova at 10 kpc. The other parameters (e.g., neutrino spectrum temperatures) are given in the text. In rows with two reactions listed, the number of events is the total for both. The middle column indicates the detected particle(s). The notation ν indicates the sum of ν_e, ν_μ, and ν_τ, though they do not contribute equally to a given reaction, and X indicates either $n+^{15}$O or $p+^{15}$N.

Events in 1 kton D_2O		
$\nu + d \to \nu + p + n$ $\bar{\nu} + d \to \bar{\nu} + p + n$	n n	485
$\nu_e + d \to e^- + p + p$ $\bar{\nu}_e + d \to e^+ + n + n$	e^- e^+nn	160
$\nu +^{16}\text{O} \to \nu + \gamma + X$ $\bar{\nu} +^{16}\text{O} \to \bar{\nu} + \gamma + X$	γ, γn γ, γn	20
$\nu +^{16}\text{O} \to \nu + n +^{15}\text{O}$ $\bar{\nu} +^{16}\text{O} \to \bar{\nu} + n +^{15}\text{O}$	n n	15
$\nu + e^- \to \nu + e^-$ $\bar{\nu} + e^- \to \bar{\nu} + e^-$	e^- e^-	10
Events in 1.4 kton H_2O		
$\bar{\nu}_e + p \to e^+ + n$	e^+	365
$\nu +^{16}\text{O} \to \nu + \gamma + X$ $\bar{\nu} +^{16}\text{O} \to \bar{\nu} + \gamma + X$	γ γ	30
$\nu + e^- \to \nu + e^-$ $\bar{\nu} + e^- \to \bar{\nu} + e^-$	e^- e^-	15

$(\nu_\mu + \bar{\nu}_\mu)$, and 41% $(\nu_\tau + \bar{\nu}_\tau)$. The flavors of the neutral-current events of course cannot be distinguished. Under our assumption that only ν_τ is massive, there is already some unavoidable dilution of $S(t)$.

In Fig. 1, $S(t)$ is shown under different assumptions about the ν_τ mass. The shape of $R(t)$ is exactly that of $S(t)$ when $m_{\nu_\tau} = 0$, though the number of events in $R(t)$ will be different. The rates $R(t)$ and $S(t)$ will be measured with finite statistics, so it is possible for statistical fluctuations to obscure the effects of a mass when there is one, or to fake the effects when there is not. We determine the mass sensitivity in the presence of the statistical fluctuations by Monte Carlo modeling. We use the Monte Carlo to generate representative statistical instances of the theoretical forms of $R(t)$ and $S(t)$, so that each run represents one supernova as seen in SNO. The best model-independent test of a ν_τ mass seems to be a test of the average arrival time $\langle t \rangle$. Any massive component in $S(t)$ will always increase $\langle t \rangle$, up to statistical fluctuations.

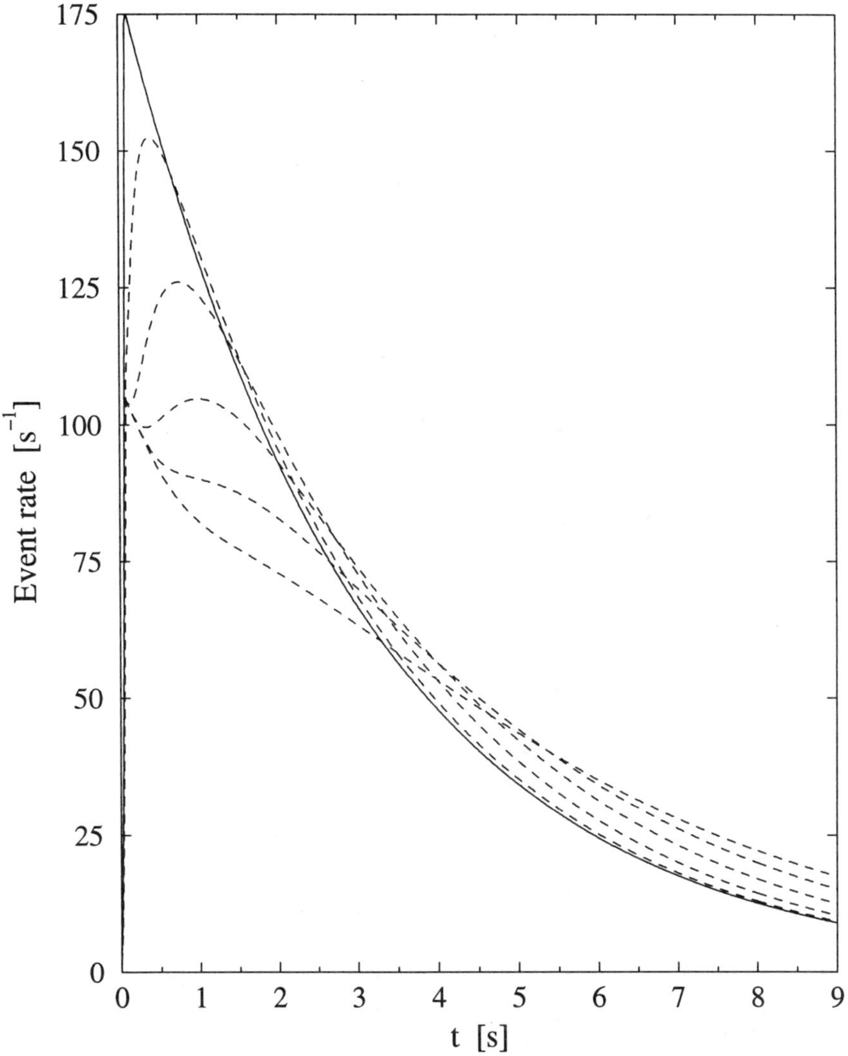

Figure 1: The expected event rate for the Signal $S(t)$ at SNO in the absence of fluctuations for different ν_τ masses, as follows: solid line, 0 eV; dashed lines, in order of decreasing height: 20, 40, 60, 80, 100 eV. Of 535 total events, 100 are massless ($\nu_e + \bar{\nu}_e$), 217.5 are massless ($\nu_\mu + \bar{\nu}_\mu$), and 217.5 are massive ($\nu_\tau + \bar{\nu}_\tau$). These totals count events at all times; in the figure, only those with $t \leq 9$ s are shown.

1.6 $\langle t \rangle$ Analysis

Given the Reference $R(t)$ (i.e., the charged-current events), the average arrival time is defined as

$$\langle t \rangle_R = \frac{\sum_k t_k}{\sum_k 1} \,, \tag{6}$$

where the sum is over events in the Reference. The effect of the finite number of counts N_R in $R(t)$ is to give $\langle t \rangle_R$ a statistical error:

$$\delta\left(\langle t \rangle_R\right) = \frac{\sqrt{\langle t^2 \rangle_R \ \langle \ t \rangle_R^2}}{\sqrt{N_R}} \,. \tag{7}$$

For a purely exponential luminosity, $\langle t \rangle_R = \sqrt{\langle t^2 \rangle_R \ \langle \ t \rangle_R^2} = \tau \simeq 3$ s. Given the Signal $S(t)$ (i.e., the neutral-current events), the average arrival time $\langle t \rangle_S$ and its error $\delta\left(\langle t \rangle_S\right)$ are defined similarly. The signal of a mass is that the measured value of $\langle t \rangle_S \ \langle \ t \rangle_R$ is greater than zero with statistical significance.

Using the Monte Carlo, we analyzed 10^4 simulated supernova data sets for a range of ν_τ masses. For each data set, $\langle t \rangle_S \ \langle \ t \rangle_R$ was calculated and its value histogrammed. These histograms are shown in the upper panel of Fig. 2 for a few representative masses. (Note that the number of Monte Carlo runs only affects how smoothly these histograms are filled out, and not their width or placement.) These distributions are characterized by their central point and their width, using the 10%, 50%, and 90% confidence levels. That is, for each mass we determined the values of $\langle t \rangle_S \ \langle \ t \rangle_R$ such that a given percentage of the Monte Carlo runs yielded a value of $\langle t \rangle_S \ \langle \ t \rangle_R$ less than that value. With these three numbers, we can characterize the results of complete runs with many masses much more compactly, as shown in the lower panel of Fig. 2. Given an experimentally determined value of $\langle t \rangle_S \ \langle \ t \rangle_R$, one can read off the range of masses that would have been likely (at these confidence levels) to have given such a value of $\langle t \rangle_S \ \langle \ t \rangle_R$ in one experiment. From the lower panel of Fig. 2, we see that SNO is sensitive to a ν_τ mass down to about 30 eV if the SK $R(t)$ is used, and down to about 35 eV if the SNO $R(t)$ is used.

We also investigated the dispersion of the event rate in time as a measure of the mass. A mass alone causes a delay, but a mass and an energy spectrum also causes dispersion (as does the separation of the massive and massless portions of the Signal). We defined the dispersion as the change in the width $\sqrt{\langle t^2 \rangle_S \ \langle \ t \rangle_S^2} \quad \sqrt{\langle t^2 \rangle_R \ \langle \ t \rangle_R^2}$. We found that the dispersion was not statistically significant until the mass was of order 80 eV or so, i.e., a plot very similar to Fig. 2 can be formed for the width, and the error band is much wider. Since the dispersion is irrelevant, the average delay is well-characterized by a single energy, which for SNO is $E_c \simeq 32$ MeV.

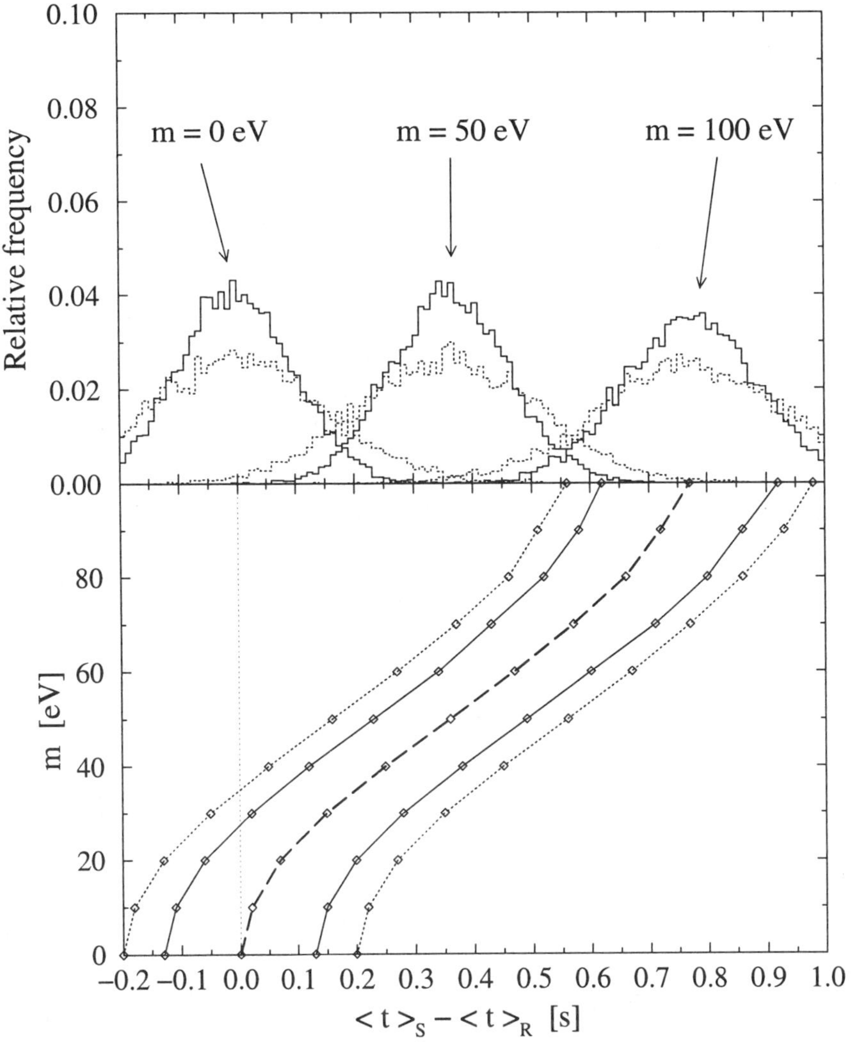

Figure 2: The results of the $\langle t \rangle$ analysis for a massive ν_τ, using the Signal $S(t)$ from SNO defined in the text. In the upper panel, the relative frequencies of various $\langle t \rangle_S \ \langle \ \ t \rangle_R$ values are shown for a few example masses. The solid line is for the results using the SK Reference $R(t)$, and the dotted line for the results using the SNO $R(t)$. In the lower panel, the range of masses corresponding to a given $\langle t \rangle_S \ \langle \ \ t \rangle_R$ is shown. The dashed line is the 50% confidence level. The upper and lower solid lines are the 10% and 90% confidence levels, respectively, for the results with the SK $R(t)$. The dotted lines are the same for the results with the SNO $R(t)$.

1.7 Analytic Results

Another nice feature of the proposed $\langle t\rangle$ analysis, besides its simplicity, is that good estimates can be made analytically. The characteristic delay is

$$\langle t\rangle_S \;\langle\; t\rangle_R \simeq \mathrm{frac}(m>0)\times 0.515\left(\frac{m}{E_c}\right)^2 D\,, \tag{8}$$

where frac$(m > 0)$ is the fraction (about 1/2) of massive events in the neutral-current Signal. The above numerical analysis shows that the mass effects occur most significantly in the delay and not the width (are characterized with a single energy, and negligible dispersion). That characteristic energy E_c is roughly at the peak of $f(E)\sigma(E)$ (see Fig. 3). This formula for the first moment is always true; the point of the Monte Carlo analysis was to show that the other moments are not changed, so this completely describes the data.

If the cross section $\sigma(E_\nu)$ depends on energy as E_ν^α ($\alpha \sim 2$ for $\nu + d$), then the characteristic energy $E_c \sim (2+\alpha)T$ and the thermally-averaged cross section is proportional to T^α, where T is the ν_μ and ν_τ temperature. For $\alpha = 2$, and a neutron detection efficiency ϵ_n, the following scaling relations hold:

$$\langle t\rangle_S \;\langle\; t\rangle_R \sim \left(\frac{m}{T}\right)^2 D\,. \tag{9}$$

$$N_S \sim \frac{1}{T}\frac{1}{D^2}T^2\epsilon_n \sim \frac{T\epsilon_n}{D^2}\,, \tag{10}$$

$$\delta\left(\langle t\rangle_S \;\langle\; t\rangle_R\right) \sim \frac{\tau}{\sqrt{N_S}} \sim \frac{\tau D}{\sqrt{T}\sqrt{\epsilon_n}}\,. \tag{11}$$

For a non-exponential luminosity, τ is more generally the width of the pulse. If zero delay is measured, then the mass limit is determined by the largest positive delay that could have fluctuated to zero, and thus

$$m_{limit} \sim T^{3/4}\,\tau^{1/2}{\epsilon_n}^{1/4}\,. \tag{12}$$

Note that is *independent* of D for any supernova distance in the Galaxy (i.e., as long as there is a reasonable number of counts). Note that the mass limit scales with the *fourth* root of the number of events (via the efficiency ϵ_n).

Besides the scaling, the value of m_{lim} at fixed values of these quantities can also be found easily [18]. The analytic results above are in excellent agreement with the full numerical results [18]. Over *reasonable* ranges of the input parameters, the results hardly change. However, some estimates of the ν_τ mass sensitivity for proposed detectors [22,23] assumed a very short duration of the supernova ($\tau \simeq 0.5$ s). If such a short duration were valid, then the SNO sensitivity would also be of order 10 eV.

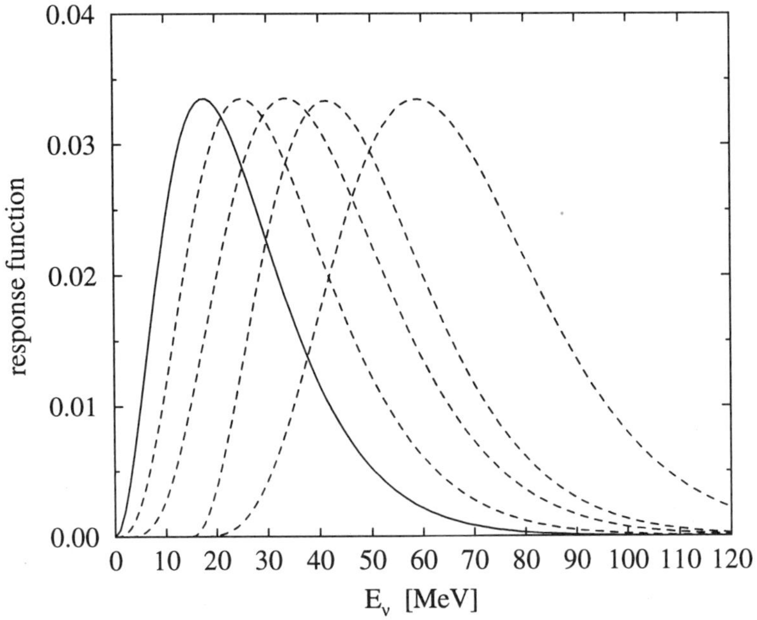

Figure 3: The *shapes* of response functions for different reactions. The solid line is a Fermi-Dirac distribution $f(E_\nu)$ with temperature 8 MeV. The dashed lines are products of that with the various neutral-current cross sections, i.e., $f(E_\nu)\sigma(E_\nu)$. The normalizations were chosen so that all curves would have the same height. From left to right, the cross sections are $\nu + e$, $\nu + d$, $\nu + ^{12}$C, $\nu + ^{16}$O. For ^{12}C, only the level at 15.11 MeV was used. For ^{16}O, a fit [17] to the full calculation [20] was used. In a given experiment, only the integral of one of these curves is measured, but each samples the underlying $f(E_\nu)$ differently. The peak positions are what is called E_c in the text.

1.8 How Model-Dependent are the Results?

Above, I discussed the proposed $\langle t \rangle$ technique for analyzing the supernova neutrino data for a ν_τ mass. This technique was designed to be as model-independent as possible. One underlying theme is to calibrate the assumed models as much as possible from the data [17,18].

While is is probably a good approximation that the luminosities of the different flavors are equal as a function of time, the normalization of the luminosity drops out in the definition of $\langle t \rangle$, and hence the results depend on the weaker assumption that the *shapes* of the luminosities are approximately equal as a function of time. The differences among the different flavors at the very early stages (for example, the higher ν_e flux) persist over a duration much less than 10 s, and hence have no effect on the analysis.

Another very important point is that the details of the shape of the luminosity as a function of time are *never used*, because the delay is measured just from the $\langle t \rangle$ of the neutral-current events minus the same for the charged-current events. Why is this enough? In general, one could describe the scattering rates of charged-current and neutral-current events by their normalizations and a series of all moments ($\langle t \rangle$, $\langle t^2 \rangle$, etc.). While this would likely require many terms, it *is* efficient to consider the *difference* in the moments between the two. The mass effects do not change the normalizations (the numbers of events), and we have shown that the higher moments (e.g., the width) are not significantly changed by a mass. That is, the only statistically significant effects of a mass on the event rate difference are in the average arrival time $\langle t \rangle$. If the details of the supernova model are not known from theory, then $\langle t \rangle$ is what is called a sufficient statistic for the mass [24].

The only aspect of the shape of the luminosity $L(t)$ that was used in the analysis was its duration, and this appears only in the error on the delay. *Thus the particular assumption of using an exponential luminosity completely disappears from the problem.* Any other reasonable model of similar duration would give a very similar result.

One also needs to know the ν_τ temperature to determine the mass limit. Since none of the neutral-current reactions measure the incoming neutrino energy, there is no direct spectral information. The ν_τ temperature must be determined from the number of events. Since different neutral-current reactions have different energy dependence, one can crudely reconstruct the underlying spectrum (see Fig. 3). The moments analysis of the rates reveals that only the average delay is significant, and thus that just a single energy (the characteristic energy E_c) is relevant, instead of the fine details of the spectral shape.

The final results for the mass sensitivity are given in Table III.

2 Supernova Location by Neutrinos

2.1 Overview

There has been great interest recently in the question of whether or not a supernova can be located by its neutrinos. If so, this may offer an opportunity to give an early warning to the astronomical community, so that the supernova light curves can be observed from the earliest possible time. An international supernova early alert network has been formed for this purpose, and the details of its implementation [25,26] were the subject of a recent workshop [27]. One of the primary motivations for such a network is to greatly reduce the false signal rate by demanding a coincidence between several different detectors. The second motivation is to locate the supernova by its neutrino signal.

The interest in the latter is driven by two considerations. First, the neutrinos leave the core at the time of collapse, while the electromagnetic radiation leaves the envelope some hours later, depending on the stellar mass. Thus the neutrino observations can give the astronomers a head start. Second, the next Galactic supernova will likely be in the disk, and hence obscured by dust, and perhaps never visible optically.

There are two classes of techniques to locate a supernova by its neutrinos. The first class of techniques is based on angular distributions of the neutrino reaction products, which can be correlated with the neutrino direction. In this case, a single experiment can independently announce a direction and its error. The second method of supernova location, triangulation, is based on the arrival-time differences of the supernova pulse at two or more widely-separated detectors. This technique would require significant and immediate data sharing among the different experiments.

2.2 Angular Distributions

Neutrino-Electron Scattering: Electron Angular Distribution

In neutrino-electron scattering, the scattered electrons have a very forward angular distribution, due to the small electron mass. At supernova neutrino energies, the angle between the incoming neutrino and the outgoing electron is about 10°, depending somewhat on neutrino energy and flavor. However, multiple scattering of the struck electron smears out its Čerenkov cone, washing out the dependence on energy and flavor, and introducing an angular resolution of about 25°.

Naively, if the one-sigma width of the electron angular distribution is 25°, then the precision with which its center (i.e., the average) can be defined given

N_S events is

$$\delta\theta \simeq \frac{25^\circ}{\sqrt{N_S}}\,, \tag{13}$$

where N_S is the number of events. For SK [17], $N_S \simeq 320$, so the cone center could be defined to within about 1.5°. For SNO [18] (using both the light and heavy water), $N_S \simeq 25$, so the cone center could be defined to about 5°. The equivalent error on the cosine is $\delta(\cos\theta) \simeq (\delta\theta)^2/2$, i.e., $3 \times 10^{\ 4}$ and $4 \times 10^{\ 3}$, respectively.

However, the neutrino-electron scattering events are indistinguishable on an event-by-event basis from other reactions with only a single electron or positron detected. The primary background in SK or the light water in SNO is thus $\bar{\nu}_e + p \to e^+ + n$ (in the heavy water in SNO, there are several background reactions).

This problem is similar to the SK solar neutrino studies, in which the neutrino-electron scattering events have to be separated from the intrinsic detector background (due to radioactivity, etc). The SK results are presented as an angular distribution in the cosine along the known direction to the Sun, and clear bump in the solar direction is seen, with signal/noise $\simeq 1/10$. Though the concept is the same, for the supernova events the signal/noise is only $\simeq 1/30$. Taking this into account [28] makes the centroiding error larger by a correction factor of $\simeq 4$ for both SK and SNO. With some cuts on energy, it may be possible to reduce this correction factor to $\simeq 2 \quad 3$.

Neutrino-Nucleus Scattering: Lepton Angular Distributions

In the charged-current reactions of ν_e and $\bar{\nu}_e$ on nucleons or nuclei, the outgoing electrons or positrons typically have angular distributions of the form

$$\frac{d\sigma}{d\cos\theta} \simeq 1 + v_e a(E_\nu)\cos\theta\,, \tag{14}$$

where θ is the angle between the neutrino and electron (or antineutrino and positron) directions in the laboratory (where the nuclear target is assumed to be at rest) and v_e is the lepton velocity in $c = 1$ units. At energies higher than for supernova neutrinos, terms proportional to higher powers of $\cos\theta$ appear, and at the highest energies, all reaction products are strongly forward simply by kinematics. Just above threshold, where the lepton velocity is small, the angular distribution obviously becomes isotropic. At supernova neutrino energies, $v_e = 1$.

Typically, the asymmetry coefficient $a(E_\nu)$ is not large, so the angular distribution is weak, but the number of events may be large (for $\bar{\nu}_e + p \to e^+ + n$, there are nearly 10^4 events expected in SK). Thus it may be possible to use these events to locate the supernova. We first consider how well one could localize the supernova assuming that a is known and constant.

Given a sample of events, one can attempt to find the axis defined by the neutrino direction. Along this axis, the distribution should be flat in the azimuthal angle ϕ and should have the form in Eq. (14) in $\cos\theta$. Along any other axis, the distribution will be a complicated function of both the altitude and azimuthal angles. We assume that the axis has been found numerically, and ask how well the statistics allow the axis to be defined. A convenient way to assess that is to define the forward-backward asymmetry as

$$A_{FB} = \frac{N_F \quad N_B}{N_F + N_B}\,, \tag{15}$$

where N_F and N_B are the numbers of events in the forward and backward hemispheres. The total number of events is $N = N_F + N_B$. Note that A_{FB} will assume an extremal value A_{FB}^{extr} along the correct neutrino direction. It can be shown [28] that

$$\delta A_{FB} = \frac{1}{\sqrt{N}}\sqrt{1 \quad \left(\frac{a}{2}\right)^2} \simeq \frac{1}{\sqrt{N}}\,, \tag{16}$$

where the error is nearly independent of a for small $|a|$, which is the case for the reactions under consideration.

In the above, the coordinate system axis was considered to be correctly aligned with the neutrino direction. Now consider what would happen if the coordinate system were misaligned. While in general, all three Euler angles would be needed to specify an arbitrary change in the coordinate system, symmetry considerations dictate that the computed value of A_{FB} depends only upon one – the angle θ between the true and the supposed neutrino axis. Thus A_{FB} is some function of θ if the axis is misaligned. Using a Legendre expansion, one can show that

$$A_{FB}(\theta) = \frac{a}{2}\cos\theta\,. \tag{17}$$

The error on the alignment is then

$$\delta(\cos\theta) = \frac{2}{|a|}\,\delta A_{FB} \simeq \frac{2}{|a|}\frac{1}{\sqrt{N}}\,. \tag{18}$$

Treating the nucleons as infinitely heavy, the coefficient a in Eq. (14) is related to the competition of the Fermi (no spin flip) and Gamow-Teller (spin flip) parts of the matrix element squared:

$$a = \frac{|M_F|^2 \quad |M_{GT}|^2}{|M_F|^2 + 3|M_{GT}|^2} \,. \tag{19}$$

Naive use of this formula for $\bar{\nu}_e + p \to e^+ + n$ gives $a = \quad 0.10$ for ($|M_{GT}/M_F| = 1.26$). One expects 10^4 events in SK, so that $\delta(\cos\theta) \simeq 0.2$. Even though the asymmetry parameter a is quite small, the number of events is large enough that this technique in SK could give a reasonable pointing error (SNO only has 400 of these events, so $\delta(\cos\theta) \simeq 1.0$, which is too large to be useful). For the charged-current reactions on deuterons, $\bar{\nu}_e + d \to e^+ + n + n$ and $\nu_e + d \to e \quad + p + p$, one would have $a = \quad 1/3$ ($M_F/M_{GT} \simeq 0$). If these channels could be combined (160 events in total), then $\delta(\cos\theta) \simeq 0.5$, which is again rather large.

In general, the coefficient $a(E_\nu)$ has energy dependence coming from recoil and weak magnetism corrections, each of order $1/M$, where M is the nucleon mass (such terms also change the total cross section [29,30,31], which has some important applications [32]). For the reaction $\bar{\nu}_e + p \to e^+ + n$, these corrections have a dramatic effect on the angular distribution, making it backward at low energies, isotropic at about 15 MeV, and forward at higher energies. Averaged over the expected $\bar{\nu}_e$ spectrum from a supernova, one obtains $\langle\cos\theta\rangle = +0.08$, i.e., a forward distribution. Using the common assumption that the angular distribution is backward (with $a = \quad 0.10$) would lead one to infer *exactly the wrong direction to the supernova.*

It is convenient to describe the angular distribution by the average cosine, weighted by the differential cross section. In the limit that Eq. (14) holds,

$$\langle\cos\theta\rangle = \frac{1}{3} v_e a(E_\nu) \,. \tag{20}$$

For $\bar{\nu}_e + p \to e^+ + n$, the variation of the average cosine can be written as (keeping only the largest terms of the full expression [31]):

$$\langle\cos\theta\rangle \simeq \quad 0.034 v_e + 2.4 \frac{E_\nu}{M} \,. \tag{21}$$

The factor 2.4 is actually 1.0 + 1.4, the first term from recoil, which always makes the angular distribution more forward, and the second from weak magnetism (which would change sign for the reaction $\nu_e + n \to e \quad + p$, if there were free neutron targets). The variation of $\langle\cos\theta\rangle$ with E_ν is shown in Fig. 4.

Since the deuteron is weakly bound, the recoil and weak magnetism corrections to $\langle\cos\theta\rangle$ can be reasonably estimated [31]. The effects of recoil and weak magnetism add in the $\bar{\nu}_e + d$ channel, and partially cancel in the $\nu_e + d$ channel, as seen in Fig. 5.

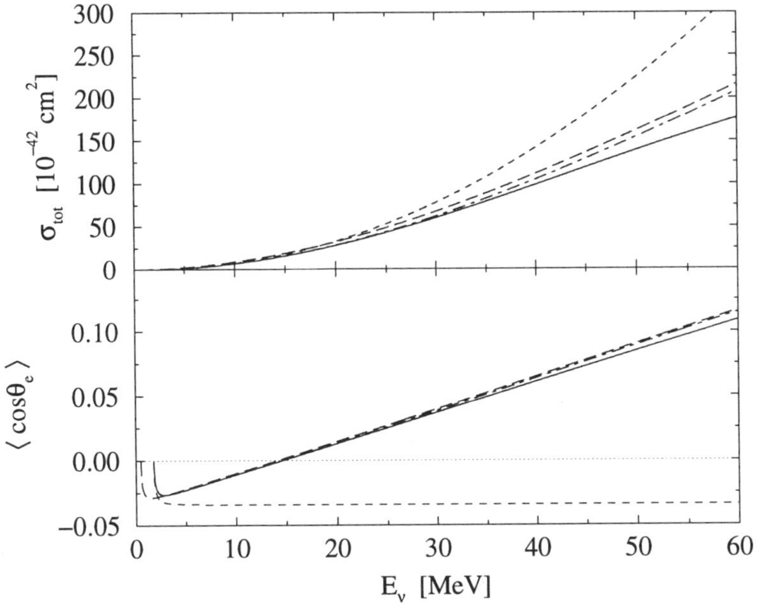

Figure 4: Upper panel: total cross section for $\bar{\nu}_e + p \to e^+ + n$; bottom panel: $\langle \cos\theta \rangle$ for the same reaction; both as a function of the antineutrino energy. The solid line is our $\mathcal{O}(1/M)$ result and the short-dashed line is the $\mathcal{O}(1)$ result. The long-dashed line is the high-energy formula of Llewellyn-Smith, and the dot-dashed line contains our threshold modifications to the same. The inner radiative corrections are included (see the text), but the outer radiative corrections are not. The long-dashed and dot-dashed lines are nearly indistinguishable in the lower panel.

Inverse Beta Decay: Positron-Neutron Separation Vector

The results above were based on lepton angular distributions as observed in water-Čerenkov detectors (light or heavy water). Some of the same reactions occur in scintillator detectors, but the lepton angular distributions are not observable due to the isotropic character of scintillation light.

Despite that, there is a way to get directional information with a scintillator detector. The idea [33] makes use of the fact that scintillator detectors can measure *final particle positions* by the relative timing between different phototubes. For $\bar{\nu}_e + p \to e^+ + n$, the positron is detected nearly at the point of creation. The neutron, however, will be detected (by its capture gamma rays), on average a few centimeters *forward* of the point of creation. The initially forward motion of the neutron is just a consequence of kinematics. Thus the

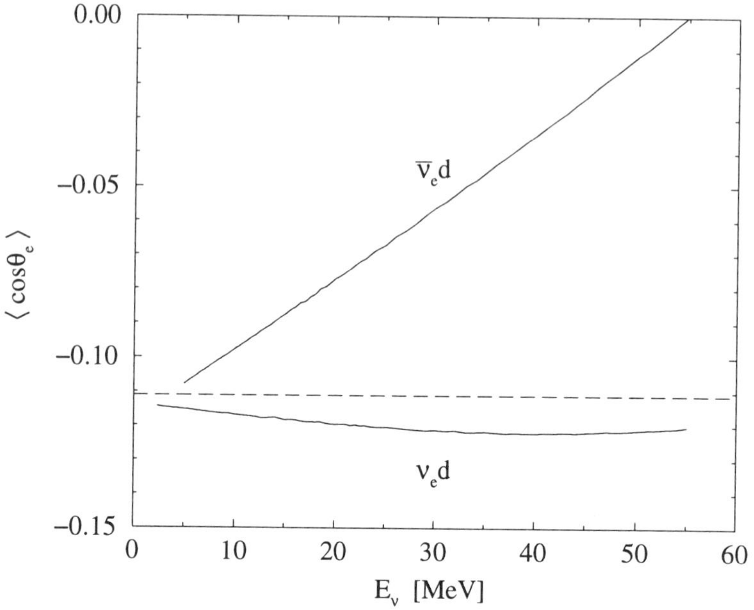

Figure 5: The average lepton cosine for the charged-current deuteron reactions, versus the neutrino or antineutrino energy, using Kubodera's calculations. We plot only from 1 MeV above threshold, so the ν_e-dependence at low energies is not shown. The small jitters are due to the coarse integration grid.

positron-neutron separation vector points in the incoming antineutrino direction.

However, before the neutron can be captured, it must be thermalized. On the first several scatterings (about half of the kinetic energy is lost on each step), the forward motion tends to be preserved. In the remaining scatterings until thermal energy is reached, the scattering is isotropic. In the last phase, the neutron wanders randomly, undergoing further scatterings at thermal energy, until it is captured. The neutron wander has the effect of degrading the significance of the initially forward motion. The average displacement and its fluctuations can be calculated by Monte Carlo simulation [31,34].

At the energies of reactor antineutrinos, the positron-neutron separation is about 1.5 cm and the statistical fluctuation several cm, depending on the neutron capture time in the scintillator [31]. There is an additional uncertainty in the position due to the gamma-ray localization error. For reactor antineutrinos, the source direction is known, so a measurement of the average positron-neutron separation vector can be used to make a background measurement, since background events will degrade the significance of the separation vector from the expectation. The precision with which the reactor can be located was evaluated by the Chooz collaboration and found to be[34] about 18°. Given the small displacement and the large error in localization, this is an impressive measurement.

At the higher energies of supernova neutrinos, the neutron kinetic energy is larger and the elastic scattering cross section smaller, so the displacement is larger. The Chooz collaboration estimates[34] that if they were to scale their detector to the size of SK (32 kton), that they would be able to locate a supernova to within about 9°, not so much worse than the neutrino-electron scattering result of about 5° in SK. The positron-neutron displacement technique is not possible in SK since neutrons are not detected, and not possible in SNO because the neutron wanders over several meters.

2.3 Triangulation

Finally, there is the technique of triangulation by arrival time differences. The idea is very simple – for two detectors, the cosine of the angle θ between the axis connecting the two detectors (with separation d) and the supernova direction is determined by the arrival-time difference Δt by

$$\cos\theta = \frac{\Delta t}{d} \,. \tag{22}$$

For detectors on opposite sides of the Earth, $d = 40$ ms, and for SK and SNO, $d = 30$ ms. Two detectors thus define a cone of allowed directions to the supernova. Since the arrival time difference has an error $\delta(\Delta t)$, the cone will have an angular thickness

$$\delta(\cos\theta) = \frac{\delta(\Delta t)}{d} \,. \tag{23}$$

The timing error is obviously the crucial point, and in order to have $\delta(\cos\theta) = 0.1$ (comparable to the neutrino-electron scattering result in SNO), one would need $\delta(\Delta t) = 3$ ms for the SK-SNO difference. This is much smaller than the duration of the supernova pulse (10 s) or even the risetime (assumed 100 ms).

In order to assess the best that triangulation could do, consider the scenario where SK and the light water portion of SNO are each perfect detectors, identical except for size. (The shape of the event rate in the heavy water in SNO will be different, due to the different reactions, and it is not clear how to correct for that). Then the scattering rates observed in SK and the light water in SNO will have the *same underlying distribution*, which depends on unknown details of the supernova models.

The observed scattering rates in SK and the light water of SNO can then differ only in normalization, statistical fluctuations, and a possible delay. The normalizations matter only in that they determine the scale of the statistical fluctuations. SK will have about 10^4 events and the light water in SNO about 400 events, so the statistical error will be dominated by the timing error in SNO. That is, the underlying distribution will be measured in SK and this template will be used in SNO to test for a delay.

In order to make estimates of the timing error, we need the shape of the event rate. We assume a short rise (in the current supernova models, this is somewhere between 0 and 200 ms, so we will very conservatively use 100 ms), followed by a long decay (10 s, as for SN1987A).

For the normalized event rate, we take

$$f(t, t_0) = \alpha_1 \times \frac{1}{\tau_1} \exp\left[+\frac{(t \quad t_0)}{\tau_1}\right], \qquad t < t_0 \tag{24}$$

$$f(t, t_0) = \alpha_2 \times \frac{1}{\tau_2} \exp\left[\ \ \frac{(t \quad t_0)}{\tau_2}\right], \qquad t > t_0\,. \tag{25}$$

where

$$\alpha_1 = \frac{\tau_1}{\tau_1 + \tau_2}, \qquad \alpha_2 = \frac{\tau_2}{\tau_1 + \tau_2}\,. \tag{26}$$

Then $f(t, t_0)$ is a normalized probability density function built out of two exponentials, and joined continuously at $t = t_0$. In what follows, we assume that this form of $f(t, t_0)$ is known to be correct and that τ_1 and τ_2 are *known*. The shape of the event rate is shown in Fig. 6.

The event rate in SNO will consist of N events sampled from $f(t, t_0)$, and the event rate in SK will consist of N' events sampled from $f(t, t_0')$. Then $\Delta t = t_0 \quad t_0'$, and $\delta(\Delta t) \simeq \delta t_0$, since the SNO error dominates. We consider only the statistical error determined by the number of counts. As noted, we want to determine the *minimal* error on the triangulation. This model, while simple, contains the essential timescales and an adjustable offset.

These considerations lead to a well-posed statistical problem: If N events are sampled from a *known* distribution $f(t, t_0)$, how well can t_0 be determined? The Rao-Cramer theorem [24] provides an answer to this question (with a possible subtlety [35]). This theorem allows one to calculate the minimum possible variance on the determination of a parameter (here t_0), by any technique whatsoever. This minimum variance can be achieved when all of the data are used as "efficiently" as possible, which is frequently possible in practice.

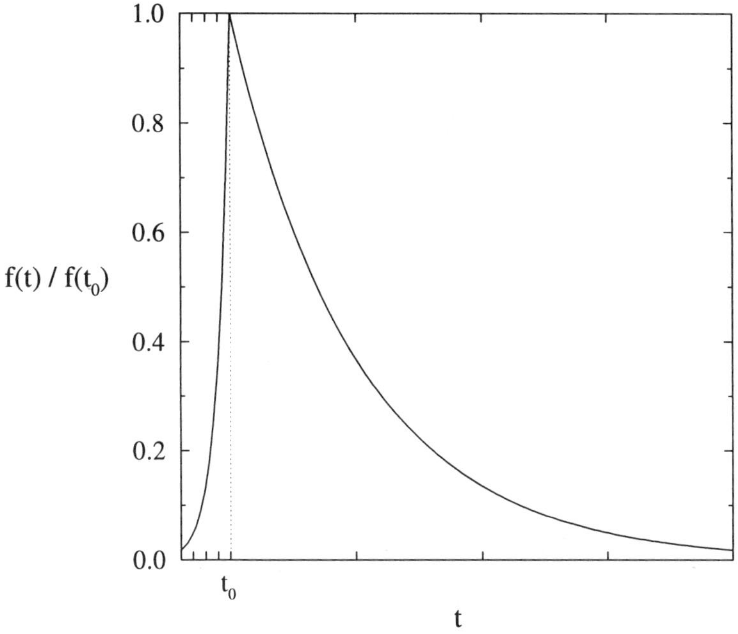

Figure 6: The schematic form of the normalized event rate $f(t, t_0)$ is shown. To the left of t_0 there is an exponential rise with time constant τ_1. To the right of t_0 there is an exponential decay with time constant τ_2. The tick marks on the t-axis are in units of the respective time constants τ_1 and τ_2. For clarity of display, we used $\tau_1/\tau_2 \simeq 10^{\ \ 1}$ in the figure, instead of the $\tau_1/\tau_2 \simeq 10^{\ \ 2}$ assumed in the analysis.

One requirement of the theorem is that the domain of positive probability must be independent of the parameter to be determined. This condition is obviously not met for a zero risetime, since then the domain is (t_0, ∞). For a nonzero risetime, the domain is technically $(\infty \ \ , \infty)$, independent of t_0, and so the theorem applies. The minimum possible variance on the determination of t_0 is:

$$\frac{1}{(\delta t_0)^2_{\rm min}} = N \times \int dt\, f(t, t_0) \left[\frac{\partial \ln f(t, t_0)}{\partial t_0}\right]^2 . \tag{27}$$

This is the general form for an arbitrary parameter t_0. When t_0 is a translation parameter, i.e., $f(t, t_0)$ depends only on $t \quad t_0$, this reduces to

$$\frac{1}{(\delta t_0)^2_{\rm min}} = N \times \int dt\, f(t, t_0) \left[\frac{\partial \ln f(t, t_0)}{\partial t}\right]^2 \tag{28}$$

$$= N \times \int dt \, \frac{\left[\partial f(t,t_0)/\partial t\right]^2}{f(t,t_0)} \,. \tag{29}$$

For the particular choice of $f(t,t_0)$ above, this reduces to

$$\frac{1}{(\delta t_0)^2_{\text{min}}} = N \times \;\; \alpha_1/\tau_1^2 + \alpha_2/\tau_2^2) \,, \tag{30}$$

and so the minimum error is

$$(\delta t_0)_{\text{min}} = \frac{\sqrt{\tau_1 \tau_2}}{\sqrt{N}} = \frac{\tau_1}{\sqrt{N_1}} \,. \tag{31}$$

Note that $N_1 = N(\tau_1/\tau_2)$ is approximately the number of events in the rising part of the pulse. Since the rise is the sharpest feature in $f(t,t_0)$, it is unsurprising that it contains almost all of the information about t_0. The total number of events N is fixed by the supernova binding energy release, so a change in the assumed total duration of the pulse, i.e., τ_2, would affect the peak event rate and hence the fraction of events in the leading edge. For a more general $f(t,t_0)$, one would replace τ_1/τ_2 by this fraction computed directly.

For SNO, $N_1 \simeq 10^{\;\;2} \times 400 \simeq 4$, so $\delta(t_0) \simeq 30$ ms$/\sqrt{4} \simeq 15$ ms. Since SK has about 25 times more events, the corresponding error would be about 3 ms. Therefore, the error on the delay is $\delta(\Delta t) \simeq 15$ ms and $\delta(\cos\theta) \simeq 0.50$ at one sigma. We have not specified the method for extracting t_0 and hence Δt from the data [28]. That is exactly the point of the Rao-Cramer theorem – that one can determine the minimum possible error without having to try all possible methods.

If the risetime were zero, which seems to be unrealistic, then one can show [28] by application of order statistics that the error becomes

$$\delta(t_0) = \frac{\tau_2}{N} \,. \tag{32}$$

Here τ_2/N is simply the spacing between events near the peak. For a more general $f(t,t_0)$, but still with a sharp edge at t_0, one would simply replace τ_2 by $1/f(t_0)$. In fact, the shape of $f(t,t_0)$ is irrelevant except for its effect on the peak rate, i.e., $f(t_0)$. So long as $f(t,t_0)$ has a sharp edge and the right total duration, allowing a more general time dependence would therefore not change the results significantly.

For the two cases, zero and nonzero risetime, we used different mathematical techniques. This may seem like an artificial distinction, and that these two cases do not naturally limit to each other. In particular, it may seem incompatible that in the first case the error $\sim 1/\sqrt{N}$, while in the second the error

$\sim 1/N$. Further, one obviously cannot take $\tau_1 \to 0$ in the first result to obtain the second. However, it can be shown [28] that the two techniques have disjoint regions of applicability (as a function of τ_1) and that the formulas match numerically at the boundary between the two. The boundary is the value of τ_1 such that the number of events in the rise is of order 1, i.e., the edge appears sharp for this or smaller τ_1.

The final results for the pointing errors are given in Table IV.

3 Concluding Remarks

The next Galactic supernova should be a bonanza for neutrino physics and astrophysics. Should, that is, provided that we can make sense of the data. The difficulty is that there are many unknown aspects of both the particle physics properties of neutrinos and the astrophysics of the expected signal. Since the expected core-collapse supernova rate in the Galaxy is about 3 per century [14], and since there are no neutrino detectors sensitive to supernovae in distant galaxies, we will not have the luxury of many observations. Of course, there is a worldwide effort to improve the numerical models of core-collapse supernovae [36]. And besides neutrinos, there is one other direct probe of the supernova core, and that is gravitational radiation. LIGO may be sensitive to supernovae out to the Virgo Cluster (at rates of order 1 per year [37]), and its observations may improve the understanding of stellar collapse.

Since there are uncertainties in the supernova models, the $\langle t \rangle$ test for the ν_μ and ν_τ masses was designed to be as model-independent as possible. The resulting sensitivities for either ν_μ or ν_τ are in Table III. If the ν_μ and ν_τ are maximally mixed with a small mass difference, then the mass test is really on $m_2 = m_3$, and the limits in Table III improve by a factor of about $\sqrt{2}$.

Table 3: Mass sensitivity for various detectors and neutral-current reactions. The mass quoted is the limit which would be placed on either the ν_μ or ν_τ mass at 90% CL if no delay were seen (equivalently, it is the smallest mass which would cause an unambiguous delay). If the ν_μ and ν_τ masses are nearly equal, then the sensitivity is improved by $\sqrt{2}$. See also [21] for the ^{12}C results.

detector	neutral-current signal	mass limit
SNO	$\nu + d$	30 eV
SK	$\nu +^{16}$ O	45 eV
SK	$\nu + e$	50 eV
Kamland	$\nu +^{12}$ C	55 eV
Borexino	$\nu +^{12}$ C	75 eV

There are also many possible signals of supernova neutrino oscillations [38]. Ideally, the current and near-term terrestrial neutrino oscillation experiments can answer some of the key questions before the next supernova. Particularly crucial is the question of whether there are sterile neutrinos (or several flavors thereof). SNO [39] may tell us whether the solar ν_e neutrinos are oscillating into sterile or active flavors (or at all). Similarly for SK [40], K2K [41], and MINOS [42] and the atmospheric ν_μ neutrinos. And MiniBoone [43] may decide whether the LSND [44] anomaly is correct (and hence whether sterile neutrinos are necessary). These questions and more are discussed thoroughly in several recent reviews [45].

Since the neutrinos leave the supernova core a few hours before the light leaves the envelope, it should be possible for the neutrino experiments to give astronomers an advance warning of a Galactic core-collapse supernova, and also to provide some guidance as to where in the sky to look. Neutrino-electron scattering in SK seems to be the best technique, with an error of order 5°, depending on distance and the suppression of backgrounds from other reactions. While 5° may sound large to an astronomer, it is about 0.1% of the sky, which will make subsequent searches with small telescopes much easier (the role of amateur astronomers is discussed in a recent article [46]). Supernova location by triangulation seems to be rather difficult at present (see Table IV).

Table 4: One-sigma errors on how well the direction to the supernova is defined by various techniques, at $D = 10$ kpc. The other parameters used are noted in the text. For neutrino-electron scattering, the most pessimistic background assumptions were used.

technique	error
$\nu + e$ forward scattering (SK)	$\delta\theta \simeq 5°$, $\delta(\cos\theta) \simeq 4 \times 10^{3}$
$\nu + e$ forward scattering (SNO)	$\delta\theta \simeq 20°$, $\delta(\cos\theta) \simeq 6 \times 10^{2}$
$\bar{\nu}_e + p$ angular distribution (SK)	$\delta(\cos\theta) \simeq 0.2$
$\bar{\nu}_e + p$ angular distribution (SNO)	$\delta(\cos\theta) \simeq 1.0$
$\nu_e + d, \bar{\nu}_e + d$ angular distributions (SNO)	$\delta(\cos\theta) \simeq 0.5$
triangulation (SK and SNO)	$\delta(\cos\theta) \simeq 0.5$

ACKNOWLEDGMENTS

I am grateful to Gabor Domokos and Susan Kovesi-Domokos for the invitation to a very interesting workshop, and to Petr Vogel for his collaboration[17,18,28,31]. This work was supported by a Sherman Fairchild fellowship at Caltech.

References

1. Ch. Weinheimer *et al.*, Phys. Lett. **B460**, 219 (1999);
   V.M. Lobashev *et al.*, Phys. Lett. **B460**, 227 (1999).
2. See the "Focus Period on Direct Measurements of the Neutrino Mass", (http://nucth.physics.wisc.edu/conferences/int.html).
3. K. Assamagan *et al.*, Phys. Rev. **D53**, 6065 (1996).
4. R. Barate *et al.*, Eur. Phys. J. **C2**, 395 (1998).
5. P. Fisher, B. Kayser and K.S. McFarland, hep-ph/9906244.
6. V. Barger, T.J. Weiler and K. Whisnant, Phys. Lett. **B442**, 255 (1998).
7. S.A. Bludman, D.C. Kennedy and P.G. Langacker, Phys. Rev. **D45**, 1810 (1992);
   M.A. Diaz, J. Ferrandis, J.C. Romao and J.W. Valle, hep-ph/9906343.
8. R.A. Malaney, G.D. Starkman and S. Tremaine, Phys. Rev. **D51**, 324 (1995);
   A. Kusenko and G. Segre, Phys. Rev. Lett. **77**, 4872 (1996);
   A. Kusenko and G. Segre, Phys. Lett. **B396**, 197 (1997);
   A.D. Dolgov, S.H. Hansen, S. Pastor and D.V. Semikoz, Nucl. Phys. **B548**, 385 (1999);
   S. Samurovic and M.M. Cirkovic, astro-ph/9907142;
   R.J. Lindebaum, G.B. Tupper and R.D. Viollier, astro-ph/9906004;
   G.C. McLaughlin, J.M. Fetter, A.B. Balantekin and G.M. Fuller, Phys. Rev. **C59**, 2873 (1999).
9. W. Hu, D.J. Eisenstein and M. Tegmark, Phys. Rev. Lett. **80**, 5255 (1998).
10. M. Fukugita, G. Liu and N. Sugiyama, hep-ph/9908450.
11. R.E. Lopez, S. Dodelson, R.J. Scherrer and M.S. Turner, Phys. Rev. Lett. **81**, 3075 (1998).
12. A.R. Cooray, Astron. Astrophys. **348**, 31 (1999).
13. R.A. Croft, W. Hu and R. Dave, Phys. Rev. Lett. **83**, 1092 (1999).
14. S. van den Bergh, Phys. Rept. **204**, 385 (1991);
    S. van den Bergh, Comments Astrophys. **17**, 125 (1993);
    G.A. Tammann, W. Löffler, A. Schröder, Astrophys. J. Suppl. **92**, 487 (1994).
15. T. Totani, Phys. Rev. Lett. **80**, 2039 (1998).

16. G.T. Zatsepin, JETP Lett. **8**, 205 (1968) [Pis'ma Zh. Eksp. Teor. Fiz. **8**, 333 (1968)];
T. Piran, Phys. Lett. **B102**, 299 (1981);
P. Reinartz and L. Stodolsky, Z. Phys. **C27**, 507 (1985).
17. J.F. Beacom and P. Vogel, Phys. Rev. **D58**, 053010 (1998).
18. J.F. Beacom and P. Vogel, Phys. Rev. **D58**, 093012 (1998).
19. H.-T. Janka, Astropart. Phys. **3**, 377 (1995).
20. K. Langanke, P. Vogel and E. Kolbe, Phys. Rev. Lett. **76**, 2629 (1996).
21. L. Cadonati, F. Calaprice, and M. Chen, in preparation.
22. C.K. Hargrove, I. Batkin, M.K. Sundaresan and J. Dubeau, Astropart. Phys. **5**, 183 (1996).
23. P.F. Smith, Astropart. Phys. **8**, 27 (1997).
24. R.V. Hogg and A.T. Craig, *Introduction to Mathematical Statistics, 4th ed.* (Macmillan, New York, 1978);
A. Stuart and J.K. Ord, *Kendall's Advanced Theory of Statistics, Vol. 2, 5th ed.* (Oxford, New York, 1991).
25. A. Habig and K. Scholberg, "An Early Supernova Detection System", unpublished.
26. A. Habig, K. Scholberg, and M. Vagins, "The International Early Supernova Alert System", poster at the Neutrino '98 Conference (June 1998).
27. First International Workshop on the Supernova Early Alert Network (Boston University, September 1998), and bound photocopied transparencies of the same.
28. J.F. Beacom and P. Vogel, Phys. Rev. **D60**, 033007 (1999).
29. P. Vogel, Phys. Rev. D **29**, 1918 (1984).
30. S. A. Fayans, Sov. J. Nucl. Phys. **42**, 590 (1985).
31. P. Vogel and J.F. Beacom, Phys. Rev. **D60**, 053003 (1999).
32. D. Seckel, hep-ph/9305311;
R.E. Lopez, M.S. Turner and G. Gyuk, Phys. Rev. **D56**, 3191 (1997);
C.J. Horowitz and G. Li, Phys. Rev. Lett. **82**, 5198 (1999).
33. C. Bemporad, in *Third International Workshop on Neutrino Telescopes*, edited by M.B. Ceolin (Padua University, Padua, Italy, 1991); C. Bemporad, Neutrino '98 Conference (June 1998).
34. M. Apollonio *et al.*, hep-ex/9906011.
35. Bob Cousins (UCLA) reminds me that what we really want is the minimum-sized confidence interval, which does not always correspond to a minimum variance (e.g., the Lorentzian distribution). However, there is apparently no such pathology here.
36. J.R. Wilson and R.W. Mayle, Phys. Rept. **227**, 97 (1993);
M. Herant, W. Benz, W.R. Hix, C.L Fryer, and S.A. Colgate, Astrophys.

J. **435**, 339 (1994);
A. Mezzacappa, A.C. Calder, S.W. Bruenn, J.M. Blondin, M.W. Guidry, M.R. Strayer and A.S. Umar, Astrophys. J. **495**, 911 (1998);
S. Yamada, H. Janka and H. Suzuki, astro-ph/9809009;
A. Burrows, T. Young, P. Pinto, R. Eastman and T. Thompson, astro-ph/9905132;
C.L. Fryer, A. Heger, astro-ph/9907433.

37. A. Abramovici *et al.*, Science **256**, 325 (1992);
B.C. Barish, gr-qc/9905026.
38. L. Wolfenstein, Phys. Rev. **D20**, 2634 (1979);
P. Reinartz and L. Stodolsky, Z. Phys. **C27**, 507 (1985);
S.P. Mikheev and A.Y. Smirnov, Sov. Phys. JETP **64**, 4 (1986);
T.P. Walker and D.N. Schramm, Phys. Lett. **B195**, 331 (1987);
T.K. Kuo and J. Pantaleone, Phys. Rev. **D37**, 298 (1988);
A.B. Balantekin and F. Loreti, Phys. Rev. **D45**, 1059 (1992);
A.Y. Smirnov, D.N. Spergel and J.N. Bahcall, Phys. Rev. **D49**, 1389 (1994);
G.M. Fuller, W.C. Haxton and G.C. McLaughlin, Phys. Rev. **D59**, 085005 (1999);
S. Choubey, D. Majumdar and K. Kar, J. Phys. G **25**, 1001 (1999);
G. Dutta, D. Indumathi, M.V. Murthy and G. Rajasekaran, hep-ph/9907372;
A.S. Dighe and A.Y. Smirnov, hep-ph/9907423.
39. J. Klein, these proceedings.
40. M. Nakahata, Nucl. Phys. Proc. Suppl. **76**, 425 (1999).
41. Y. Oyama, hep-ex/9803014.
42. B. Choudhary, these proceedings; M. Goodman, *ibid.*
43. A.O. Bazarko, hep-ex/9906003.
44. C. Athanassopoulos et al., Phys. Rev. C **54**, 2685 (1996).
45. K. Zuber, Phys. Rept. **305**, 295 (1998);
S.M. Bilenkii, C. Giunti and W. Grimus, hep-ph/9812360;
P. Fisher, B. Kayser and K.S. McFarland, hep-ph/9906244;
A.Y. Smirnov, hep-ph/9901208;
A.B. Balantekin and W.C. Haxton, nucl-th/9903038;
G.G. Raffelt, hep-ph/9902271;
and several articles in this proceedings.
46. L.J. Robinson, Sky & Telescope, Aug. 1999, p. 30.

NEUTRINOS IN THE NEXT MILLENIUM - SUMMARY TALK

LINCOLN WOLFENSTEIN
Carnegie Mellon University
Pittsburgh, Pennsylvania

1 Natural Sources of Neutrinos

The first twenty years of neutrino experiments involved man-made neutrinos starting with the reactor experiments of Fred Reines and culminating with the discovery of neutral currents using the neutrinos from the CERN proton synchrotron. For the last twenty years, the most exciting discoveries involved natural neutrinos: atmospheric, solar, and supernova.

It is interesting to compare and contrast these natural sources. Solar neutrinos are ν_e produced in nuclear reactions at a typical energy of 20 kev. Supernova neutrinos come in all flavors produced primarily by weak neutral current reactions at energies of 10-100 Mev. Atmospheric neutrinos consist of ν_μ and ν_e, the products of the decay of muons and pions produced in the collisions of cosmic ray protons of 1 to 100 Gev.

As we study these natural sources, there arises the interesting question of whether we are trying to understand the source or the neutrino; are we doing astrophysics or particle physics? The first detection of solar neutrinos in the pioneering experiment of Ray Davis provided a wonderful confirmation of the theory that nuclear reactions are the source of solar energy. However, today our study of solar neutrinos is designed to learn about neutrino oscillations with the belief that our understanding of the sun is much better than that of neutrinos.

The observation of neutrinos from supernova 1987 was one of the great events of modern science. It confirmed our picture that the energy of gravitational collapse is emitted in the form of some 10^{58} neutrinos, 19 of which were detected. It is clear that neutrinos play an essential role in the energetics of supernovae, in understanding the optical burst and explosive nucleosynthesis. It is essential that as many detectors as possible are always ready to detect neutrinos from a supernova in our own galaxy. It is to be hoped that other experiments will determine the oscillation parameters so that these can be applied to supernova theory and better predict the neutrino fluxes to be expected. The confirmation of these predictions or their failure may then help us to better understand supernovae.

The study of atmospheric neutrinos is unlikely to tell us more about cosmic rays than we can determine from detailed muon measurements. To many people's surprise, however, this study has provided us with the most convincing evidence of neutrino oscillations.

Over the next twenty years or more, the great challenge is the study of high-energy neutrinos originating outside our galaxy. Two possible sources are active galactic nuclei (AGN) and gamma-ray bursts (GRB). A number of AGN models predict neutrino fluxes that exceed that produced by cosmic rays for energies greater than 10^5 Gev, however, the predictions differ by more than an order of magnitude. These neutrinos, mainly ν_μ, can arise from the decay of pions produced by photoproduction when high-energy protons collide with photons in the AGN jet. They may also come from reactions near the central black hole. The predictions in the case of GRB's are even more uncertain. Since neutrinos can escape from regions from which no other particles can, it is hoped that if the fluxes are large enough, they will constrain the models of AGN and GRB. Even more exotic sources of very high energy neutrinos have been suggested. There is one interesting aspect of neutrino physics that may also be explored: oscillations of ν_μ into ν_τ for very small values of Δm^2. High-energy ν_τ can penetrate the earth by a succession of interaction $\nu_\tau + N \to \tau + X$ and decays $\tau \to \nu_\tau + X$ so that upward going ν_τ can be a signal of oscillations.

2 Evidence for Neutrino Oscillations

The most compelling evidence for neutrino oscillations comes from the up-down asymmetry of multi-Gev ν_μ from the atmosphere observed in the Super Kamiokande experiment. The flux of upcoming ν_μ ($\cos\theta = -0.2$ to -1) that have traveled between 2500 and 13000 km is just about half that downgoing ($\cos\theta = 0.2$ to 1) that have gone less than or about 100 km. The expectation that these fluxes should be the same does not depend on detailed calculations, but just on the observed isotrophy of the cosmic rays. (Geomagnetic corrections are small.) This is a 5σ effect. It is supported by the observed deficiency of ν_μ relative to ν_e as compared with detailed calculations.

Since no excess in upgoing ν_e is observed, the most natural conclusion is that ν_μ have oscillated to ν_τ which cannot be directly detected. The best fit parameters have Δm^2 =3.5x10^{-3}ev^{2}with the mixing $\sin^2 2\theta$ close to 1. Whereas there is no doubt about the effect the possible range of Δm^2 depends on the details of the analysis; Super K now quotes at 99% C.L., a range of 8x10^{-4}ev^2 to 10^{-2}ev^2. There is also the possibility that the oscillation is to a hypothetical sterile neutrino. In principle, these can be distinguished at Super

K by the neutral current reactions of ν_τ or by identifying matter effects that enter $\nu_\mu - \nu_s$ oscillations, but not $\nu_\mu - \nu_\tau$. If one assumes $\sin^2 2\theta = 1$ maximal mixing, then as a result of the different matter interaction of ν_μ and ν_s the actual mixing will be reduced for the case of ν_s. Some evidence against ν_s was presented from the data on through-going muons produced by high-energy ν_μ interactions in rock.

Evidence for neutrino oscillations comes also from solar neutrinos. This is a different oscillation since it involves the disappearance of ν_e and a lower value of Δm^2. The most compelling evidence, I believe, now comes from two gallium experiments (Gallex and Sage) which measure a flux about 57% of that calculated; both of these experiments have been calibrated with a neutrino source. What is striking is that the flux observed is less than that required by the luminosity constraint if there were only neutrinos from the pp reaction and leaves no room for neutrinos from $e^- + {}^7$Be, which constitute almost 30% of the calculated flux. A similar ^{7}Be problem exists when one compares the result of the chlorine (Davis) experiment with the result from Super Kamiokande. There are different oscillation solutions with values of Δm^2 between 10^{-11} and 10^{-4}ev^2.

There exists one more possible sign of oscillations from the LSND experiment at Los Alamos. The source here is stopping π^+ from the LAMPF accelerator. What is observed is the appearance of $\bar{\nu}_e$ presumably from $\bar{\nu}_\mu$to $\bar{\nu}_e$oscillations. What is required is a value of Δm^2 of order 1 ev^2 or greater with a small mixing angle. Given the small statistics, it is hard to make detailed checks for systematic errors so that it is important to have an independent confirmation of this possible oscillation.

When the LSND result is combined with atmospheric and solar, three very different Δm^2 values are required and, therefore, the existence of a fourth neutrino, usually referred to as a sterile neutrino, must be assumed.

3 Masses and Mixings - Theory?

From a theoretical point of view, the major motivation for neutrino masses comes from the apparent symmetry between leptons and quarks. They come in three generations; the charged leptons have a mass hierarchy like the quarks, and they have the same weak interactions. Thus, it is very natural to imagine that this symmetry is exact at some scale and is spontaneously broken. In this case, the right-handed neutrino ν_R must exist and there should be Dirac mass terms M_D linking ν_L to ν_R comparable to the quark masses. These considerations, in the context of SO(10), led to the "see-saw" idea: when the symmetry is broken at a high scale, the ν_R obtain a large majorana mass M and then the masses of the observed neutrinos are given by

$$m_\nu = M_D M^{-1} M_D \backsim m^2_{quark}/M$$

Within this general scenario, there are many possibilities, most of which are declared to be "natural" or "well-motivated" by their authors. The Dirac matrices for the neutrinos and the charged leptons should be related to those of the quarks, but they clearly are not identical. There are many ways of introducing the difference, particularly as one goes to the first and second generations. The matrix M is a completely new feature with no guide for its generation structure.

It seems natural to have a mass hierarchy among the three neutrino masses, but models with two or even all three neutrinos almost degenerate with small splittings are possible. Recently , many authors have found "natural" models in which mixings are large, even though it was previously thought that neutrino mixings should be not too large. Unfortunately, the large number of theoretical papers with different models is a testimony not to our understanding, but to our ignorance.

4 Future Experiments

For the value of Δm^2 $\left(10^{-2} to 10^{-3} ev^2\right)$ that fits the atmospheric neutrinos, a number of questions need to be answered:

1. Is the oscillation to ν_τ or a sterile ν?

2. Is there a small, but significant, probability of oscillation to ν_e?

3. What are the more precise values of Δm^2 and $\sin^2 2\theta$?

While all these answers will be sought by Super Kamiokande, definite conclusions probably require the long baseline experiments being planned. The first, which begins this year, will send a ν_μ neutrino beam from KEK to Super K at a distance of 250 km. In the U.S., the MINOS experiment, planned to start in late 2002, involves ν_μ beams from Fermilab to the SOUDAN mine at a distance of 750 km. While the broad band beam involves neutrino energies from 8 to 30 Gev, the possibility of a beam centered at 4 Gev is being planned to cover the lower Δm^2 values. Similar experiments are being discussed at CERN, with a beam going to the Gran Sasso laboratory.

A number of experiments are in progress or being prepared to further study solar neutrinos. The most important goal is to discover a "smoking gun," that is, direct evidence of oscillations independent of any calculated flux. Possibilities are:

1. A spectral distortion of the ^{8}B neutrinos.
2. A day-night difference in the flux. This could occur as a result of the regeneration of ν_e in the earth, giving a bigger flux at night due to the effect of matter on the mixing angle.
3. A seasonal variation in the flux over and above that due to $(1/r^2)$.
4. The ratio of neutral current events to charged currents. The observations, so far, are primarily sensitive to ν_e. If oscillations to ν_μ or ν_τ are taking place, then one can detect the total flux $(\nu_e + \nu_\mu + \nu_\tau)$ by detecting neutral current interactions on nucleons.

Each of these effects occurs only for certain values of $(\Delta m^2, \sin^2 2\theta)$ so the observation would also constrain these values.

The spectral distortion may be very hard to detect because any distortion is likely to be greatest at high energies where the flux is small or low energies where the background is large. Both Super Kamiokande and SNO will be measuring the spectrum of ^{8}B neutrinos. Both will also be searching for a day-night effect; in fact, Super Kamiokande reports a value of (6 ± 3) % which is interesting, but not decisive. The possibility for an anomalous seasonal variation occurs for the vacuum solution for the monoenergetic ^{7}Be neutrinos. Observing these is one goal of the proposed Borexino experiment in the Gran Sasso and perhaps also in the future of the Kamland experiment in Japan.

The best hope is the discovery of neutral current reactions due to the ν_μ or ν_τ to which the ν_e has oscillated. This is a major goal of SNO, a liquid deuterium Cerenkov detector located in Sudbury, Ontario, which has just started taking data. The neutral current reaction

$$\nu_x + d \to n + p + \nu_x$$

has the same cross-section for all flavors of neutrinos and can be directly compared to the charged current reaction

$$\nu_e + d \to p + p + e^-$$

In the case of oscillation to ν_μ or ν_τ (but not sterile neutrinos) the ratio of the neutral current reaction to the charged current would be at least twice as large as in the case of no oscillations.

It is of great importance to confirm (or refute) the LSND result since the existence of a sterile neutrino would be of great significance. To check the LSND result, there is the continuing Karman experiment in England, but it will take some time to get a definitive result given their estimated background. An experiment Miniboone is being prepared at Fermilab, a short baseline experiment using neutrinos of 1 Gev or less produced using the 8 Gev booster, to explore this parameter range. Since this experiment uses a single detector at one distance to detect ν_e, the background subtraction has to be carefully evaluated. It is intended to follow this up using a second detector at a different distance.

The detection of high energy neutrinos from extra-galactic sources requires a very large volume detector. The minimum requirement is generally thought to be a (kilometer)^3 detector. This is to be accomplished by instrumenting a target existing in nature. Projects being planned include the Ice Cube project, using ice at the South Pole and the Nestor project, using water in the Aegean Sea. The forerunner to Ice Cube is the Amanda detector which has instrumented .01 km^2 of ice and will be expanded to 0.1 km^2 next year.

The possibility of detecting extremely high energy neutrinos greater than 10^{18}ev might require other techniques. One possibility, called Rice, involves antennae in ice to detect Cerenkov radiation for large areas. The Auger project, designed to detect very large cosmic ray showers initiated by protons, could also be sensitive to ultra-high energy neutrinos that produce horizontal showers.

5 Conclusion

Neutrinos are very important for particle physics, astrophysics and cosmology. The discovery of neutrino oscillations and neutrino mass has implications yet to be determined for particle theory, supernovae, and dark matter. The detection of solar neutrinos and neutrinos from supernova 1987a marked the beginning of neutrino astronomy. Major efforts are now in progress to move to extra-galactic neutrino astronomy.

This research has been supported, in part, by the U.S. Department of Energy under grant DE-FG02-91-ER-40682.